ROWAN UNIVERSITY
CAMPBELL LIBRARY
201 MULLICA HILL RD.
GLASSBORO, NJ 08028-1701

*Principles of
Random Signal Analysis
and Low Noise Design*

Principles of Random Signal Analysis and Low Noise Design

The Power Spectral Density and its Applications

Roy M. Howard
Curtin University of Technology
Perth, Australia

A JOHN WILEY & SONS, INC., PUBLICATION

This is printed on acid-free paper. ⊗

Copyright © 2002 by John Wiley & Sons, Inc., New York. All rights reserved.

Published simultaneously in Canada.

No part of this publication may be reproduced, stored in a retrieval system or transmitted in any form or by any means, electronic, mechanical, photocopying, recording, scanning or otherwise, except as permitted under Sections 107 or 108 of the 1976 United States Copyright Act, without either the prior written permission of the Publisher, or authorization through payment of the appropriate per-copy fee to the Copyright Clearance Center, 222 Rosewood Drive, Danvers, MA 01923, (978) 750-8400, fax (978) 750-4744. Requests to the Publisher for permission should be addressed to the Permissions Department, John Wiley & Sons, Inc., 605 Third Avenue, New York, NY 10158-0012, (212) 850-6011, fax (212) 850-6008, E-Mail: PERMREQ@WILEY.COM.

For ordering and customer service, call 1-800-CALL-WILEY.

Library of Congress Cataloging-in-Publication Data is available.

ISBN: 0-471-22617-3

Printed in the United States of America.
10 9 8 7 6 5 4 3 2 1

Contents

Preface   ix

About the Author   xi

1. Introduction   1

2. Background: Signal and System Theory   3

    2.1 Introduction / 3
    2.2 Background Theory / 3
    2.3 Functions, Signals and Systems / 7
    2.4 Signal Properties / 12
    2.5 Measure and Lebesgue Integration / 23
    2.6 Signal Classification / 35
    2.7 Convergence / 36
    2.8 Fourier Theory / 38
    2.9 Random Processes / 44
    2.10 Miscellaneous Results / 45
          Appendix 1: Proof of Theorem 2.11 / 46
          Appendix 2: Proof of Theorem 2.13 / 47
          Appendix 3: Proof of Theorem 2.17 / 47
          Appendix 4: Proof of Theorem 2.27 / 49
          Appendix 5: Proof of Theorem 2.28 / 50
          Appendix 6: Proof of Theorem 2.30 / 52
          Appendix 7: Proof of Theorem 2.31 / 53
          Appendix 8: Proof of Theorem 2.32 / 56

3. The Power Spectral Density   59

3.1 Introduction / 59
3.2 Definition / 60
3.3 Properties / 65
3.4 Random Processes / 67
3.5 Existence Criteria / 73
3.6 Impulsive Case / 74
3.7 Power Spectral Density via Autocorrelation / 78
Appendix 1: Proof of Theorem 3.4 / 84
Appendix 2: Proof of Theorem 3.5 / 85
Appendix 3: Proof of Theorem 3.8 / 88
Appendix 4: Proof of Theorem 3.10 / 89

4. Power Spectral Density Analysis   92

4.1 Introduction / 92
4.2 Boundedness of Power Spectral Density / 92
4.3 Power Spectral Density via Signal Decomposition / 95
4.4 Simplifying Evaluation of Power Spectral Density / 98
4.5 The Cross Power Spectral Density / 102
4.6 Power Spectral Density of a Sum of Random Processes / 107
4.7 Power Spectral Density of a Periodic Signal / 112
4.8 Power Spectral Density — Periodic Component Case / 119
4.9 Graphing Impulsive Power Spectral Densities / 122
Appendix 1: Proof of Theorem 4.2 / 123
Appendix 2: Proof of Theorem 4.4 / 126
Appendix 3: Proof of Theorem 4.5 / 128
Appendix 4: Proof of Theorem 4.6 / 128
Appendix 5: Proof of Theorem 4.8 / 130
Appendix 6: Proof of Theorem 4.10 / 132
Appendix 7: Proof of Theorem 4.11 / 134
Appendix 8: Proof of Theorem 4.12 / 136

5. Power Spectral Density of Standard Random Processes — Part 1   138

5.1 Introduction / 138
5.2 Signaling Random Processes / 138
5.3 Digital to Analogue Converter Quantization / 152
5.4 Jitter / 155
5.5 Shot Noise / 160
5.6 Generalized Signaling Processes / 166
Appendix 1: Proof of Theorem 5.1 / 168
Appendix 2: Proof of Theorem 5.2 / 171
Appendix 3: Proof of Equation 5.73 / 173
Appendix 4: Proof of Theorem 5.3 / 174

Appendix 5: Proof of Theorem 5.4 / 176
Appendix 6: Proof of Theorem 5.5 / 177

6. Power Spectral Density of Standard Random Processes — Part 2 — 179

- 6.1 Introduction / 179
- 6.2 Sampled Signals / 179
- 6.3 Quadrature Amplitude Modulation / 185
- 6.4 Random Walks / 192
- 6.5 1/f Noise / 198
  Appendix 1: Proof of Theorem 6.1 / 200
  Appendix 2: Proof of Theorem 6.2 / 201
  Appendix 3: Proof of Theorem 6.3 / 202
  Appendix 4: Proof of Equation 6.39 / 204

7. Memoryless Transformations of Random Processes — 206

- 7.1 Introduction / 206
- 7.2 Power Spectral Density after a Memoryless Transformation / 206
- 7.3 Examples / 211
  Appendix 1: Proof of Theorem 7.1 / 223
  Appendix 2: Fourier Results for Raised Cosine Frequency Modulation / 224

8. Linear System Theory — 229

- 8.1 Introduction / 229
- 8.2 Impulse Response / 230
- 8.3 Input-Output Relationship / 232
- 8.4 Fourier and Laplace Transform of Output / 232
- 8.5 Input-Output Power Spectral Density Relationship / 238
- 8.6 Multiple Input-Multiple Output Systems / 243
  Appendix 1: Proof of Theorem 8.1 / 246
  Appendix 2: Proof of Theorem 8.2 / 248
  Appendix 3: Proof of Theorem 8.3 / 249
  Appendix 4: Proof of Theorem 8.4 / 251
  Appendix 5: Proof of Theorem 8.6 / 252
  Appendix 6: Proof of Theorem 8.7 / 253
  Appendix 7: Proof of Theorem 8.8 / 255

9. Principles of Low Noise Electronic Design — 256

- 9.1 Introduction / 256
- 9.2 Gaussian White Noise / 259
- 9.3 Standard Noise Sources / 264
- 9.4 Noise Models for Standard Electronic Devices / 266
- 9.5 Noise Analysis for Linear Time Invariant Systems / 269

9.6 Input Equivalent Current and Voltage Sources / 278
9.7 Transferring Noise Sources / 282
9.8 Results for Low Noise Design / 285
9.9 Noise Equivalent Bandwidth / 285
9.10 Power Spectral Density of a Passive Network / 287
   Appendix 1: Proof of Theorem 9.2 / 291
   Appendix 2: Proof of Theorem 9.4 / 294
   Appendix 3: Proof of Conjecture for Ladder Structure / 296

Notation 300

References 302

Index 307

Preface

This book gives a systematic account of the Power Spectral Density and details the application of this theory to Communications and Electronics. The level of the book is suited to final year Electrical and Electronic Engineering students, post-graduate students and researchers.

This book arises from the author's research experience in low noise amplifier design and analysis of random processes.

The basis of the book is the definition of the power spectral density using results directly from Fourier theory rather than the more popular approach of defining the power spectral density in terms of the Fourier transform of the autocorrelation function. The difference between use of the two definitions, which are equivalent with an appropriate definition for the autocorrelation function, is that the former greatly facilitates analysis, that is, the determination of the power spectral density of standard signals, as the book demonstrates. The strength, and uniqueness, of the book is that, based on a thorough account of signal theory, it presents a comprehensive and straightforward account of the power spectral density and its application to the important areas of communications and electronics.

The following people have contributed to the book in various ways. First, Prof. J. L. Hullett introduced me to the field of low noise electronic design and has facilitated my career at several important times. Second, Prof. L. Faraone facilitated and supported my research during much of the 1990s. Third, Prof. A. Cantoni, Dr. Y. H. Leung and Prof. K. Fynn supported my research from 1995 to 1997. Fourth, Mr. Nathanael Rensen collaborated on a research project with me over the period 1996 to early 1998. Fifth, Prof. A. Zoubir has provided collegial support and suggested that I contact Dr. P. Meyler from John Wiley & Sons with respect to publication. Sixth, Dr. P. Meyler,

Ms. Melissa Yanuzzi and staff at John Wiley have supported and efficiently managed the publication of the book. Finally, several students—undergraduate and postgraduate—have worked on projects related to material in the book and, accordingly, have contributed to the final outcome.

I also wish to thank the staff at Kiribilli Cafe, the Art Gallery Cafe and the King Street Cafe for their indirect support whilst a significant level of editing was in progress. Finally, family and friends will hear a little less about 'The Book' in the future.

ROY M. HOWARD

December 2001

About the Author

Roy M. Howard has been awarded a BE and a Ph.D. in Electrical Engineering, as well as a BA (Mathematics and Philosophy), by the University of Western Australia. Currently he is a Senior Lecturer in the School of Electrical and Computer Engineering at Curtin University of Technology, Perth, Australia. His research interests include signal theory, stochastic modeling, $1/f$ noise, low noise amplifier design, nonlinear systems, and nonlinear electronics.

1
Introduction

Random phenomena have their basis in the nature of the physical order (e.g., the nature of electron movement) and limit the performance of many systems including electronic and communication systems. For example, the minimum sensitivity of an amplifier and the distance a signal can be transmitted and recovered, are both limited by random signal variations. On the other hand, there are applications where introduced randomness will enhance aspects of system performance. One example is where a low level randomly varying waveform is added to a repetitive signal to improve the resolution in signal values obtained by an analogue to digital converter, and, after averaging (Potzick, 1999; Gray, 1993). Further, in recent years there has been increasing interest in stochastic resonance which occurs when the system response to a weak periodic signal is enhanced by an increase in the level of random variations associated with the system (Luchinsky, 1999; Hanggi, 2000).

The importance of random phenomena has led to an increasing number of theoretical results as can be found in books such as, Gardner (1990), Papoulis (2002), and Taylor (1998). In communications and electronics a standard way of characterizing random phenomenon is through a power spectral density which, for example, facilitates derivation of the signal to noise ratio of a system operating under prescribed conditions. There are two standard approaches for defining the power spectral density. First, there is a direct Fourier approach. Second, and more commonly, an approach based on the Fourier transform of an autocorrelation function.

With the direct Fourier approach the power spectral density of a single signal x, for the interval $[0, T]$, is defined as

$$G(T, f) = \frac{|X(T, f)|^2}{T} \tag{1.1}$$

where X is the Fourier transform of x evaluated over the interval $[0, T]$. The alternative approach is to determine the autocorrelation of the signal, defined as

$$R(T, t, \tau) = \begin{cases} x(t)x^*(t-\tau) & t \in [0, T], t-\tau \in [0, T] \\ 0 & \text{elsewhere} \end{cases} \quad (1.2)$$

and then take a time average to form an averaged autocorrelation function:

$$\bar{R}(T, \tau) = \begin{cases} \dfrac{1}{T} \displaystyle\int_0^{T+\tau} R(T, t, \tau)\, dt & \tau < 0 \\ \dfrac{1}{T} \displaystyle\int_\tau^{T} R(T, t, \tau)\, dt & \tau > 0 \end{cases} \quad (1.3)$$

Finally, the Fourier transform of this function is taken to obtain the power spectral density, that is,

$$G(T, f) = \int_{-T}^{T} \bar{R}(T, \tau) e^{-j2\pi f \tau}\, d\tau \quad (1.4)$$

These two approaches lead to identical power spectral density functions where the definitions can be readily generalized for random processes and the infinite time interval. Analytically, the Fourier approach is more direct and leads directly to the interpretation of the power spectral density, at a given frequency, as being proportional to the power in the constituent sinusoidal signal with that frequency. Further, the direct nature of the Fourier approach facilitates the derivation of the power spectral density of signals and random processes.

The following chapters give a systematic account of the theory related to the direct Fourier approach to defining and evaluating the power spectral density. This theory is applied to the derivation of the power spectral density of the random processes commonly encountered in communications and electronics, noise analysis in linear electronic systems, and memoryless transformations of random processes.

Chapter 2 gives appropriate background theory for this book, while Chapter 3 gives a detailed discussion of the two alternative ways the power spectral density can be defined and the equivalence between them. Chapter 4 gives important results that facilitate the derivation of the power spectral density. Chapter 5 and 6 detail the derivation of the power spectral density of standard random processes encountered in communications and electronics. Chapter 7 details an approach for ascertaining the power spectral density of random processes after a nonlinear memoryless transformation. Chapter 8 discusses the relationship between the input and output signals, and input and output power spectral densities of a linear time invariant system. This chapter gives the necessary background material for Chapter 9, which details the characterization of standard noise signals that occur in electronic devices, and how analysis of such noise signals can be carried out to quantify, and hence, minimize the noise of a linear electronic system.

2

Background: Signal and System Theory

2.1 INTRODUCTION

The power spectral density arises from signal analysis of deterministic signals, and random processes, and is required to be evaluated over both the finite and infinite time intervals. While signal analysis for the finite case, for example, the integral on a finite interval of a finite summation of bounded signals, causes few problems, signal analysis for the infinite case is more problematic. For example, it can be the case that the order of the integration and limit operators cannot be interchanged. With the infinite case, careful attention to detail and a reasonable knowledge of underlying mathematical theory is required. Clarity is best achieved for integration, for example, through measure theory and Lebesgue integration.

This chapter gives the necessary mathematical background for the development and application, of theory related to the power spectral density that follows in subsequent chapters. First, a review of fundamental results from set theory, real and complex analysis, signal theory and system theory is given. This is followed by an overview of measure and Lebesgue integration, and associated results. Finally, consistent with the requirements of subsequent chapters, results from Fourier theory and a brief introduction to random process theory are given.

2.2 BACKGROUND THEORY

2.2.1 Set Theory

Set theory is fundamental to mathematical analysis, and the following results from set theory are consistent with subsequent analysis. Useful references for set theory include Sprecher (1970), Lipschutz (1998), and Epp (1995).

DEFINITION: SET   A set is a collection of distinct entities.

The notation $\{\alpha_1, \alpha_2, \ldots, \alpha_N\}$ is used for the set of distinct entities $\alpha_1, \alpha_2, \ldots, \alpha_N$. The notation $\{x: f(x)\}$ is used for the set of elements x for which the property $f(x)$ is true. The notation $x \in S$ means that the entity denoted x is an element of the set S. The empty set $\{\,\}$ is denoted by $\emptyset$. The complement of a set S, denoted S^C, is defined as $S^C = \{x: x \notin S\}$, where S is usually a subset of a large set — often the "universal set." The union and intersection of two sets are defined as follows:

$$A \cup B = \{x: x \in A \text{ or } x \in B\}$$
$$A \cap B = \{x: x \in A \text{ and } x \in B\} \tag{2.1}$$

DEFINITION: CHARACTERISTIC FUNCTION OF A SET   The characteristic function of a set S is defined according to

$$\chi_S(x) = \begin{cases} 1 & x \in S \\ 0 & x \notin S \end{cases} \tag{2.2}$$

DEFINITION: ORDERED PAIR AND CARTESIAN PRODUCT   An ordered pair, denoted (x_1, x_2), where $x_1 \in A$ and $x_2 \in B$, is the set $\{x_1, \{x_1, x_2\}\}$. This definition clearly indicates, for example, that $(x_1, x_2) \neq (x_2, x_1)$ when $x_1 \neq x_2$. The Cartesian product of two sets A and B, denoted $A \times B$, is defined as the set of all possible ordered pairs from these sets, that is,

$$A \times B = \{(x, y): x \in A, y \in B\} \tag{2.3}$$

DEFINITION: SUPREMUM AND INFIMUM   The supremum of a set A of real numbers, denoted $\sup\{A\}$, is the least upper bound of that set. The infimum of a set A of real numbers, denoted $\inf(A)$, is the greatest lower bound of that set. Formally, $\sup(A)$ is such that (Marsden, 1993 p. 45)

$$\sup(A) \geq x \quad \forall x \in A$$
$$\forall \varepsilon > 0 \quad \exists x \in A \quad \text{s.t.} \quad \sup(A) - x < \varepsilon \tag{2.4}$$

Similarly, $\inf(A)$ is such that

$$\inf(A) \leq x \quad \forall x \in A$$
$$\forall \varepsilon > 0 \quad \exists x \in A \quad \text{s.t.} \quad x - \inf(A) < \varepsilon \tag{2.5}$$

DEFINITION: PARTITION   The set $\{I_1, \ldots, I_N\}$, where $I_i \cap I_j = \emptyset$ for $i \neq j$ and $\bigcup_{i=1}^{N} I_i = I$, is a partition of the set I.

An equivalent relationship generates a partition of a set (Sprecher, 1970 p. 14; Epp, 1995 p. 558).

Finally, set theory is not without its problems. For example, associated with set theory is Russell's paradox and Cantor's paradox (Epp, 1995 p. 268; Lipschutz, 1998 p. 222).

2.2.2 Real and Complex Analysis

The following, gives a review of real and complex analysis consistent with the development of subsequent theory. Useful references for real analysis include Sprecher (1970) and Marsden (1993), while useful references for complex analysis include Marsden (1987) and Brown (1995).

Real analysis has its basis in the natural numbers, denoted N and defined as

$$N = \{1, 2, 3, \ldots\} \tag{2.6}$$

To this set can be added the number zero and the negative of all the numbers in N to form the set of integers, denoted Z, that is,

$$Z = \{\ldots, -3, -2, -1, 0, 1, 2, 3, \ldots\} \tag{2.7}$$

The set of positive integers Z^+ is defined as being equal to N. The set of rational numbers, denoted Q, readily follows:

$$Q = \{p/q : p, q \in Z, q \neq 0, \gcd(p, q) = 1\} \tag{2.8}$$

where gcd is the greatest common divisor function. The set of rational numbers, however, is not "complete", in the sense that it does not include useful numbers such as the length of the hypotenuse of a right triangle whose sides have unity length, or the area of a circle of unit radius, etc. "Completing" the set of rational numbers to yield the familiar set of real numbers, denoted R, can be achieved in two ways. First, through the limit of sequences of rational numbers. Consistent with this approach, a real number can be considered to be the limit of a sequence of rational numbers that converge. For example, the real number 2 is the limit of the sequence $\{2, 2, 2, \ldots\}$, while $\sqrt{2}$ is the limit of the sequence $\{1, 7/5, 141/100, 707/500, \ldots\}$ and so on. Strictly speaking, a real number is an equivalence class associated with a Cauchy sequence of rational numbers (Sprecher, 1970 Ch. 3). Second, through use of a partition (Dedekind cut) of the set of rational numbers into two sets (Dedekind sections). The point of partition is associated with a real number (Ball, 1973 p. 22). For example, the partition of Q according to

$$\{\{x : x \in Q, x \leqslant 0 \text{ or } x^2 < 2\}, \{x : x \in Q, x > 0 \text{ and } x^2 > 2\}\} \tag{2.9}$$

defines the real number $\sqrt{2}$.

Algebra on the real numbers is defined through axioms that are of two types (Sprecher, 1970 p. 37; Marsden, 1993 p. 26). First, there are "field" axioms that specify the arithmetic operations of addition and multiplication and appropriate additive and multiplicative identity elements. Second, there are "order" axioms that specify the order qualities of real numbers, such as equality, greater than, and less than. The set of real numbers is an "ordered field."

The set of complex numbers, denoted C, is the set of possible ordered pairs that can be generated from real numbers, that is,

$$C = \{(\alpha, \beta): \alpha, \beta \in R\} \qquad (2.10)$$

When representing a complex number in the plane the notation $(x, y) = x + jy$ is used where $j = (0, 1)$. The algebra of complex numbers is governed by the rules of vector addition and scalar multiplication, that is,

$$(x_1, y_1) + (x_2, y_2) = (x_1 + x_2, y_1 + y_2)$$
$$a(x_1, y_1) = (ax_1, ay_1) \quad a \in R \qquad (2.11)$$
$$(x_1, y_1)(x_2, y_2) = (x_1 x_2 - y_1 y_2, x_1 y_2 + y_1 x_2)$$

From these definitions, the familiar result of $j^2 = -1$, or $j = \sqrt{-1}$, follows.

The conjugate of a complex number (x, y), by definition, is $(x, -y)$.

DEFINITION: COUNTABLE AND UNCOUNTABLE SETS   A set is a countable set if each element of the set can be associated, uniquely, with an element of N (Sprecher, 1970 p. 29). If such an association is not possible, then the set is an uncountable set.

The sets N, Z, and Q are countable sets. The sets R and C are uncountable sets.

DEFINITION. INTERVALS   If α and β are distinct real numbers with $\alpha < \beta$, then the following sets of points of R, denoted intervals, can readily be defined:

$$\begin{aligned}
[\alpha, \beta] &= \{x: \alpha \leqslant x \leqslant \beta\} & \text{closed interval} \\
(\alpha, \beta) &= \{x: \alpha < x < \beta\} & \text{open interval} \\
[\alpha, \beta) &= \{x: \alpha \leqslant x < \beta\} & \text{closed/open interval} \\
(\alpha, \beta] &= \{x: \alpha < x \leqslant \beta\} & \text{open/closed interval}
\end{aligned} \qquad (2.12)$$

DEFINITION: NEIGHBORHOOD   A neighborhood (NBHD) of a point $x \in R$ is the open interval $(x - \delta, x + \delta)$ where $\delta > 0$ (Sprecher, 1970 p. 79).

DEFINITION: A CONTIGUOUS PARTITION   The set of intervals $\{I_1, \ldots, I_N\}$ is a contiguous partition of the interval I if $\{I_1, \ldots, I_N\}$ is a partition of I and the intervals are ordered such that

$$t \in I_i \Rightarrow t < t_x \qquad \forall t_x \in I_{i+1},\ i \in \{1, \ldots, N-1\} \qquad (2.13)$$

2.3 FUNCTIONS, SIGNALS, AND SYSTEMS

Signal and system theory form the basis for a significant level of subsequent analysis. Appropriate definitions and discussion follows. A useful reference for signal theory is Franks (1969).

DEFINITION: FUNCTION OR MAPPING A function, f, is a mapping from a set D, the domain, to a set R, the range, such that only one element in the range is associated with each element in the domain. Such a function is written as $f: D \to R$. If $y \in R$ and $x \in D$ with x mapping to y under f, then the notation $y = f(x)$ is used (Sprecher, 1970 p. 16).

Note, a function is a special type of relationship between elements from two sets. A "relation," for example, is a more general relationship (Smith, 1990 ch. 3; Polimeni, 1990 ch. 4).

DEFINITION: SIGNAL A real and continuous signal is a function from R, or a subset of R, to R, or a subset of R. A real and discrete signal is a function from Z, or a subset of Z, to R, or a subset of R.

The term "continuous" used here is not related to the concept of continuity. A continuous signal can be represented, for example diagrammatically, as shown in Figure 2.1. Commonly, a real function is implicitly defined by its graph which is a display, for the continuous case, of the set of points $\{(t, f(t)): t \in R\}$. In many instances the variable t denotes time.

A complex signal is a mapping from R, or a subset of R, to C, or a subset of C.

DEFINITION: SYSTEM In the context of engineering, a system is an entity which produces an output signal, usually in response to an input signal which is transformed in some manner. An autonomous system is one which produces an output signal when there is no input signal. Chaotic systems and oscillators are examples of autonomous systems.

DEFINITION: OPERATOR A system which produces an output signal in response to an input signal can be modeled by an operator, F, as illustrated in

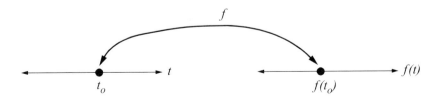

Figure 2.1 Mapping involved in a continuous real function.

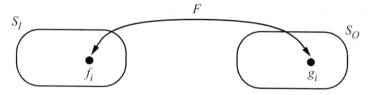

Figure 2.2 Mapping produced by a system.

Figure 2.2. In this figure, S_I is the set of possible input signals, and S_O is the set of possible output signals. Hence, the operator is a mapping from S_I to S_O, that is, $F: S_I \to S_O$.

DEFINITION: CONJUGATION OPERATOR   A conjugation operator, F_C, is a mapping from the set of complex signals $\{f: R \to C\}$ to the same set of complex signals, and is defined according to $F_C[f] = f^*$, where $f^*(t) = x(t) - jy(t)$ when $f(t) = x(t) + jy(t)$. Here, the signals x and y are real signals, that is, mappings from R to R.

2.3.1 Disjoint and Orthogonal Signals

DEFINITION: DISJOINT SIGNALS   Two signals $f_1: R \to C$ and $f_2: R \to C$ are disjoint on the interval I, if

$$\forall t \in I \quad f_1(t) f_2(t) = 0 \tag{2.14}$$

DEFINITION: SET OF DISJOINT SIGNALS   A set of real or complex signals $\{f_1, \ldots, f_N\}$ is a set of disjoint signals on the interval I, if they are pairwise disjoint, that is,

$$\forall t \in I, \; i \neq j \quad f_i(t) f_j(t) = 0 \tag{2.15}$$

DEFINITION: ORTHOGONALITY   Two signals $f_1: R \to C$ and $f_2: R \to C$ are orthogonal on an interval I, if

$$\int_I f_1(t) f_2^*(t) \, dt = 0 \tag{2.16}$$

Clearly, disjointness implies orthogonality. Note, orthogonality is defined, in general, via an inner product on elements of an "inner product space" or a Hilbert space (Debnath, 1999 ch. 3; Kresyzig, 1978 ch. 3).

DEFINITION: ORTHOGONAL SET A set of signals $\{f_i: \mathbf{R} \to \mathbf{C},\ i \in \mathbf{Z}^+\}$ is an orthogonal set on an interval I, if the signals are pairwise orthogonal, that is,

$$\int_I f_i(t) f_j^*(t)\, dt = 0 \qquad i \neq j \tag{2.17}$$

The most widely used orthogonal sets for an interval $[\alpha, \beta]$ are the sets

$$\left\{1,\ \cos(2\pi i f_o t),\ \sin(2\pi i f_o t):\ i \in \mathbf{Z}^+,\ f_o = \frac{1}{\beta - \alpha}\right\} \tag{2.18}$$

$$\left\{e^{j2\pi i f_o t}:\ i \in \mathbf{Z},\ f_o = \frac{1}{\beta - \alpha}\right\} \tag{2.19}$$

THEOREM 2.1. SIGNAL DECOMPOSITION Any signal $f: I \to \mathbf{C}$ can be written as the sum of disjoint waveforms, from a disjoint set $\{f_1, \ldots, f_N\}$, according to

$$f(t) = \sum_{i=1}^{N} f_i(t) \qquad \text{where} \qquad f_i(t) = \begin{cases} f(t) & t \in I_i \\ 0 & \text{elsewhere} \end{cases} \tag{2.20}$$

and $\{I_1, \ldots, I_N\}$ is a partition of I.

Proof. The proof of this result follows directly from the definition of a partition, the definition of set of disjoint waveforms, and by construction.

Signal decomposition using orthogonal basis sets is widely used. A common example is signal decomposition to generate the Fourier series of a signal. Such decomposition is best formulated through use of an inner product on a Hilbert space (Kreyszig, 1978 ch. 3; Debnath, 1999 ch. 3).

2.3.2 Types of Systems and Operators

The following paragraphs define several types of systems commonly encountered in engineering. In terms of notation, the ith input signal is denoted f_i and the corresponding output signal is denoted g_i.

(a) In general, there may not be an explicit rule defining the mapping between input and output signals produced by a system. In such a case, the relationship between input and output signals can be explicitly stated in a one-to-one manner according to

$$f_1 \to g_1 \qquad f_2 \to g_2 \ldots \tag{2.21}$$

(b) *Linear systems.* A linear system is one that can be characterized by an operator L which exhibits the properties of superposition and homogeneity, that is,

$$L[\alpha f_i(t) + \beta f_j(t)] = \alpha L[f_i(t)] + \beta L[f_j(t)] \quad (2.22)$$

(c) *Memoryless systems*. A memoryless system is one where the relationship between the input and output signals can be explicitly defined by an operator F, such that

$$g_i = F[f_i] \quad (2.23)$$

An example of such a system is one defined by $F(f) = f^2$ that implies $g_i(t) = f_i^2(t)$.

(d) *Argument altering systems*. Another class of systems is where the relation between input and output signals can be explicitly written in the form

$$g_i(t) = f_i(G[t]) \quad (2.24)$$

for some function G. An example of such a system is a delay system, defined by the operator F according to $F[f(t)] = f[G(t)] = f(t - t_d)$, where $G(t) = t - t_d$. Consistent with such a definition $g_i(t) = f_i(t - t_d)$.

(e) Combining the memoryless and argument operators, another class of system can be defined, using an operator F and a function G, according to

$$g_i(t) = F[f_i(G[t])] \quad (2.25)$$

An example of such a system is one where $g_i(t) = f_i^2(t - t_d)$.

(f) A generalization of the memoryless but argument altering system, is one where

$$g_i(t) = \sum_{j=1}^{N} F_j[f_i(G_j[t])] \quad (2.26)$$

An example of such a system is one described by the convolution operator according to

$$g_i(t) = \int_0^t f_i(\lambda) h(t - \lambda)\, d\lambda = \int_0^t f_i(t - \lambda) h(\lambda)\, d\lambda \quad (2.27)$$

As the integral is the limit of a sum, it follows that

$$g_i(t) = \lim_{\Delta t \to 0} \Delta t \sum_{j=1}^{\lfloor t/\Delta t \rfloor} f_i(t - j\Delta t) h(j\Delta t) \quad (2.28)$$

Hence, the convolution can be written as

$$g_i(t) = \lim_{\Delta t \to 0} \Delta t \sum_{j=1}^{\lfloor t/\Delta t \rfloor} F_j[f_i(G_j[t])] \quad (2.29)$$

where $G_j[t] = t - j\Delta t$ and $F_j[f_i] = h(j\Delta t) f_i$.

(g) *Implicitly characterized systems.* Systems characterized by, for example, differential equations result in implicit operator definitions. For example, consider the system defined by the differential equation

$$\frac{dg_i(t)}{dt} + G[g_i(t)] = F[f_i(t)] \quad (2.30)$$

With D denoting the differentiation operator, the system can be defined as

$$(D + G)(g_i) = F(f_i) \quad (2.31)$$

2.3.3 Defining Output Signal from a Memoryless System

Consider, as shown in Figure 2.3, a memoryless system defined by the operator F. Such a operator can be written in terms of a set of disjoint operators according to

$$F(f) = \sum_{i=1}^{N} F_i(f) \quad \text{where} \quad F_i(f) = \begin{cases} F(f) & f \in [f_{i-1}, f_i) \\ 0 & \text{elsewhere} \end{cases} \quad (2.32)$$

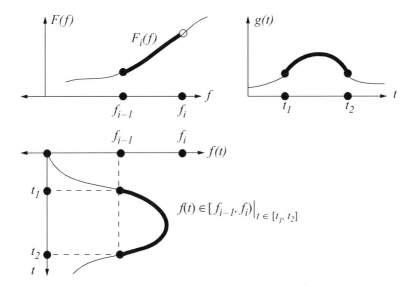

Figure 2.3 Input and output signal of a memoryless system.

The output signal, g, of such a system, in response to an input signal f, can then be determined, consistent with the illustration in Figure 2.3, according to

$$g(t) = F(f(t)) = \sum_{i=1}^{N} F_i(f(t)) \qquad (2.33)$$

or in terms of specific time intervals:

$$g(t) = \begin{cases} F_1(f(t)) & t \in I_1 \quad I_1 = \{t: f(t) \in [f_0, f_1)\} \\ F_2(f(t)) & t \in I_2 \quad I_2 = \{t: f(t) \in [f_1, f_2)\} \\ \vdots & \end{cases} \qquad (2.34)$$

Such a characterization is well-suited to a piecewise linear memoryless system.

2.3.3.1 Decomposition of Output Using Time Partition The input signal, f, to a memoryless nonlinear system can be written, over an interval I, as a summation of disjoint waveforms, that is,

$$f(t) = \sum_{i=1}^{N} f_i(t) \qquad f_i(t) = \begin{cases} f(t) & t \in I_i \\ 0 & \text{elsewhere} \end{cases} \qquad (2.35)$$

where $\{I_1, \ldots, I_N\}$ is a partition of I. It then follows, by using this partition of I, that the output signal can be written as a summation of disjoint waveforms according to

$$g(t) = \sum_{i=1}^{N} g_i(t) \qquad g_i(t) = \begin{cases} g(t) & t \in I_i \\ 0 & t \notin I_i \end{cases} \qquad (2.36)$$

The relationship between the ith disjoint output waveform and the input waveform is

$$g_i(t) = \begin{cases} F(f_i(t)) & t \in I_i \\ 0 & t \notin I_i \end{cases} \qquad (2.37)$$

This result is easily proved by noting the following:

$$g_i(t) = \begin{cases} g(t) & t \in I_i \\ 0 & t \notin I_i \end{cases} = \begin{cases} F(f(t)) & t \in I_i \\ 0 & t \notin I_i \end{cases} = \begin{cases} F(f_i(t)) & t \in I_i \\ 0 & t \notin I_i \end{cases} \qquad (2.38)$$

2.4 SIGNAL PROPERTIES

To establish precise criteria for the validity of various signal relationships related to the power spectral density, precise definitions for basic signal properties such as continuity, differentiability, piecewise smoothness, boundedness, bounded variation, and absolute continuity are required. These properties are detailed in this section. First, however, definitions for signal energy and signal power are given.

DEFINITION: SIGNAL ENERGY AND SIGNAL POWER The energy and average power of a signal $f: \mathbf{R} \to \mathbf{C}$ on an interval $[\alpha, \beta]$, respectively, are defined as

$$E = \int_\alpha^\beta |f(t)|^2 \, dt \qquad \bar{P} = \frac{1}{\beta - \alpha} \int_\alpha^\beta |f(t)|^2 \, dt \qquad (2.39)$$

2.4.1 Piecewise Continuity and Continuity

DEFINITION: LEFT AND RIGHT HAND CONTINUITY AT A POINT A function is right continuous at a point t_o if the right limit, $f(t_o^+)$, defined as follows, exists:

$$f(t_o^+) = \lim_{\delta \to 0} f(t_o + \delta) \qquad \delta > 0 \qquad (2.40)$$

Similarly, a function is left continuous at a point t_o if the left limit, $f(t_o^-)$, defined as follows, exists:

$$f(t_o^-) = \lim_{\delta \to 0} f(t_o - \delta) \qquad \delta > 0 \qquad (2.41)$$

DEFINITION: PIECEWISE CONTINUITY AT A POINT A function f is piecewise continuous at a point t_o if the left and right limits, $f(t_o^-)$ and $f(t_o^+)$, exist, that is,

$$\forall \varepsilon > 0 \quad \exists \delta_o > 0 \quad \text{s.t.} \quad 0 < \delta < \delta_o \Rightarrow |f(t_o + \delta) - f(t_o^+)| < \varepsilon \qquad (2.42)$$

$$\forall \varepsilon > 0 \quad \exists \delta_o > 0 \quad \text{s.t.} \quad 0 < \delta < \delta_o \Rightarrow |f(t_o^-) - f(t_o - \delta)| < \varepsilon \qquad (2.43)$$

and $f(t_o) \in \{f(t_o^-), f(t_o^+)\}$. Here, s.t. is an abbreviation for "such that." The last requirement excludes functions, such as

$$f(t) = \begin{cases} \infty & t = t_o \\ k & t \neq t_o \end{cases} \quad \text{or} \quad f(t) = \begin{cases} k_o & t = t_o \\ k & t \neq t_o, k \neq k_o \end{cases} \qquad (2.44)$$

from being piecewise continuous at t_o.

DEFINITION: PIECEWISE CONTINUITY ON AN INTERVAL A function f is piecewise continuous over an interval I, if it is piecewise continuous at all points in the interval I. For a closed interval $[\alpha, \beta]$ right continuity is required at α while left continuity is required at β.

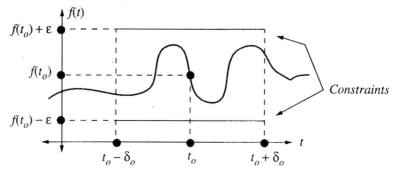

Figure 2.4 Constraints on a function imposed by continuity.

DEFINITION: CONTINUITY AT A POINT A function $f: \mathbf{R} \to \mathbf{C}$ is continuous at a point t_o if it is both left and right continuous at that point, and the left and right limits are equal to the function at the point (Jain, 1986 p. 12), that is,

$$\forall \varepsilon > 0 \quad \exists \delta_o > 0 \quad \text{s.t.} \quad \forall |\delta| < \delta_o \quad |f(t_o + \delta) - f(t_o)| < \varepsilon \quad (2.45)$$

or

$$\forall \varepsilon > 0 \quad \exists \delta_o > 0 \quad \text{s.t.} \quad \forall |\delta| < \delta_o \quad f(t_o) - \varepsilon < f(t_o + \delta) < f(t_o) + \varepsilon \quad (2.46)$$

Consistent with this last equation, continuity implies the function f is constrained around t_o, as shown in Figure 2.4.

DEFINITION: POINTWISE CONTINUITY ON AN INTERVAL A function f is pointwise continuous over an interval I, if it is continuous at all points in the interval I. For a closed interval $[\alpha, \beta]$, right continuity is required at α, while left continuity is required at β with $f(\alpha^+) = f(\alpha)$, and $f(\beta^-) = f(\beta)$.

DEFINITION: UNIFORM CONTINUITY ON AN INTERVAL A function is uniformly continuous over an interval I if (Jain, 1986 p. 13)

$$\forall \varepsilon > 0 \ \exists \delta_o > 0 \quad \text{s.t.} \quad \forall |\delta| < \delta_o \quad |f(t_o + \delta) - f(t_o)| < \varepsilon \quad (2.47)$$

where δ_o is independent of the value of $t_o \in I$ and, close to the end points of the interval, δ is such that $t_o + \delta \in I$.

THEOREM 2.2. UNIFORM AND POINTWISE CONTINUITY *Uniform continuity implies pointwise continuity but the converse is not true. For a closed interval $[\alpha, \beta]$, pointwise continuity on (α, β), right continuity at α and left continuity at β imply uniform continuity on $[\alpha, \beta]$.*

Proof. It is clear from the definition of uniform continuity that it implies pointwise continuity. To illustrate why the converse is not true, consider the function $f(t) = 1/t$ which is pointwise continuous, but not uniformly continuous, on the interval $(0, 1)$.

To prove the second result, consider a fixed $\varepsilon > 0$. Pointwise continuity on the interval implies that it is possible to choose N numbers $\delta_1, \ldots, \delta_N$, and N points $t_1, \ldots, t_N$, where $t_1 = \alpha$, $t_{i+1} > t_i$ and $t_N = \beta$, such that $t_i + \delta_i > t_{i+1} - \delta_{i+1}$, and it is the case that

$$|f(t_i + \delta) - f(t_i)| < \varepsilon \quad \begin{cases} \forall |\delta| < \delta_i, t_i + \delta \in [\alpha, \beta] \\ i \in \{1, \ldots, N\} \end{cases} \quad (2.48)$$

Appropriate left- and right-hand limits are assumed for $t_1 = \alpha$ and $t_N = \beta$. The intervals $[t_1, t_1 + \delta_1)$, $(t_i - \delta_i, t_i + \delta_i)$ for $i \in \{2, \ldots, N-1\}$, and $(t_N - \delta_N, t_N]$ "cover" the interval $[\alpha, \beta]$, and with the definition $\delta_{min} = \inf\{\delta_1, \ldots, \delta_N\}$ it follows that

$$\forall |\delta| < \delta_{min}, \; t + \delta \in [\alpha, \beta] \quad |f(t + \delta) - f(t)| < \varepsilon \quad (2.49)$$

which implies uniform continuity as required.

2.4.2 Differentiability and Piecewise Smoothness

DEFINITION: DIFFERENTIABILITY A function f is differentiable at t_o iff

$$\lim_{\delta \to 0} \left[\frac{f(t_o + \delta) - f(t_o)}{\delta} \right]$$

exists. This limit is denoted $f'(t_o)$ and exists if $f'(t_o)$ is such that

$$\forall \varepsilon > 0 \; \exists \delta_o > 0 \; \text{s.t.} \; 0 < |\delta| < \delta_o \Rightarrow \left| \frac{f(t_o + \delta) - f(t_o)}{\delta} - f'(t_o) \right| < \varepsilon \quad (2.50)$$

The requirement of differentiability constrains a function for the interval $(t_o - \delta, t_o + \delta)$ such that, as shown in Figure 2.5, it lies between the lines f_1 and f_2 defined according to

$$f_1(t) = f(t_o) + (t - t_o)[f'(t_o) + \varepsilon] \quad (2.51)$$
$$f_2(t) = f(t_o) + (t - t_o)[f'(t_o) - \varepsilon] \quad (2.52)$$

These constraining lines arise from writing the inequality in Eq. (2.50) in the form

$$|f(t_o + \delta) - f(t_o) - \delta f'(t_o)| < \varepsilon |\delta| \quad (2.53)$$

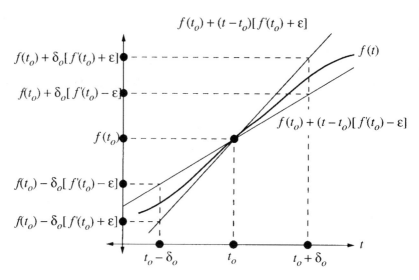

Figure 2.5 Constraints on a function consistent with differentiability at a point t_o.

and, equivalently, as

$$f(t_o) + \delta[f'(t_o) \mp \varepsilon] < f(t_o + \delta) < f(t_o) + \delta[f'(t_o) \pm \varepsilon] \quad (2.54)$$

where the choice of $\pm$ depends on whether $\delta < 0$ or $\delta > 0$. With $\delta = t - t_o$ the required result follows.

Clearly, differentiability when compared with continuity, places a higher degree of constraint on the variation of a function around a point t_o. Further, provided $f'(t_o)$ is nonzero, it is possible to choose ε, such that $\varepsilon \ll |f'(t_o)|$ whereupon it follows for $t_o - \delta < t < t_o + \delta$, that the function f can be approximated by the first-order Taylor series expansion:

$$f(t) \approx f(t_o) + (t - t_o)f'(t_o) \quad (2.55)$$

DEFINITION: PIECEWISE DIFFERENTIABILITY OR PIECEWISE SMOOTHNESS A function f is piecewise differentiable, or piecewise smooth at t_o iff the left- and right-hand derivatives defined according to (Champeney, 1987 p. 42)

$$\begin{aligned} f'(t_o^+) &= \lim_{\delta \to 0} \left[\frac{f(t_o + \delta) - f(t_o^+)}{\delta} \right] \quad \delta > 0 \\ f'(t_o^-) &= \lim_{\delta \to 0} \left[\frac{f(t_o^-) - f(t_o - \delta)}{\delta} \right] \quad \delta > 0 \end{aligned} \quad (2.56)$$

exist. The assumption in these definitions is that left- and right-hand limits $f(t_o^+)$ and $f(t_o^-)$ also exist. As for the case of piecewise continuity, the additional constraint $f(t_o) \in \{f(t_o^-), f(t_o^+)\}$ is included in the definition.

Piecewise smoothness at a point t_o constrains a function for the case where $f'(t_o^-)$ and $f'(t_o^+)$ are nonzero, such that it can be approximated by the first-order Taylor series expansions either side of the point; that is,

$$f(t) \approx f(t_o^+) + (t - t_o)f'(t_o^+) \qquad t_o < t < t_o + \delta, \delta > 0 \qquad (2.57)$$

$$f(t) \approx f(t_o^-) + (t - t_o)f'(t_o^-) \qquad t_o - \delta < t < t_o, \delta > 0 \qquad (2.58)$$

Clearly, if $f(t_o^+) = f(t_o^-)$ and $f'(t_o^+) = f'(t_o^-)$ then f is differentiable at t_o.

DEFINITION: PIECEWISE SMOOTHNESS ON AN INTERVAL A function f, is piecewise smooth on an interval I, iff f is piecewise smooth at all points in the interval. Appropriate left and right limits apply for the end points of a closed interval.

2.4.3 Boundedness, Bounded Variation, and Absolute Continuity

Absolute continuity is important because it is a sufficient condition to guarantee that a function is the indefinite integral of its derivative. Furthermore, absolute continuity is a sufficient condition to guarantee that integration by parts will be valid (Champeney, 1987 p. 22; Jain 1986 p. 197). Associated with absolute continuity is the concept of bounded variation and a related concept is that of signal pathlength. These signal properties are defined below, after the concept of boundedness is defined.

DEFINITION: BOUNDEDNESS A signal $f: I \to C$ is bounded on the interval I, if there exists a constant f_o, such that $|f(t)| < f_o$ for all $t \in I$.

DEFINITION: SIGNAL PATHLENGTH Over the interval $[\alpha, \beta]$ the signal pathlength of a real piecewise smooth signal, f, with discontinuities at points $\{t_1, \ldots\}$, is defined according to

$$\int_\alpha^{t_1} \sqrt{1 + (f'(t))^2}\, dt + \int_{t_1}^{t_2} \sqrt{1 + (f'(t))^2}\, dt + \cdots + \sum_i |f(t_i^+) - f(t_i^-)| \quad (2.59)$$

This result readily follows from the definition of a derivative as shown in Figure 2.6.

By considering the interval $[\Delta, \beta]$, as $\Delta \to 0$, it can be readily shown that the signal $t \cos(1/t)$, while bounded, has infinite signal pathlength over any neighborhood of $t = 0$.

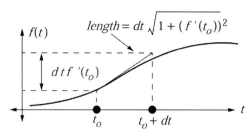

Figure 2.6 Illustration of the signal pathlength of a function between two closely spaced points.

DEFINITION: BOUNDED VARIATION   A signal $f: \mathbf{R} \to \mathbf{C}$ is of bounded variation on a closed interval $[\alpha, \beta]$, if there exists a constant $k_o > 0$ such that, for every set of numbers $\{t_0, \ldots, t_N\}$, where $\alpha \leq t_0 < t_1 < \cdots < t_N \leq \beta$, it is the case that (Champeney, 1987 p. 39)

$$\sum_{i=0}^{N-1} |f(t_{i+1}) - f(t_i)| < k_o \tag{2.60}$$

The signal $\cos(1/t)$, while bounded, is not of bounded variation on any closed interval that includes the point $t = 0$. To establish that the signal $f(t) = t\cos(1/t)$ is not of bounded variation over any interval that includes $t = 0$, note that a sequence of times $1/\pi, 2/3\pi, 1/2\pi, 2/5\pi, 1/3\pi, 2/7\pi, \ldots$ yields the corresponding function values $-1/\pi, 0, 1/2\pi, 0, -1/3\pi, \ldots$ and the summation of the numbers $|f(t_{i+1}) - f(t_i)|$ for $i \in \mathbf{Z}^+$ does not converge.

THEOREM 2.3. FINITE SIGNAL PATHLENGTH IMPLIES BOUNDED VARIATION   A real and piecewise smooth signal with a finite signal pathlength on a closed interval $[\alpha, \beta]$, has bounded variation on this interval.

Proof. As shown in Figure 2.6, it follows that if a signal is real, piecewise smooth, and with a finite pathlength over $[\alpha, \beta]$, then dt can be chosen, such that, over any interval $[t_o, t_o + dt]$ the signal pathlength is closely approximated by $dt\sqrt{1 + (f'(t_o^+))^2} + |f(t_o^+) - f(t_o^-)|$. Now, as

$$dt\sqrt{1 + (f'(t_o^+))^2} > dt|f'(t_o^+)| \approx |f(t_o + dt) - f(t_o^+)| \tag{2.61}$$

and $|f(t_o^+) - f(t_o^-)|$ is finite, it follows that the signal has bounded variation over $[t_o, t_o + dt]$. The required result readily follows.

DEFINITION: ABSOLUTE CONTINUITY ON AN INTERVAL   A function $f: \mathbf{R} \to \mathbf{C}$ is absolutely continuous on an interval I if $\forall \varepsilon > 0$ there exists a $\delta_o > 0$, such that

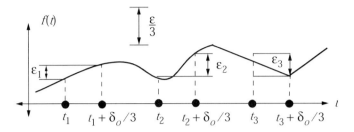

Figure 2.7 Illustration of the requirement of absolute continuity. The case shown is for three disjoint intervals of equal length.

(Titchmarsh, 1939 p. 364; Jain, 1986 p. 192)

$$\sum_{i=1}^{N} |f(t_i + \delta_i) - f(t_i)| < \varepsilon \tag{2.62}$$

for every set of nonoverlapping intervals $(t_i, t_i + \delta_i) \subseteq I$ where $\Sigma_{i=1}^{N} \delta_i < \delta_o$. For a closed interval $[\alpha, \beta]$, the intervals $[\alpha, \alpha + \delta_1)$ and $(\beta - \delta_N, \beta]$ are to be considered.

This criterion is illustrated in Figure 2.7. Absolute continuity states that for any $\varepsilon > 0$ there exists a δ_o, such that the variation in the function f is less than ε over any subset of the interval I, whose length, or "measure," is less than δ_o. As the signal variation of $t \cos(1/t)$ over any neighborhood of $t = 0$ is infinite, then this function is not absolutely continuous over any interval that includes $t = 0$.

2.4.4 Relationships between Signal Properties

The following theorems state important relationships between the above defined signal properties.

THEOREM 2.4. CONTINUITY IMPLIES BOUNDEDNESS *If f is piecewise continuous on the closed and finite interval I, then f is bounded on I. The converse is not true. If I is an open interval, then f may be unbounded at either or both ends of the interval.*

Proof. Piecewise continuity implies that for any point $t_o \in I$ the left- and right-hand limits, according to Eqs. (2.42) and (2.43), exist, and that

$$f(t_o) \in \{f(t_o^-), f(t_o^+)\}$$

Hence, the definition excludes the function being unbounded at any point of I. It does not preclude the function being unbounded as its argument becomes unbounded. To show the converse does not hold, consider the function f

defined as being unity if its argument is rational, and zero if its argument is irrational. Such a function is clearly bounded but is not piecewise continuous at any point.

To illustrate the potential unboundedness of a continuous function on an open interval, consider the function $1/t$ that is continuous on the interval $(0, 1)$, but is unbounded as t approaches zero.

THEOREM 2.5. CONTINUITY IMPLIES FINITE NUMBER OF MAXIMA AND MINIMA
If f is piecewise continuous at a point t_o, then for all $\varepsilon > 0$ there exists a neighborhood of t_o, such that in this neighborhood f has a finite number of local maxima and minima, where the difference between adjacent maxima and minima is greater than ε.

Proof. Consider the contrapositive form: If there exists a $\varepsilon > 0$, such that f has an infinite number of local maxima and minima in all neighborhoods of t_o, where the difference between adjacent maxima and minima is greater than ε, then f is not piecewise continuous at t_o.

Assume that in all neighborhoods of a point t_o, the function f has an infinite number of local maxima and minima, where the difference between a maxima and minima is greater than a fixed number ε. It then follows, for any chosen $f(t_o^+)$, that

$$\forall \delta_o > 0 \quad \exists \delta < \delta_o \quad \text{s.t.} \quad |f(t_o + \delta) - f(t_o^+)| > \varepsilon/2 \quad \delta > 0 \qquad (2.63)$$

which implies that f is not right-hand continuous at t_o. The lack of left-hand continuity can be similarly proved.

For example, the function $\cos(1/t)$ is not piecewise continuous at $t = 0$.

2.4.4.1 Continuity and Infinite Pathlength
Continuity at a point can be consistent with infinite signal pathlength in the neighborhood of the point in question. The function $t\cos(1/t)$, which is uniformly continuous on all neighborhoods of $t = 0$, demonstrates this point.

2.4.4.2 Continuity and Infinite Number of Discontinuities
Continuity and piecewise continuity at a point, can be consistent with an infinite number of discontinuities in the neighborhood at that point. Consider a function defined by

$$f(t) = \begin{cases} k & t \leq 0, t > 1 \\ k + \dfrac{1}{(n+1)^p} & t \in \left(\dfrac{1}{n+1}, \dfrac{1}{n}\right], \quad n \text{ even} \\ k - \dfrac{1}{(n+1)^p} & t \in \left(\dfrac{1}{n+1}, \dfrac{1}{n}\right], \quad n \text{ odd} \end{cases} \qquad (2.64)$$

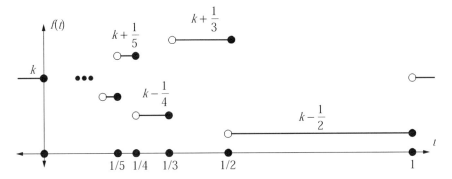

Figure 2.8 Function which has an infinite number of discontinuities in all neighborhoods of $t = 0$ but it continuous at this point.

for the case where $p = 1$. The graph of this function is shown in Figure 2.8. Clearly, f is such that $|f(\delta) - k| < \delta$ for positive δ. Hence, for any $\varepsilon > 0$ it is the case, for all $|\delta|$ less than ε, that $|f(\delta) - k| < \varepsilon$ which implies continuity at $t = 0$.

2.4.4.3 Piecewise Smoothness and Infinite Number of Discontinuities

As with piecewise continuity, it is the case that piecewise smoothness can be consistent with an infinite number of discontinuities in the neighborhood of a point. To illustrate this, consider the function f defined by Eq. (2.64) and shown in Figure 2.8 for the case where $p = 1$. Given that to the right of the point $t_o = 0$, the function alternates between being above and below k, the obvious choice for $f'(t_o^+)$, and $f'(t_o^-)$ is zero, whereupon, it follows, for $\delta \in [1/(n+1), 1/n)$, that

$$\frac{f(t_o + \delta) - f(t_o^+)}{\delta} = \frac{k \pm \frac{1}{(n+1)^p} - k}{\delta} = \pm \frac{1}{(n+1)^p \delta} \qquad (2.65)$$

Since, on $[1/(n+1), 1/n)$ the minimum and maximum value of δ, respectively, are $1/(n+1)$ and $1/n$ it follows that

$$\frac{n}{(n+1)^p} < \left| \frac{f(t_o + \delta) - f(t_o^+)}{\delta} \right| < \frac{1}{(n+1)^{p-1}} \qquad (2.66)$$

Thus, when $p = 1$, $(f(t_o + \delta) - f(t_o^+))/\delta$ does not converge as δ decreases, and n increases, which implies f is not right differentiable at t_o. However, when $p = 2$, $(f(t_o + \delta) - f(t_o^+))/\delta$ does converge as δ decreases, which implies f is right differentiable at $t_o = 0$.

THEOREM 2.6. PIECEWISE SMOOTHNESS IMPLIES PIECEWISE CONTINUITY *If f is piecewise smooth on an interval, then f is piecewise continuous over that interval. The converse is not necessarily true.*

Proof. Piecewise differentiability to the right of a point t_o, implies there exists a $f'(t_o^+)$, such that

$$\forall \varepsilon > 0 \quad \exists \delta_o > 0 \quad \text{s.t.} \quad 0 < \delta < \delta_o \Rightarrow |f(t_o + \delta) - f(t_o^+) - \delta f'(t_o^+)| < \delta \varepsilon$$

(2.67)

This implies

$$\forall \varepsilon > 0 \quad \exists \delta_o > 0 \quad \text{s.t.} \quad 0 < \delta < \delta_o \Rightarrow |f(t_o + \delta) - f(t_o^+)| < \delta[|f'(t_o^+)| + \varepsilon]$$

(2.68)

which is consistent with continuity, for example, let $\delta_0 = \varepsilon/(|f'(t_o^+)| + \varepsilon)$ when $f'(t_o^+) \neq 0$.

Jain (1986 pp. 232f) and Burk (1998 pp. 279f) give examples of functions that are continuous everywhere, but which are not differentiable at any point.

THEOREM 2.7. PIECEWISE SMOOTHNESS IMPLIES BOUNDED VARIATION *If f is piecewise smooth on a closed interval $[\alpha, \beta]$, then f has bounded variation on this interval. The converse is not true.*

Proof. First, piecewise smoothness implies $|f(t_i^+) - f(t_i^-)| < \infty$ for all $t_i \in [\alpha, \beta]$. Fix $\varepsilon > 0$. As in the proof of Theorem 2.6, piecewise differentiability at an arbitrary point t_i implies there exists $\delta_i > 0$, $f'(t_i^+)$, and $f'(t_i^-)$ such that

$$\begin{aligned} 0 < \delta < \delta_i \Rightarrow |f(t_i + \delta) - f(t_i^+)| < \delta[|f'(t_i^+)| + \varepsilon] \\ 0 < \delta < \delta_i \Rightarrow |f(t_i^-) - f(t_i - \delta)| < \delta[|f'(t_i^-)| + \varepsilon] \end{aligned}$$

(2.69)

Thus, over the interval $(t_i - \delta_i, t_i + \delta_i)$ the signal pathlength is finite. For any fixed ε there will be a finite number of intervals $[\alpha, \alpha + \delta_1), (t_i - \delta_i, t_i + \delta_i)$ and $(\beta - \delta_N, \beta]$ which "cover" the interval $[\alpha, \beta]$, and the theorem is then proved. To prove that the converse is not true, consider the function $f(t) = \sqrt{t}$ for $t > 0$ and $f(t) = 0$ for $t \leq 0$, which has bounded variation on all neighborhoods of zero but is not piecewise smooth at $t = 0$.

THEOREM 2.8. ABSOLUTE CONTINUITY IMPLIES CONTINUITY AND BOUNDED VARIATION *If f is absolutely continuous on an interval I, then f is uniformly continuous, and of bounded variation, on this interval* (Jain, 1986 pp. 192–3). *Uniform continuity does not necessarily imply absolute continuity. Bounded variation does not necessarily imply absolute continuity.*

Proof. Setting $N = 1$ in the definition of absolute continuity [Eq. (2.62)] shows that f is uniformly continuous. The proof of bounded variation also follows in a direct manner from the definition of absolute continuity. The function $t\cos(1/t)$, which is uniformly continuous in a neighborhood of $t = 0$, is not absolutely continuous over such a neighborhood. Any signal with bounded variation, but with a discontinuity, is not absolutely continuous.

THEOREM 2.9. CONTINUITY AND PIECEWISE SMOOTHNESS YIELDS ABSOLUTE CONTINUITY *If a function f is continuous at all points in $[\alpha, \beta]$, and is piecewise smooth on the same interval, then it is absolutely continuous on $[\alpha, \beta]$* (Champeney 1987 p. 22). *If f is differentiable at all points in $[\alpha, \beta]$, then it is absolutely continuous on $[\alpha, \beta]$.*

Proof. A straightforward application of the definitions for continuity, piecewise smoothness, and absolute continuity yields the required result.

Continuity is consistent with infinite pathlength of a function in the neighborhood of a point, and piecewise continuity is consistent with discontinuities in a function. Both conditions are inconsistent with absolute continuity. The combination of continuity and piecewise smoothness ensures that a first-order Taylor series approximation to the function can be made either side of any point in the interval of interest. This implies that the signal pathlength and signal variation of the function can be made arbitrarily small over all intervals whose total length or "measure" is appropriately chosen. This, in turn, implies absolute continuity.

THEOREM 2.10. ABSOLUTE CONTINUITY IMPLIES DIFFERENTIABILITY ALMOST EVERYWHERE *If a function f is absolutely continuous over $[\alpha, \beta]$, then it is differentiable everywhere except, at most, on a set of countable points of $[\alpha, \beta]$, that is, it is differentiable "almost everywhere"* (Champeney, 1987 p. 22; Jain, 1986 p. 193).

Proof. See Jain (1986 p. 193).

The function $f(t) = \sqrt{t}$ for $t > 0$ and $f(t) = 0$ for $t \leq 0$, shows why absolute continuity does not guarantee the existence of a derivative, or even the existence of both left- and right-hand derivatives, at all points. This function is absolutely continuous in all neighborhoods of $t = 0$ but $f'(0^+)$ does not exist.

2.5 MEASURE AND LEBESGUE INTEGRATION

The following subsections give a brief introduction to measure theory and Lebesgue integration.

2.5.1 Measure and Measurable Sets

The measure of a set of real numbers is a generalization of the notion of length and, broadly speaking, is the length of the intervals comprising the set. The simplest example is an interval $I = [\alpha, \beta]$ whose measure is $\beta - \alpha$. The measure of a set E is denoted $M(E)$ where M is the measure operator (strictly speaking an outer measure operator). Consistent with our understanding of length, it follows that the measure of two disjoint sets is the sum of their individual measures. Thus, if $E_1, \ldots, E_N$ are disjoint sets, then

$$M\left(\bigcup_{i=1}^{N} E_i\right) = \sum_{i=1}^{N} M(E_i) \tag{2.70}$$

A detailed discussion of measure can be found in books such as Jain (1986 ch. 3), Burk (1998 ch. 3), and Titchmarsh (1939 ch. 10).

The first issue that needs to be clarified is whether all sets of real numbers are, in fact, measurable. For the purposes of this book the following definition will suffice (Jain, 1986 p. 80).

DEFINITION: MEASURABLE SET A set E of real numbers is a measurable set, if it can be approximated arbitrarily closely by an open set and a closed set, that is, if $\forall \varepsilon > 0$, there exists an open set O and a closed set C, such that

$$E \subseteq O \quad\quad C \subseteq E \tag{2.71}$$

and

$$M(O \cap E^C) < \varepsilon \quad\quad M(E \cap C^C) < \varepsilon \tag{2.72}$$

These relationships imply

$$M(O) - M(E) < \varepsilon \quad\quad M(E) - M(C) < \varepsilon \tag{2.73}$$

It is difficult, but possible, to construct a set which is nonmeasurable (Jain, 1986 pp. 83f).

DEFINITION: ZERO MEASURE A set E is said to have zero measure if $M(E) = 0$.

Note, the measure of a countable set of points has zero measure. For example, $M(Q) = 0$.

DEFINITION: ALMOST EVERYWHERE (a.e.) A property is said to hold "almost everywhere" if it holds everywhere except on a set of points that have zero measure.

2.5.2 Measurable Functions

The importance of a function being measurable is that measurability is a prerequisite for Lebesgue integrability. A detailed discussion of measurable functions can be found in Jain (1986 ch. 4) and Burk (1998 ch. 4). For subsequent discussion, the following definition will suffice (Jain, 1986 p. 93).

DEFINITION: MEASURABLE FUNCTION A function $f: R \to C$ is a measurable function if for any open set, O, of C the inverse image defined by $f^{-1}(O) = \{t: f(t) \in O\}$ is a measurable set.

2.5.3 Lebesgue Integration

A detailed discussion of Lebesgue integration can be found in such books as Burk (1998 ch. 5), Jain (1986 ch. 5), Titchmarsh (1939 pp. 332f), and Debnath (1999 ch. 2). The following is a brief overview of Lebesgue integration: Consider a bounded measurable function $f: R \to R$ on an interval (α, β), where the function is bounded according to

$$f_L \leq f(t) \leq f_U \qquad t \in (\alpha, \beta) \tag{2.74}$$

The range of f is partitioned by the $N + 1$ numbers $f_0, f_1, \ldots, f_N$ such that

$$f_L = f_0 < f_1 < \cdots < f_{N-1} < f_N = f_U \tag{2.75}$$

and the sets $E_0, E_1, \ldots, E_N$ are then defined according to

$$\begin{aligned} E_i &= \{t: f_i \leq f(t) < f_{i+1}\} \qquad i \in \{0, \ldots, N-1\} \\ E_N &= \{t: f(t) = f_N\} \end{aligned} \tag{2.76}$$

Note that it is the measurability of f that guarantees the existence of the sets $E_0, \ldots, E_N$. As illustrated in Figure 2.9, the area under the function f over the

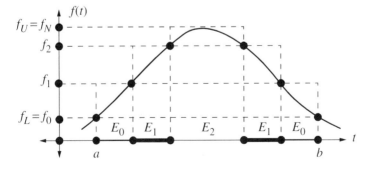

Figure 2.9 Illustration of the partition of the range of f, and the sets partitioning the domain of f, for the case where N = 3.

interval (α, β) can be approximated by the lower and upper sums defined by

$$S_L = \sum_{i=0}^{N-1} f_i M(E_i) \qquad S_U = \sum_{i=0}^{N-1} f_{i+1} M(E_i) \qquad (2.77)$$

Clearly, $S_L < S_U$. As the number of points, $N + 1$, demarcating the range of f increases in a manner, such that $f_{i+1} - f_i$ tends towards zero for $i \in \{0, \ldots, N-1\}$, then S_L and S_U converge to the same number and this number is defined as the Lebesgue integral of the function f over the interval (α, β). The Lebesgue integral of a function f over a set E is written as

$$\int_E f \qquad (2.78)$$

The Lebesgue integral is defined for a larger class of functions than a Riemann integral. For example, the function defined as being unity when its argument is irrational and zero otherwise is Lebesgue integrable on a finite interval but not Riemann integrable. If a function is bounded on $[\alpha, \beta]$, and is Riemann integrable over this interval, then it is also Lebesgue integrable and the two integrals are equal (Burk, 1998 pp. 181–182; Jain, 1986 p. 136). For bounded functions that are continuous almost everywhere on a finite interval, the Riemann integral exists and is equal to the Lebesgue integral (Burk, 1998 p. 182; Jain, 1986 p. 229), that is,

$$\int_{[\alpha, \beta]} f = \int_\alpha^\beta f(x)\, dx \qquad (2.79)$$

It is useful to use both the integral notations shown in this equation for Lebesgue integrals, and both forms are used in subsequent analysis.

2.5.4 Lebesgue Integrable Functions

The following definitions find widespread use in analysis (Jain, 1986 p. 205):

DEFINITION: SET OF LEBESGUE INTEGRABLE FUNCTIONS If $f: R \to C$ is a measurable function, and the Lebesgue integral of $|f|^p$ ($p > 0$) over a set E is finite, then f is said to be p integrable over E. The set of p integrable functions over E is denoted $L^p(E)$, that is,

$$L^p(E) = \left\{ f: E \to C, \int_E |f|^p < \infty \right\} \qquad (2.80)$$

For the case of integration over $(-\infty, \infty)$ the simpler notation

$$L^p = \left\{ f: \mathbf{R} \to \mathbf{C}, \int_{(-\infty,\infty)} |f|^p < \infty \right\} \tag{2.81}$$

is used, and when $p = 1$, the superscript on L is omitted. For the case of integration over the interval $[\alpha, \beta]$ notation, as follows, is used:

$$L^p[\alpha, \beta] = \left\{ f:[\alpha, \beta] \to \mathbf{C}, \int_{[\alpha,\beta]} |f|^p < \infty \right\} \tag{2.82}$$

Again, when $p = 1$, the superscript is omitted.

DEFINITION: LOCALLY INTEGRABLE If a function is an element of $L[\alpha, \beta]$, for all finite $\alpha, \beta \in \mathbf{R}$, then it is said to be "locally integrable."

2.5.5 Properties of Lebesgue Integrable Functions

2.5.5.1 Basic Properties The following are some basic results for a Lebesgue integrable function (Jain, 1986 p. 151). First, the integral of a function over a set of zero measure is zero, that is,

$$M(E) = 0 \implies \int_E f = 0 \tag{2.83}$$

Thus, if

$$\lim_{n \to \infty} f_n(t) = \begin{cases} \text{undefined} & t = t_o \\ 0 & t \neq t_o \end{cases} \tag{2.84}$$

and for all n

$$\int_{-\infty}^{\infty} f_n(t)\, dt = k \tag{2.85}$$

then

$$\lim_{n \to \infty} \int_{-\infty}^{\infty} f_n(t)\, dt = k \quad \text{but} \quad \int_{-\infty}^{\infty} \lim_{n \to \infty} f_n(t)\, dt = 0 \tag{2.86}$$

THEOREM 2.11. AREA ASSOCIATED WITH THE TAIL AND A NEIGHBORHOOD If $f \in L$, then the area under the tail of f, the area associated with the neighborhood of any point, and the area under f in the neighborhood of a point where f is unbounded, can be made arbitrarily small, that is, $\forall \varepsilon > 0$, there exists $T_o > 0$, $\delta > 0$, $f_o > 0$, such that

$$\int_{T_o}^{\infty} |f(t)| \, dt < \varepsilon, \quad \int_{-\infty}^{-T_o} |f(t)| \, dt < \varepsilon, \quad \int_{t_o-\delta}^{t_o+\delta} |f(t)| \, dt < \varepsilon \quad t_o \in \mathbf{R}$$

$$\int_E |f(t)| \, dt < \varepsilon \qquad E = \{t : |f(t)| > f_o\} \tag{2.87}$$

Proof. The proof of the last of these results is detailed in Appendix 1. The proof of the other results follow in a similar manner.

THEOREM 2.12. LIMITS ON UNBOUNDEDNESS FOR INTEGRABILITY *If $f \in L$, then the measure of the set over which f is unbounded is zero. Formally, if $f \in L$, then $\forall \varepsilon > 0$ there exists a constant $f_o > 0$, such that*

$$M\{t : |f(t)| > f_o\} < \varepsilon \tag{2.88}$$

Proof. This result is readily proved by considering the contrapositive form of the Theorem.

THEOREM 2.13. FINITE ENERGY IMPLIES ABSOLUTE INTEGRABILITY *If $f \in L^2[\alpha, \beta]$, then $f \in L[\alpha, \beta]$. It is not necessarily the case that if $f \in L^2$ then $f \in L$, or if $f \in L$ then $f \in L^2$.*

Proof. The proof of the first part of this theorem is detailed in Appendix 2.

Figure 2.10 shows the results stated in the second part of the theorem. For example, consider f_1 and f_2 defined according to

$$f_1(\delta, t) = \begin{cases} 0 & t \leq 0, t > 1 \\ \dfrac{1}{t^\delta} & 0 < t \leq 1 \end{cases} \qquad f_2(\delta, t) = \begin{cases} 0 & t \leq 1 \\ \dfrac{1}{t^\delta} & t > 1 \end{cases} \tag{2.89}$$

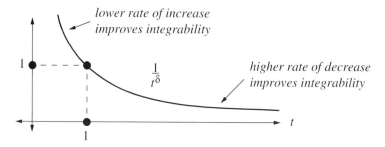

Figure 2.10 Illustration of how rate of increase, or rate of decrease, affects integrability.

From elementary integration results it follows, for $0.5 \leq \delta < 1$, that $f_1(\delta, t) \in L$, but $f_1(\delta, t) \notin L^2$. Also, for $0.5 < \delta \leq 1$, it follows that $f_2(\delta, t) \in L^2$, but $f_2(\delta, t) \notin L$.

THEOREM 2.14. BOUNDEDNESS AND INTEGRABILITY *If f is bounded and $f \in L$, then $f \in L^2$.*

Proof. Assume $|f(t)| \leq f_{max}$ for $\forall t \in \mathbf{R}$. It then follows that $|f(t)|/f_{max} < 1$ for $\forall t \in \mathbf{R}$ which implies $0 < |f|^2/f_{max}^2 < |f|/f_{max}$. From this result, it follows that $\int |f|^2 < f_{max} \int |f| < \infty$, and hence, $f \in L^2$.

2.5.5.2 Schwarz Inequality

Schwarz's inequality is a general relationship that applies to any two elements of an inner product space (Debnath, 1999 p. 90; Kreyszig, 1978 p. 137). The specific forms relevant to the development of theory in later chapters, are detailed in the following theorem.

THEOREM 2.15. SCHWARZ INEQUALITY *If $f, g \in L^2[\alpha, \beta]$, then*

$$\left| \int_\alpha^\beta f(t) g^*(t) \, dt \right| \leq \sqrt{\int_\alpha^\beta |f(t)|^2 \, dt} \sqrt{\int_\alpha^\beta |g(t)|^2 \, dt} \qquad (2.90)$$

If $f, g \in L^2$, then

$$\left| \int_{-\infty}^\infty f(t) g^*(t) \, dt \right| \leq \sqrt{\int_{-\infty}^\infty |f(t)|^2 \, dt} \sqrt{\int_{-\infty}^\infty |g(t)|^2 \, dt} \qquad (2.91)$$

If $\sum_{i=1}^\infty |f_i|^2$ and $\sum_{i=1}^\infty |g_i|^2$ are finite, then

$$\left| \sum_{i=1}^\infty f_i g_i^* \right| \leq \sqrt{\sum_{i=1}^\infty |f_i|^2} \sqrt{\sum_{i=1}^\infty |g_i|^2} \qquad (2.92)$$

Proof. First, for the case where $f, g \in L^2[\alpha, \beta]$, equality holds if either, or both, of f, g are zero almost everywhere. Second, assume g is nonzero on a set of nonzero measure and consider the following inequality

$$0 \leq \int_\alpha^\beta |f(t) - k g(t)|^2 \, dt$$

$$= \int_\alpha^\beta |f(t)|^2 \, dt + |k|^2 \int_\alpha^\beta |g(t)|^2 \, dt - k^* \int_\alpha^\beta f(t) g^*(t) \, dt - k \int_\alpha^\beta f^*(t) g(t) \, dt$$

$$(2.93)$$

which is valid for any $k \in C$. For the case where $k = \int_\alpha^\beta f(t)g^*(t)\, dt / \int_\alpha^\beta |g(t)|^2\, dt$ it follows that

$$0 \leq \int_\alpha^\beta |f(t)|^2\, dt - \frac{\left|\int_\alpha^\beta f(t)g^*(t)\, dt\right|^2}{\int_\alpha^\beta |g(t)|^2\, dt} \qquad (2.94)$$

and the required result follows directly. The proof for the other two forms follow in an analogous manner.

2.5.5.3 Approximation by a Simple Function

DEFINITION: SIMPLE FUNCTION A simple function $\psi: I \to C$ is defined as

$$\psi(x) = \sum_{i=1}^N a_i \chi_{E_i}(x) \qquad \chi_{E_i}(x) = \begin{cases} 1 & x \in E_i \\ 0 & x \notin E_i \end{cases} \qquad (2.95)$$

where $\{E_1, \ldots, E_N\}$ is a partition of I and χ_{E_i} is the characteristic function of E_i.

THEOREM 2.16 *If $f \in L$, then for all $\varepsilon > 0$ there exists a simple function $\psi: R \to C$, such that*

$$\left| \int_R f - \sum_{i=1}^N a_i M(E_i) \right| < \varepsilon \qquad (2.96)$$

where the measure of each set E_i is finite.

Proof. The proof of this result is implicit in the definition of the Lebesgue integral [see, for example, Jain (1986 pp. 130f) and Titchmarsh (1939 pp. 332f)].

2.5.5.4 Continuous Approximation to a Lebesgue Integrable Function

It is plausible that a Lebesgue integrable function can be closely approximated by a continuous function. Figures 2.11 and 2.12 show two cases, where a continuous function cannot approximate a Lebesgue integrable function at all points. The following theorem formulates precisely the ability of a continuous function to approximate a Lebesgue integrable function. Appropriate references are Titchmarsh (1939 p. 376) and Jain (1986 p. 116).

THEOREM 2.17. CONTINUOUS APPROXIMATION TO A MEASURABLE FUNCTION *If $f: R \to C$ is a measurable function on a finite interval $[\alpha, \beta]$ and $f \in L[\alpha, \beta]$, then there exists an absolutely continuous function $\phi: R \to C$ which approximates f arbitrarily closely, except on a set of arbitrarily small measure, that is*

$$\forall \varepsilon, \delta > 0 \quad \exists \phi \quad \text{s.t.} \quad M\{t \in [\alpha, \beta]: |f(t) - \phi(t)| > \delta\} < \varepsilon \qquad (2.97)$$

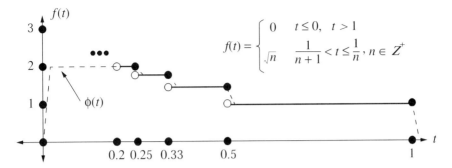

Figure 2.11 Approximating a function, which has an infinite number of discontinuities that diverge at the point $t = 0$, by a continuous function.

That is, $|f(t) - \phi(t)| < \delta$ for $t \in [\alpha, \beta]$ except on a set whose measure is less than ε.

Proof. The proof is detailed in Appendix 3.

The following theorem details the implication of the existence of an absolutely continuous function that closely approximates a measurable function (Titchmarsh, 1939 pp. 376–7).

THEOREM 2.18. INTEGRATED ERROR IN CONTINUOUS APPROXIMATION *If $f: \mathbf{R} \to \mathbf{C}$ is Lebesgue integrable on $[\alpha, \beta]$, that is, $f \in L[\alpha, \beta]$, then, $\forall \varepsilon > 0$, there exists an absolutely continuous function $\phi: \mathbf{R} \to \mathbf{C}$ such that $\phi \in L[\alpha, \beta]$, and*

$$\int_\alpha^\beta |f(t) - \phi(t)|\, dt < \varepsilon \tag{2.98}$$

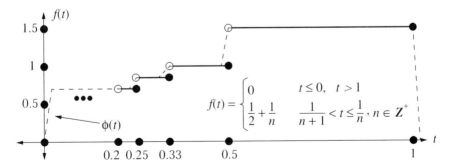

Figure 2.12 Approximating a function, which has an infinite number of discontinuities that converge at the point $t = 0$, by a continuous function.

Proof. If $f \in L[\alpha, \beta]$, then according to Theorem 2.11 it is possible to find a constant f_o and a set $E = \{t : |f(t)| > f_o\}$, such that $\int_E |f| < \varepsilon/3$. Define $F = [\alpha, \beta] \cap E^C$.

Clearly, $f(t)$ is bounded within the range $[-f_o, f_o]$ for $t \in F$. According to Theorem 2.17, there exists an absolutely continuous function, ϕ, that approximates f, within an arbitrary δ, for all $t \in F$ except on a set whose measure is ε_1. It then follows that

$$\int_\alpha^\beta |f(t) - \phi(t)| \, dt < \int_E |f(t)| \, dt + \int_F |f(t) - \phi(t)| \, dt < \frac{\varepsilon}{3} + \delta M(F) + \varepsilon_1 f_o$$

(2.99)

For any given value of $\varepsilon > 0$, it is the case that f_o and $M(F)$ are fixed. For such fixed values δ and ε_1 can be made arbitrarily small, such that $\delta M(F) < \varepsilon/3$ and $\varepsilon_1 f_o < \varepsilon/3$. The required result then follows.

COROLLARY *If $f: \mathbf{R} \to \mathbf{C}$ is a Lebesgue integrable function on the infinite interval $(-\infty, \infty)$, that is, $f \in L$, then $\forall \varepsilon > 0$ there exists an absolutely continuous function $\phi: \mathbf{R} \to \mathbf{C}$ such that*

$$\int_{-\infty}^\infty |f(t) - \phi(t)| dt < \varepsilon$$

(2.100)

Proof. If $f \in L$, then according to Theorem 2.11 there exists a constant T_o, such that

$$\int_{T_o}^\infty |f(t)| \, dt < \frac{\varepsilon}{3} \qquad \int_{-\infty}^{-T_o} |f(t)| \, dt < \frac{\varepsilon}{3}$$

(2.101)

It is then possible to define a function ϕ, that is zero on the intervals $(-\infty, -T_o)$ and (T_o, ∞), and on the interval $[-T_o, T_o]$, according to Theorem 2.18, is such that

$$\int_{-T_o}^{T_o} |f(t) - \phi(t)| \, dt < \frac{\varepsilon}{3}$$

(2.102)

2.5.5.5 Step Approximation to a Lebesgue Integrable Function
Often, it is useful to employ a step approximation to a function where the step function approximates the function arbitrarily closely. To illustrate the problem of step values being chosen, such that the step approximation "poorly" approximates the function of interest, consider the function

$$f(t) = \begin{cases} 0 & t \in \mathbf{Q} \\ 1 & t \notin \mathbf{Q} \end{cases}$$

(2.103)

A step approximation, f_S to f could be defined according to $f_S(t) = f(\lfloor t/\Delta \rfloor \Delta)$ for a suitably small $\Delta \in Q^+$. However, this step approximation is such that

$$\int_0^1 |f(t) - f_S(t)|\, dt = 1 \qquad (2.104)$$

for all $\Delta \in Q^+$. Intuition suggests that, if the signal pathlength of a function is finite, then a step approximation to the function can be achieved over a finite interval with arbitrarily small integrated error. This is the case and is formally stated in the following theorem.

THEOREM 2.19. STEP APPROXIMATION TO A LEBESGUE INTEGRABLE FUNCTION
If $f \in L[\alpha, \beta]$ is real and has bounded variation on $[\alpha, \beta]$, then $\forall \varepsilon > 0$, there exists a constant $\Delta_o > 0$ such that for $0 < \Delta < \Delta_o$ it is the case that

$$\int_\alpha^\beta |f(t) - f_S(t)|\, dt < \varepsilon \qquad (2.105)$$

where f_S is a step approximation to f, with uniform step width Δ, defined as

$$f_S(t) = \sum_{i=0}^N f(\alpha + i\Delta)\chi_{I_i}(t) = \begin{cases} f(\alpha + \lfloor (t-\alpha)/\Delta \rfloor \Delta) & t \in [\alpha, \beta] \\ 0 & \text{elsewhere} \end{cases} \qquad (2.106)$$

Here

$$N = \left\lfloor \frac{\beta - \alpha}{\Delta} \right\rfloor \qquad \chi_{I_i}(t) = \begin{cases} 1 & t \in I_i \\ 0 & \text{elsewhere} \end{cases} \qquad (2.107)$$

$$I_i = \begin{cases} [\alpha + i\Delta, \alpha + (i+1)\Delta) & 0 \leq i < N-1 \\ [\alpha + N\Delta, \beta] & i = N \end{cases}$$

If $f \in L$, is real and has bounded variation on all finite closed intervals, then this result is valid for the infinite interval.

Proof. Consider the intervals I_i defined in the theorem, and the maximum and minimum values of the function on these intervals defined according to $M_i = \sup\{f(t): t \in I_i\}$ and $m_i = \inf\{f(t): t \in I_i\}$. On the ith interval, it follows that $|M_i - f(\alpha + i\Delta)| \leq M_i - m_i$ and $|f(\alpha + i\Delta) - m_i| \leq M_i - m_i$. Since f is real, and has bounded variation on the interval $[\alpha, \beta]$, it follows that there exists a constant k_o, such that

$$\sum_{i=0}^N M_i - m_i < k_o \qquad (2.108)$$

where k_o is independent of Δ and N, that is, bounded variation implies that the difference between maxima and minima on an interval, on average, will decrease as the interval measure is decreased. Thus,

$$\int_\alpha^\beta |f(t) - f_S(t)|\, dt = \sum_{i=0}^N \int_{I_i} |f(t) - f(\alpha + i\Delta)|\, dt \leq \Delta \sum_{i=0}^N M_i - m_i < \Delta k_o \quad (2.109)$$

As Δ can be made arbitrarily small the required result follows.

The result for the infinite interval follows from bounded variation on all intervals of the form $[-T_o, T_o]$, the definition $f_S(t) = 0$ on $[-\infty, -T_o]$ and $[T_o, \infty]$, and the fact that $f \in L$ implies, according to Theorem 2.11, that there exists a constant $T_o > 0$, such that $\int_{T_o}^\infty |f(t)|\, dt < \varepsilon$ and $\int_{-\infty}^{-T_o} |f(t)|\, dt < \varepsilon$.

Note that f_S may not converge to f at a given point as the step size decreases. For example, consider a function defined according to $f(t) = 1$ for $t \geq \pi$ and that is zero elsewhere. Clearly, $f_S(\pi) = 0$ for all rational step sizes while $f(\pi) = 1$.

2.5.5.6 Interchanging Integration Order and Summation Order
It is often the case, that the order of integration in multiple integrals needs to be interchanged. The Fubini and Tonelli theorems, summarized below, specify when this interchange is valid (Champeney, 1987 p. 18).

THEOREM 2.20. FUBINI–TONELLI THEOREM *If $f: \mathbf{R}^2 \to \mathbf{C}$ is measurable and defined almost everywhere, and one of the integrals*

$$\iint |f(t, \tau)|\, dt\, d\tau \qquad \iint |f(t, \tau)|\, d\tau\, dt \qquad (2.110)$$

can be shown to be finite, then f is integrable over the t, τ plane, and

$$\int f(t, \tau)\, d(t, \tau) = \iint f(t, \tau)\, dt\, d\tau = \iint f(t, \tau)\, d\tau\, dt \qquad (2.111)$$

This result can be generalized to higher dimension in a straightforward manner.

An analogous result follows for interchanging the order of summation in double summations (Hirschman, 1962 pp. 119–121).

THEOREM 2.21. ABSOLUTE CONVERGENCE IMPLIES SUMMATION INTERCHANGE *If $\sum_{i=1}^\infty \sum_{j=1}^\infty |x_{ij}|$ is finite, then the following summations are equal:*

$$\sum_{i=1}^\infty \left[\sum_{j=1}^\infty x_{ij} \right] = \sum_{j=1}^\infty \left[\sum_{i=1}^\infty x_{ij} \right] \qquad (2.112)$$

2.6 SIGNAL CLASSIFICATION

Figure 2.13 shows a partial classification of measurable signals in terms of integrability, continuity, differentiability, and boundedness. This diagram is based on the following results. (1) Differentiability implies piecewise smoothness and pointwise continuity implies piecewise continuity; (2) piecewise smoothness is consistent with a discontinuity at a point, and thus, piecewise smoothness does not necessarily imply pointwise continuity. According to Theorem 2.6, piecewise smoothness implies piecewise continuity; (3) absolute continuity implies continuity as per Theorem 2.8; (4) piecewise continuity implies boundedness from Theorem 2.4; (5) boundedness implies local integrability, that is, if $f: \boldsymbol{R} \to \boldsymbol{C}$ is measurable and bounded, then $f \in L[\alpha, \beta]$. It is not necessarily the case that $f \in L$. This result follows because boundedness on a finite interval implies a finite integral. The function $f(t) = |\sin(2\pi f_c t)|$ shows the potential lack of integrability of a bounded function on the infinite interval $(-\infty, \infty)$; (6) integrability implies local integrability, that is, if $f \in L$, then $f \in L[\alpha, \beta]$ for $\alpha, \beta \in \boldsymbol{R}$. The converse is not true; (7) boundedness and an appropriate level of signal decay implies integrability on the infinite interval.

THEOREM 2.22. BOUNDEDNESS AND DECAY IMPLY INTEGRABILITY *If $f: \boldsymbol{R} \to \boldsymbol{C}$ is bounded and f decays at a rate greater than $k/|t|$ as $|t| \to \infty$, that is, $\exists \delta > 0$, $\exists t_o > 0$, such that $\forall |t| > t_o$ it is the case that $|f(t)| < k/|t|^{1+\delta}$, then $f \in L$.*

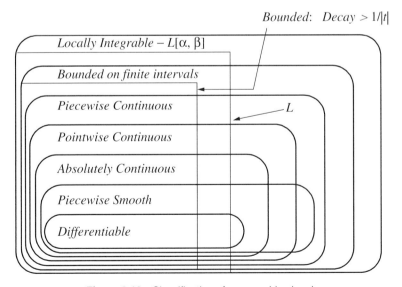

Figure 2.13 Classification of measurable signals.

Proof. Consider the interval $(0, \infty)$. Finiteness of the integral of f over this interval is demonstrated according to

$$\int_0^\infty |f| = \int_0^{t_o} |f| + \int_{t_o}^\infty |f| < \int_0^{t_o} |f| + \int_{t_o}^\infty \frac{k}{t^{1+\delta}} dt$$
$$= \int_0^{t_o} |f| + k \frac{-1}{\delta t^\delta} \bigg|_{t_o}^\infty = \int_0^{t_o} |f| + \frac{k}{\delta t_o^\delta} \tag{2.113}$$

2.7 CONVERGENCE

The standard forms of convergence for a sequence of functions include pointwise convergence, uniform convergence, and convergence in the mean. Two important associated results related to the interchange of limit and integration operations, are the monotone and dominated convergence theorems. In terms of notation a sequence of functions $\{f_n: \mathbf{R} \to \mathbf{C}\}_{n=1}^\infty$ is written as $\{f_n\}$ for convenience. A useful reference is Champeney (1987 ch. 4).

DEFINITION: POINTWISE AND UNIFORM CONVERGENCE OF FUNCTIONS A sequence of functions $\{f_n\}$ converges pointwise to a function $f: \mathbf{R} \to \mathbf{C}$ on a set E, written $\lim_{n \to \infty} f_n(t) = f(t)$, if for all $t \in E$ it is the case that

$$\forall \varepsilon > 0 \quad \exists N_o > 0 \quad \text{s.t.} \quad \forall n > N_o \quad |f_n(t) - f(t)| < \varepsilon \tag{2.114}$$

where N_o, in general, depends on t and ε. For the case where N_o is independent of the value of $t \in E$ the convergence is said to be uniform over E.

DEFINITION: CONVERGENCE IN MEAN A sequence of functions $\{f_n\}$ converges to a function f "in the mean" on a set E, if

$$\lim_{n \to \infty} \int_E |f_n(t) - f(t)| \, dt = 0 \tag{2.115}$$

that is,

$$\forall \varepsilon > 0 \quad \exists N_o > 0 \quad \text{s.t.} \quad \forall n > N_o \quad \int_E |f_n(t) - f(t)| \, dt < \varepsilon \tag{2.116}$$

2.7.1 Dominated and Monotone Convergence

Two important results that give sufficient conditions for the interchange of limit and integral operations, are the monotone and dominated convergence theorems (Champeney, 1987 p. 26).

THEOREM 2.23. MONOTONE CONVERGENCE THEOREM *If $\{f_n\}$ is a sequence of nonnegative functions, such that $f_n \in L$, $f_n(t) \leq f_{n+1}(t)$ for all n and all t, except on a set of zero measure, and $\int f_n < A$ for all n, then there will exist a function $f \in L$ such that*

$$\lim_{n \to \infty} f_n(t) = f(t) \quad \text{pointwise a.e.} \tag{2.117}$$

$$\lim_{n \to \infty} \int f_n(t) = \int \lim_{n \to \infty} f_n(t) = \int f \tag{2.118}$$

THEOREM 2.24. DOMINATED CONVERGENCE THEOREM *If $\{f_n\}$ is a sequence of functions, such that $f_n \in L$, $\lim_{n \to \infty} f_n(t) = f(t)$ pointwise almost everywhere, and $|f_n| < |g|$ for all n almost everywhere where $g \in L$, then*

$$\lim_{n \to \infty} \int f_n(t) = \int \lim_{n \to \infty} f_n(t) = \int f \tag{2.119}$$

THEOREM 2.25. DOMINATED CONVERGENCE IMPLIES CONVERGENCE IN MEAN *Consider a sequence of functions $\{f_n\}$, where $f_n \in L$ and $\lim_{n \to \infty} f_n(t) = f(t)$ pointwise almost everywhere on a set E. Assume dominated convergence of $\{f_n\}$, that is, assume there exists a function $g \in L$, such that $|f_n| < g$ for all $n \in \mathbf{Z}^+$ almost everywhere. Dominated convergence of this sequence implies, for all subsets $E_i \subseteq E$, that*

$$\lim_{n \to \infty} \int_{E_i} f_n(t) = \int_{E_i} f \tag{2.120}$$

It then follows, that the sequence of functions $\{f_n\}$ converges to f "in the mean" on the set E, that is,

$$\lim_{n \to \infty} \int_E |f_n(t) - f(t)| \, dt = 0 \tag{2.121}$$

Proof. Assume dominated convergence, such that, for all $\varepsilon > 0$ there exists a constant N_o, and for all $n > N_o$ and for all subsets E_i of E, it is the case that

$$\left| \int_{E_i} [f_n(t) - f(t)] \, dt \right| < \varepsilon/2 \tag{2.122}$$

For all $n > N_o$ the set E can be partitioned according to

$$E_1 = \{t \in E : f_n(t) - f(t) \geq 0\} \quad E_2 = \{t \in E : f_n(t) - f(t) < 0\} \tag{2.123}$$

whereupon it follows that

$$\int_{E_1} [f_n(t) - f(t)] \, dt < \varepsilon/2 \qquad \int_{E_2} -[f_n(t) - f(t)] \, dt < \varepsilon/2 \qquad (2.124)$$

The required result then follows, namely

$$\int_E |f_n(t) - f(t)| \, dt < \varepsilon \qquad (2.125)$$

2.8 FOURIER THEORY

This section details relevant Fourier results including Parseval's theorem, which is the fundamental result that is used in defining the power spectral density function. Other important results include the Riemann–Lebesgue theorem and the definition of a Dirichlet point. The definition of a Dirichlet point facilitates the definition of the integral properties of the Dirac delta function and the definition of the inverse Fourier transform. This section starts with a brief overview of Fourier series and the Fourier transform.

2.8.1 Fourier Series and Fourier Transform

DEFINITION: FOURIER SERIES A Fourier series, x_F, on the interval $[\alpha, \beta]$ for a signal $x \in L[\alpha, \beta]$ is defined according to

$$\begin{aligned} x_F(t) &= a_o + \sum_{i=1}^{\infty} a_i \cos(2\pi i f_o t) + b_i \sin(2\pi i f_o t) \\ &= \sum_{i=-\infty}^{\infty} c_i e^{j 2\pi i f_o t} \qquad t \in [\alpha, \beta] \end{aligned} \qquad (2.126)$$

where $f_o = 1/(\beta - \alpha)$ is the fundamental period and

$$a_o = \frac{1}{\beta - \alpha} \int_\alpha^\beta x(t) \, dt \qquad a_i = \frac{2}{\beta - \alpha} \int_\alpha^\beta x(t) \cos(2\pi i f_o t) \, dt \qquad (2.127)$$

$$b_i = \frac{2}{\beta - \alpha} \int_\alpha^\beta x(t) \sin(2\pi i f_o t) \, dt \qquad c_i = \frac{1}{\beta - \alpha} \int_\alpha^\beta x(t) e^{-j 2\pi i f_o t} \, dt \qquad (2.128)$$

In general, all the coefficients are complex, and from their definitions it follows that $a_{-i} = a_i$ and $b_{-i} = -b_i$. Further, it is readily shown that

$$c_i = \begin{cases} a_0 & i = 0 \\ 0.5(a_i - jb_i) & i \neq 0 \end{cases} \qquad (2.129)$$

and it then follows that

$$|c_i|^2 = \begin{cases} |a_0|^2 & i = 0 \\ 0.25(|a_i|^2 + |b_i|^2 - 2Im[a_i b_i^*]) & i \neq 0 \end{cases} \qquad (2.130)$$

The power in the sinusoidal components of the signal x_F with a frequency if_o, $i \neq 0$, is given by $(|a_i|^2 + |b_i|^2)/2$. From Eq. 2.130, and the results $a_{-i} = a_i$ and $b_{-i} = -b_i$, the important result

$$|c_{-i}|^2 + |c_i|^2 = (|a_i|^2 + |b_i|^2)/2 \qquad (2.131)$$

follows. In general $|c_{-i}|^2 \neq |c_i|^2$ but when the signal is real the simpler result $|c_{-i}|^2 = |c_i|^2$ holds.

THEOREM 2.26. CONVERGENCE OF A FOURIER SERIES *If $x \in L[\alpha, \beta]$ and is piecewise smooth, or has bounded variation, on $[\alpha, \beta]$, then*

$$x_F(t) = \begin{cases} x(t) & x \text{ is continuous at } t, \quad t \in (\alpha, \beta) \\ 0.5[x(t^+) + x(t^-)] & x \text{ is discontinuous at } t, \quad t \in (\alpha, \beta) \\ 0.5[x(\alpha^+) + x(\beta^-)] & t = \alpha, t = \beta \end{cases} \qquad (2.132)$$

Proof. See Champeney (1987 pp. 156–7) and Tolstov (1962 pp. 75f).

Consistent with this theorem, a standard way of viewing a Fourier series is as a decomposition of the underlying signal x, into its constituent component signals which are from one of the following orthogonal sinusoidal sets:

$$\left\{ 1, \cos(2\pi i f_o t), \sin(2\pi i f_o t), i \in \mathbf{Z}^+, f_o = \frac{1}{\beta - \alpha} \right\} \qquad (2.133)$$

$$\left\{ e^{j2\pi i f_o t}, i \in \mathbf{Z}, f_o = \frac{1}{\beta - \alpha} \right\} \qquad (2.134)$$

The general theory related to signal decomposition is best formulated on a Hilbert space which is vector space with an appropriately defined inner product (Kreyszig, 1978 ch. 3; Debnath, 1999 ch. 3).

DEFINITION: FOURIER TRANSFORM The Fourier transform of a signal x, denoted X and evaluated over the intervals $[\alpha, \beta]$ and $(-\infty, \infty)$ are, respectively, defined according to

$$\chi X([\alpha, \beta], f) = \int_{\alpha}^{\beta} x(t)e^{-j2\pi ft} \, dt \qquad X(f) = \int_{-\infty}^{\infty} x(t)e^{-j2\pi ft} \, dt \qquad (2.135)$$

In the following chapters, the interval $[0, T]$ is primarily considered and the Fourier transform of a signal x, over this interval, is denoted $X(T, f)$. The following properties of the Fourier transform are given with respect to this interval. (1) A requirement for the existence of X defined over $[0, T]$ is that $x \in L[0, T]$; (2) when x is real it follows that

$$X(T, -f) = X^*(T, f) \qquad (2.136)$$

(3) a link between the ith Fourier coefficient c_i in the exponential Fourier series of a signal, and the Fourier transform of the same signal, follows from Eq. (2.128) according to

$$c_i = \frac{X(T, if_o)}{T} \qquad (2.137)$$

2.8.2 Riemann–Lebesgue Theorem

If the variation in a function f is small over the period of a sinusoid $\cos(2\pi\lambda t)$, then the integral of $f(t)\cos(2\pi\lambda t)$ over such a period will be close to zero. A generalization of this result is given by the Riemann–Lebesgue theorem. This theorem is used, for example, to prove the existence of Dirichlet points for piecewise smooth signals. Appropriate reference are Champeney (1987 p. 23), Titchmarsh (1939 p. 403), and Tolstov (1962 p. 70).

THEOREM 2.27. RIEMANN–LEBESGUE THEOREM *If $f \in L$ then*

$$\lim_{\lambda \to \infty} \int_{-\infty}^{\infty} f(t) \cos(2\pi\lambda t) \, dt = 0 \qquad \lim_{\lambda \to \infty} \int_{-\infty}^{\infty} f(t) \sin(2\pi\lambda t) \, dt = 0 \qquad (2.138)$$

Proof. The proof of this theorem is detailed in Appendix 4.

2.8.3 Dirichlet Points

Consider a function g that is "impulsive" at a point t_o, and has unit area in a neighborhood $(t_o - \delta, t_o + \delta)$ of this point. If a function f is "smooth" over $(t_o - \delta, t_o + \delta)$ then it is expected that

$$\int_{t_o - \delta}^{t_o + \delta} f(t)g(t) \, dt \approx f(t_o) \qquad (2.139)$$

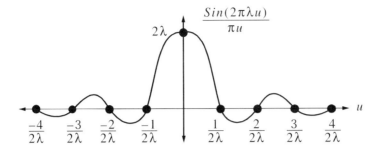

Figure 2.14 Illustration of "impulsive" function. The area under the graph equals unity.

A precise statement of this concept is given through the definition of a Dirichlet point and a Dirichlet value (Champeney, 1987 ch. 5). The existence of Dirichlet points for a piecewise smooth function, is fundamental to proving Parseval's theorem and the relationship between a function and the inverse Fourier transform of its Fourier transform.

DEFINITION: DIRICHLET POINT A point t_o is a Dirichlet point of a locally integrable function, $f: \mathbf{R} \to \mathbf{C}$, if $\exists \delta_o > 0$, such that for $0 < \delta < \delta_o$ it is the case that

$$\lim_{\lambda \to \infty} \int_{-\delta}^{\delta} f(t_o + u) \cdot \frac{\sin(2\pi\lambda u)}{\pi u} \, du \qquad (2.140)$$

is finite.

Note that the function $\sin(2\pi\lambda u)/\pi u$, as shown in Figure 2.14, is "impulsive."

DEFINITION: DIRICHLET VALUE If t_o is a Dirichlet point of a locally integrable function, $f: \mathbf{R} \to \mathbf{C}$, then the value defined by the limit in Eq. (2.140) is called the Dirichlet value of f at t_o.

2.8.3.1 Existence of Dirichlet Points
There is no known necessary and sufficient conditions for the existence of a Dirichlet point (Champeney, 1987 p. 43). In fact, on any finite interval (α, β) there exists functions that are locally integrable, but which have no Dirichlet points on that interval (Champeney, 1987 p. 38). The following two theorems give sufficient conditions for the existence of a Dirichlet point (Champeney, 1987 p. 41).

THEOREM 2.28. DIFFERENTIABILITY IMPLIES DIRICHLET POINT *If f is differentiable at a point then that point is a Dirichlet point of f. Further, the Dirichlet value* of f at t_o is $f(t_o)$, that is,

$$f(t_o) = \lim_{\lambda \to \infty} \int_{-\delta}^{\delta} f(t_o + u) \cdot \frac{\sin(2\pi\lambda u)}{\pi u} du \qquad \delta > 0 \qquad (2.141)$$

Proof. The proof of this theorem is detailed in Appendix 5.

THEOREM 2.29. PIECEWISE SMOOTHNESS IMPLIES DIRICHLET POINT *If f is piecewise smooth at a point, then that point is a Dirichlet point of f. Further, the Dirichlet value of f at t_o is $[f(t_o^+) + f(t_o^-)]/2$ (Champeney, 1987 pp. 39, 42), that is,*

$$\frac{f(t_o^+) + f(t_o^-)}{2} = \lim_{\lambda \to \infty} \int_{-\delta}^{\delta} f(t_o \pm u) \cdot \frac{\sin(2\pi\lambda u)}{\pi u} du \qquad \delta > 0 \qquad (2.142)$$

$$\frac{f(t_o^+) + f(t_o^-)}{2} = \lim_{\lambda \to \infty} \int_{t_o-\delta}^{t_o+\delta} f(u) \cdot \frac{\sin[2\pi\lambda(u - t_o)]}{\pi(u - t_o)} du \qquad \delta > 0 \quad (2.143)$$

Proof. The proof of this theorem is similar to the proof of the previous theorem. A change of variable yields the alternative forms stated in this theorem.

2.8.3.2 Continuity and Existence of a Dirichlet Point

It is reasonable to assume that if f is continuous at a point t_o, then that point is a Dirichlet point of f. However, this is not the case, and one example is given in Champeney (1987 p. 37). Thus, as noted above, it is not surprising that, if $f \in L[\alpha, \beta]$ then it can be the case that f does not possess any Dirichlet points.

2.8.4 Inverse Fourier Transform

The definition of a Dirichlet point facilitates the statement of conditions related to the existence of the inverse Fourier transform as detailed in the following theorem.

THEOREM 2.30. EXISTENCE OF INVERSE FOURIER TRANSFORM *If $x \in L[0, T]$, the Fourier transform of x on $[0, T]$, denoted $X(T, f)$, is Lebesgue integrable on $(-\infty, \infty)$, and x is piecewise smooth at a point t_o, then the inverse Fourier transform at t_o exists and is given by*

$$\int_{-\infty}^{\infty} X(T, f) e^{j2\pi f t_o} df = \begin{cases} 0 & t_o \notin [0, T] \\ x(0^+)/2 & t_o = 0 \\ \dfrac{x(t_o^+) + x(t_o^-)}{2} & t_o \in (0, T) \\ x(T^-)/2 & t_o = T \end{cases} \qquad (2.144)$$

Proof. The proof is detailed in Appendix 6.

If $X \notin L$, then indirect approaches for establishing the inverse Fourier transform for specific classes of signals can be used. A useful reference is Champeney (1987 ch. 8f).

2.8.5 Dirac Delta Function

The Dirac delta function, denoted δ, is widely used in analysis. However, its definition is problematic and best formulated using generalized function theory [see for example, Champeney (1987 ch. 12)]. For the material that follows, the following definition will suffice.

DEFINITION: DIRAC DELTA FUNCTION Consider a sequence of functions $\{\delta_i\}_{i=1}^{\infty}$ such that $\delta_i(t) = \delta_i(-t)$,

$$\lim_{i \to \infty} \delta_i(t) = \begin{cases} \infty & t = 0 \\ 0 & t \neq 0 \end{cases} \tag{2.145}$$

and

$$\lim_{i \to \infty} \int_{-\infty}^{\infty} \delta_i(t)\, dt = 1 \tag{2.146}$$

By definition

$$\delta(t) = \lim_{i \to \infty} \delta_i(t) \tag{2.147}$$

2.8.5.1 Integral Properties of Dirac Delta Function
First, as δ is non-zero only on a set of zero measure, it follows [see Eq. (2.86)] that

$$\int_{-\infty}^{\infty} \delta(t)\, dt = 0 \tag{2.148}$$

Second, if t_o is a Dirichlet point of $f \in L$, then

$$\lim_{i \to \infty} \int_{-\infty}^{\infty} \delta_i(t - t_o) f(t)\, dt = \frac{f(t_o^+) + f(t_o^-)}{2} \tag{2.149}$$

but

$$\int_{-\infty}^{\infty} \delta(t - t_o) f(t)\, dt = 0 \tag{2.150}$$

However, for notational convenience, $\lim_{i \to \infty} \int_{-\infty}^{\infty} \delta_i(t - t_o) f(t)\, dt$ is written as $\int_{-\infty}^{\infty} \delta(t - t_o) f(t)\, dt$.

2.8.6 Parseval's Theorem

Parseval's theorem is fundamental to defining the power spectral density. A formal statement of this theorem is as follows (Champeney, 1987 p. 72; Titchmarsh, 1948 pp. 50, 69–76).

THEOREM 2.31. PARSEVAL'S THEOREM *Denote the Fourier transform of x as X. If $x \in L$ and $x \in L^2$, then $X \in L^2$, and*

$$\int_{-\infty}^{\infty} |x(t)|^2 \, dt = \int_{-\infty}^{\infty} |X(f)|^2 \, df \qquad (2.151)$$

If $x \in L^2[\alpha, \beta]$ then $x \in L[\alpha, \beta]$, and

$$\int_{\alpha}^{\beta} |x(t)|^2 \, dt = \int_{-\infty}^{\infty} |X([\alpha, \beta], f)|^2 \, df \qquad \int_{0}^{T} |x(t)|^2 \, dt = \int_{-\infty}^{\infty} |X(T, f)|^2 \, df \qquad (2.152)$$

Proof. The proof of this relationship is detailed in Appendix 7.

2.9 RANDOM PROCESSES

The theory of random processes is vast, as books by Papoulis (2002), and Grimmett (1992) attest. The theory related to random processes that is required in subsequent chapters is not great, but the following concepts are fundamental.

A strict definition of a random process (RP) is that it is a set of random variables [see for example, Grimmett (1992 ch. 8)]. In engineering, however, an alternative, but consistent way of defining a random process is as a set of signals that are governed by a probability of occurrence. Consider a random process $X(T)$ that defines a set of signals on the interval $[0, T]$, called an ensemble, and denoted $E_X(T)$. This set has the form

$$E_X(T) = \{x : S_X \times [0, T] \to \mathbf{C}\} \qquad (2.153)$$

where S_X is an index set to distinguish between waveforms in the ensemble. For the case of a countable ensemble $S_X \subseteq \mathbf{Z}^+$, and for an uncountable ensemble $S_X \subseteq \mathbf{R}$. For the countable case it is usual, for general analysis, to assume that $S_X = \mathbf{Z}^+$. For notational convenience, and for the countable case, a subscript rather than an argument is used for the ith signal according to $x(i, t) = x_i(t)$.

The probability associated with each waveform defines the probability space, P_X, when the ensemble has a countable number of waveforms, that is,

$$P_X = \left\{ p_i : p_i = P[x(i, t)], \ i \in \mathbf{Z}^+, \ \sum_{i=1}^{\infty} p_i = 1 \right\} \qquad (2.154)$$

where $P : E_X(T) \to [0, 1]$ is the probability operator.

For the uncountable case, a density function f_X characterizes the probability of waveforms in the ensemble according to

$$P[x(\lambda, t)|_{\lambda \in [\lambda_o, \lambda_o + d\lambda]}] = f_X(\lambda_o) \, d\lambda \qquad f_X(\lambda) \geq 0 \tag{2.155}$$

$$\int_{-\infty}^{\infty} f_X(\lambda) \, d\lambda = 1$$

For notational convenience the argument T is dropped and X rather than $X(T)$, and E_X rather than $E_X(T)$, is used.

2.10 MISCELLANEOUS RESULTS

The following theorem details results used in later chapters for the development of the theory related to the power spectral density.

THEOREM 2.32. MISCELLANEOUS RESULTS

$$\sum_{i=1}^{N} e^{-j2\pi i f/f_o} = e^{-j2\pi f/f_o} \sum_{i=0}^{N-1} e^{-j2\pi i f/f_o} = \begin{cases} e^{-j2\pi f/f_o} \left[\dfrac{1 - e^{-j2\pi N f/f_o}}{1 - e^{-j2\pi f/f_o}} \right] \\ N \qquad f = k f_o, \, k \in \mathbf{Z} \end{cases} \tag{2.156}$$

$$\left| \sum_{i=1}^{N} e^{-j2\pi i f/f_o} \right|^2 = \left| \sum_{i=0}^{N-1} e^{-j2\pi i f/f_o} \right|^2 = \begin{cases} \dfrac{\sin^2(\pi N f/f_o)}{\sin^2(\pi f/f_o)} & -\infty < f < \infty \\ N^2 & f = k f_o, \, k \in \mathbf{Z} \end{cases}$$

$$= N \left[1 + 2 \sum_{i=1}^{N-1} \left(1 - \frac{i}{N} \right) \cos(2\pi i f/f_o) \right] \tag{2.157}$$

$$\lim_{N \to \infty} \frac{1}{N} \cdot \frac{\sin^2(\pi N f/f_o)}{\sin^2(\pi f/f_o)} = \sum_{i=-\infty}^{\infty} f_o \delta(f - i f_o) \tag{2.158}$$

$$\lim_{T \to \infty} T \operatorname{sinc}^2(fT) = \delta(f) \qquad \operatorname{sinc}(x) = \frac{\sin(\pi x)}{\pi x} \tag{2.159}$$

Proof. The proof of these results is detailed in Appendix 8.

The following theorem details the Fourier transforms of signals that occur several times in subsequent chapters.

46 BACKGROUND: SIGNAL AND SYSTEM THEORY

THEOREM 2.33. FOURIER TRANSFORMS OF COMMON SIGNALS

$$p(t) = \begin{cases} 1 & 0 \leq t < T \\ 0 & \text{elsewhere} \end{cases} \leftrightarrow P(T, f) = T \operatorname{sinc}(fT) e^{-j\pi fT} \qquad (2.160)$$

$$\begin{cases} c_k e^{j2\pi k f_i t} & 0 \leq t < T \\ 0 & \text{elsewhere} \end{cases} \leftrightarrow c_k T e^{-j\pi(f - kf_i)T} \operatorname{sinc}((f - kf_i)T) \qquad (2.161)$$

Proof. The proof of these results follows directly from the definition of the Fourier transform.

APPENDIX 1: PROOF OF THEOREM 2.11

The contrapositive form of the last result in Theorem 2.11 is: If $\exists \varepsilon > 0$, such that $\forall f_o > 0$ it is the case that

$$\int_E |f| > \varepsilon \qquad E = \{t : |f(t)| > f_o\} \qquad (2.162)$$

then $f \notin L$. To prove this contrapositive form, fix $\varepsilon > 0$. Choose $f_1 > 0$ such that

$$\int_{E_1} |f| > \varepsilon \qquad E_1 = \{t : |f(t)| > f_1\} \qquad (2.163)$$

Then choose a sufficiently large number $f_2 > f_1$, such that

$$\int_{E_2} |f| > \varepsilon \qquad \int_{E_1 \cap E_2^C} |f| > \frac{\varepsilon}{2} \qquad E_2 = \{t : |f(t)| > f_2\} \qquad (2.164)$$

Clearly, $E_2 \subseteq E_1$ and $E_2 \cap (E_1 \cap E_2^C) = \emptyset$. A sequence of numbers $\{f_i\}$, and associated sets $\{E_i\}$, can be constructed such that

$$E_1 = \bigcup_{i=1}^{\infty} (E_i \cap E_{i+1}^C) \qquad \int_{E_i \cap E_{i+1}^C} |f| > \frac{\varepsilon}{i+1} \qquad (2.165)$$

and

$$\int_{E_1} |f| = \sum_{i=1}^{\infty} \int_{E_i \cap E_{i+1}^C} |f| > \sum_{i=1}^{\infty} \frac{\varepsilon}{i+1} = \infty \qquad (2.166)$$

which implies $f \notin L$.

APPENDIX 2: PROOF OF THEOREM 2.13

Assume $f \in L^2[\alpha, \beta]$. With $f(t) = x(t) + jy(t)$, it follows that

$$\int_\alpha^\beta [x^2(t) + y^2(t)]\, dt < \infty \qquad (2.167)$$

Clearly, $x, y \in L^2[\alpha, \beta]$. Define the sets E_1 and E_2 according to

$$\begin{aligned} E_1 &= \{t : |x(t)| \le 1\} & M(E_1) &\le \beta - \alpha \\ E_2 &= \{t : |x(t)| > 1\} & M(E_2) &\le \beta - \alpha \end{aligned} \qquad (2.168)$$

It then follows that

$$\begin{aligned} x^2(t) &\le |x(t)| \le 1 & t \in E_1 \\ 1 &< |x(t)| < x^2(t) & t \in E_2 \end{aligned} \qquad (2.169)$$

and hence,

$$\int_{E_2} |x(t)|\, dt < \int_{E_2} x^2(t)\, dt < \infty \qquad \int_{E_1} |x(t)|\, dt \le \int_{E_1} dt = M(E_1) < \infty \qquad (2.170)$$

which proves that $x \in L[\alpha, \beta]$. A similar argument can be used to show $y \in L[\alpha, \beta]$ and hence, $f \in L[\alpha, \beta]$.

APPENDIX 3: PROOF OF THEOREM 2.17

First, consider $\varepsilon > 0$ and $\delta > 0$ to be a set. Second, consistent with Theorem 2.12 a range $[-f_o, f_o]$ can be defined such that the function f is outside this range only on a set of measure less than $\varepsilon/2$. For subsequent parts of the proof, it is convenient to choose f_o such that $f_o = N\delta$ for $N \in Z^+$ and $M\{t : |f(t)| > f_o\} < \varepsilon/2$. Third, demarcate the range $[-f_o, f_o]$ into subranges separated by δ. Fourth, define the $2N + 1$ sets that partition the interval $[\alpha, \beta]$, apart from the set of points where $|f| > f_o$, according to

$$E_i = f^{-1}(L_i) = \{t : f(t) \in L_i\}$$

$$L_i = [i\delta, (i+1)\delta) \quad \text{for} \quad i \in \{-N, \ldots, 0, \ldots, N-1\} \qquad L_N = \{f_o\}$$

$$(2.171)$$

Now, the ith set E_i may consist of an infinite number of intervals and/or an infinite number of points. In some instances, the separation between these intervals or points is negligibly small. For example, consider the function

$$f(t) = \begin{cases} 0 & t \in Q \\ 1 & t \notin Q \end{cases} \qquad (2.172)$$

However, since f is a measurable function, and L_i is a measurable set, it follows that E_i is a measurable set. Hence, according to the definition of a measurable set, there exists an open set F_i such that

$$E_i \subseteq F_i \qquad M(F_i) - M(E_i) < \varepsilon \qquad (2.173)$$

Now, an open set of $\mathbf{R}$ can be written as a union of a countable number of disjoint open intervals (Sprechter, 1970 pp. 123, 136–137). Thus,

$$F_i = \bigcup_{j=1}^{\infty} I_{ij} \qquad I_{ij} \cap I_{ik} = \varnothing \quad \text{when} \quad j \neq k \qquad (2.174)$$

With the definition of the set of disjoint intervals $\{J_{ij} = I_{ij} \cap E_i\}_{j=1}^{\infty}$, it follows that $E_i = \bigcup_{j=1}^{\infty} J_{ij}$, and hence, there exists a number N_i, such that

$$M(E_i) - \sum_{j=1}^{N_i} M(J_{ij}) < \frac{\varepsilon}{2} \cdot \frac{1}{2N+1} \qquad (2.175)$$

Then, it is possible to define a function φ according to

$$\varphi(t) = \sum_{i=-N}^{N} i\delta \cdot \chi_i(t) \qquad \chi_i(t) = \begin{cases} 1 & t \in \bigcup_{j=1}^{N_i} J_{ij} \\ 0 & \text{otherwise} \end{cases} \qquad (2.176)$$

This function is a simple function defined on $\sum_{i=-N}^{N} N_i$ disjoint intervals, and takes on values from the set $\{-N\delta, \ldots, N\delta\}$ for $t \in [\alpha, \beta]$. It is such that $|f(t) - \varphi(t)| > \delta$ on a set of measure less than ε, and this occurs when $|f(t)| > f_o$, or when the finite union of disjoint sets $\bigcup_{j=1}^{N_i} J_{ij}$ does not equal E_i. Hence, the measure of the set of points, such that $|f(t) - \varphi(t)| > \delta$, by construction, is less than ε. By drawing straight line segments of finite slope between the value of φ at the end point of one interval and the start of another, a continuous and piecewise smooth function ϕ, can be created, which apart from a set whose measure is less than $\varepsilon + \varepsilon'$, approximates the function f with an error less than δ. The term ε' accounts for the measure taken up by the straight line segments. This can be made arbitrarily small by increasing the slope of the line segments.

By construction, the function ϕ is continuous and piecewise smooth over $[\alpha, \beta]$. Hence, according to Theorem 2.9, ϕ is absolutely continuous.

APPENDIX 4: PROOF OF THEOREM 2.27

The goal is to show that

$$\forall \varepsilon > 0 \quad \exists \lambda_o \quad \text{s.t.} \quad \forall \lambda > \lambda_o \quad \left| \int_{-\infty}^{\infty} f(t) \begin{Bmatrix} \cos(2\pi\lambda t) \\ \sin(2\pi\lambda t) \end{Bmatrix} dt \right| < \varepsilon \quad (2.177)$$

The proof will consider the integral of $f(t) \cos(2\pi\lambda t)$. The proof for the integral of $f(t) \sin(2\pi\lambda t)$ follows in an analogous manner. First, since $f \in L$

$$\left| \int_{-\infty}^{\infty} f(t) \cos(2\pi\lambda t) \, dt \right| \leq \int_{-\infty}^{\infty} |f(t) \cos(2\pi\lambda t)| \, dt < \int_{-\infty}^{\infty} |f(t)| \, dt < \infty \quad (2.178)$$

and from Theorem 2.11, it follows $\forall \varepsilon > 0$ that there exists a $T > 0$, such that

$$\left| \int_{T}^{\infty} f(t) \cos(2\pi\lambda t) \, dt \right| \leq \int_{T}^{\infty} |f(t)| \, dt < \varepsilon$$

$$\left| \int_{-\infty}^{-T} f(t) \cos(2\pi\lambda t) \, dt \right| < \varepsilon \quad (2.179)$$

It then remains to show for any given value of $\varepsilon > 0$, that

$$\exists \lambda_o \quad \text{s.t.} \quad \forall \lambda > \lambda_o \quad \left| \int_{-T}^{T} f(t) \cos(2\pi\lambda t) \, dt \right| < \varepsilon \quad (2.180)$$

To this end, note from Theorem 2.18, that $\forall \varepsilon > 0$, there exists an absolutely continuous function $\phi: \mathbf{R} \to \mathbf{C}$, such that

$$\int_{-T}^{T} |f(t) - \phi(t)| \, dt < \varepsilon \quad (2.181)$$

It then follows from

$$\int_{-T}^{T} f(t) \cos(2\pi\lambda t) \, dt = \int_{-T}^{T} [f(t) - \phi(t)] \cos(2\pi\lambda t) \, dt + \int_{-T}^{T} \phi(t) \cos(2\pi\lambda t) \, dt$$

$$(2.182)$$

and $|A + B| \leq |A| + |B|$, that

$$\int_{-T}^{T} f(t) \cos(2\pi\lambda t)\, dt \leq \left| \int_{-T}^{T} [f(t) - \phi(t)] \cos(2\pi\lambda t)\, dt \right| + \left| \int_{-T}^{T} \phi(t) \cos(2\pi\lambda t)\, dt \right|$$

$$\leq \int_{-T}^{T} |f(t) - \phi(t)|\, dt + \left| \int_{-T}^{T} \phi(t) \cos(2\pi\lambda t)\, dt \right| \quad (2.183)$$

$$\leq \varepsilon + \left| \int_{-T}^{T} \phi(t) \cos(2\pi\lambda t)\, dt \right|$$

Since ϕ is absolutely continuous, integration by parts can be used (Jain, 1986 p. 199) in this equation to yield

$$\int_{-T}^{T} f(t) \cos(2\pi\lambda t)\, dt \leq \varepsilon + \left| \left[\frac{\phi(t) \sin(2\pi\lambda t)}{2\pi\lambda} \right]_{-T}^{T} - \frac{1}{2\pi\lambda} \int_{-T}^{T} \phi'(t) \sin(2\pi\lambda t)\, dt \right|$$
(2.184)

Absolute continuity for ϕ ensures that ϕ is bounded and ϕ' is Lebesgue integrable (Jain, 1986 p. 197), on $[-T, T]$. It then follows that

$$\lim_{\lambda \to \infty} \int_{-T}^{T} f(t) \cos(2\pi\lambda t)\, dt \leq \varepsilon \quad (2.185)$$

which completes the proof.

APPENDIX 5: PROOF OF THEOREM 2.28

The point t_o is a Dirichlet point of a differentiable function f, and the Dirichlet value of f at t_o is $f(t_o)$, if there exists a constant $\delta_o > 0$ such that, for all $0 < \delta < \delta_o$, it is the case that

$$\forall \varepsilon > 0 \quad \exists \lambda_o > 0 \quad \text{s.t.} \quad \forall \lambda > \lambda_o \quad \left| \int_{-\delta}^{\delta} f(t_o + u) \cdot \frac{\sin(2\pi\lambda u)}{\pi u}\, du - f(t_o) \right| < \varepsilon$$
(2.186)

First, it is required to show that the area under the "impulsive" function $\sin(2\pi\lambda u)/\pi u$ approaches unity as $\lambda \to \infty$. To this end define $I(\lambda)$ as

$$I(\lambda) = \int_0^{\delta} \frac{\sin(2\pi\lambda u)}{\pi u}\, du \quad (2.187)$$

A change of variable $\zeta = \lambda u$ and the definite integral (Spiegel, 1968 p. 96)

$$I(\lambda) = \int_0^{\lambda\delta} \frac{\sin(2\pi\zeta)}{\pi\zeta} d\zeta \qquad \int_0^{\infty} \frac{\sin(p\zeta)}{\zeta} d\zeta = \frac{\pi}{2} \quad p > 0 \qquad (2.188)$$

yields the required result, namely

$$\lim_{\lambda \to \infty} I(\lambda) = \int_0^{\infty} \frac{\sin(2\pi\zeta)}{\pi\zeta} d\zeta = \frac{1}{2} \qquad (2.189)$$

Second, the following result is required: With the definitions for the integrals I_1 and I_2 according to

$$I_1 = \int_0^{\delta} [\alpha + \beta_1 u] \frac{\sin(2\pi\lambda u)}{\pi u} \cdot du \qquad I_2 = \int_0^{\delta} [\alpha + \beta_2 u] \frac{\sin(2\pi\lambda u)}{\pi u} \cdot du \qquad (2.190)$$

it follows that

$$\beta_1 < \beta_2 \Rightarrow I_1 < I_2 \qquad (2.191)$$

To prove these results consider the integral I defined according to

$$I = \int_0^{\delta} [\alpha + \beta u] \frac{\sin(2\pi\lambda u)}{\pi u} du = \alpha \int_0^{\delta} \frac{\sin(2\pi\lambda u)}{\pi u} du + \frac{\beta}{2\pi^2 \lambda} [1 - \cos(2\pi\lambda\delta)] \qquad (2.192)$$

The stated inequalities follow because $0 \leq 1 - \cos(2\pi\lambda\delta) \leq 2$. Hence, I increases with β if $\beta > 0$ and decreases as β becomes more negative when $\beta < 0$.

Third, differentiability of f implies bounds on the nature of the signal around a point t_o, as indicated in Figure 2.5, according to

$$f(t_o) + u[f'(t_o) - \varepsilon] < f(t_o + u) < f(t_o) + u[f'(t_o) + \varepsilon] \qquad (2.193)$$

for $0 < u < \delta$ where δ is consistent with the chosen value of $\varepsilon > 0$. Further, $f(t_o + u) \approx f(t_o) + uf'(t_o)$ for $0 < u < \delta$. Thus, for fixed values of ε and δ consistent with this inequality, it follows from Eq. (2.191) that

$$\int_0^{\delta} [f(t_o) + u[f'(t_o) - \varepsilon]] \cdot \frac{\sin(2\pi\lambda u)}{\pi u} du < \int_0^{\delta} [f(t_o + u)] \cdot \frac{\sin(2\pi\lambda u)}{\pi u} du$$

$$< \int_0^{\delta} [f(t_o) + u[f'(t_o) + \varepsilon]] \cdot \frac{\sin(2\pi\lambda u)}{\pi u} du \qquad (2.194)$$

and hence,

$$f(t_o) \int_0^\delta \frac{\sin(2\pi\lambda u)}{\pi u} du + \frac{[f'(t_o) - \varepsilon]}{2\pi^2\lambda}[1 - \cos(2\pi\lambda\delta)] < \int_0^\delta f(t_o + u) \cdot \frac{\sin(2\pi\lambda u)}{\pi u} du$$

$$< f(t_o) \int_0^\delta \frac{\sin(2\pi\lambda u)}{\pi u} du + \frac{[f'(t_o) + \varepsilon]}{2\pi^2\lambda}[1 - \cos(2\pi\lambda\delta)] \quad (2.195)$$

This result holds for all $\lambda > 0$, and as $\lambda \to \infty$, it follows from Eq. (2.189) that

$$\frac{f(t_o)}{2} \leq \lim_{\lambda \to \infty} \int_0^\delta f(t_o + u) \cdot \frac{\sin(2\pi\lambda u)}{\pi u} du \leq \frac{f(t_o)}{2} \quad (2.196)$$

which is the required result.

APPENDIX 6: PROOF OF THEOREM 2.30

If $x \in L[0, T]$ and $X \in L$, then the following integral is finite:

$$y(t) = \lim_{\lambda \to \infty} \int_{-\lambda}^\lambda X(T, f) e^{j2\pi ft} df \quad (2.197)$$

Using the definition for the Fourier transform, it follows that

$$y(t) = \lim_{\lambda \to \infty} \int_{-\lambda}^\lambda \left[\int_0^T x(\tau) e^{-j2\pi f\tau} d\tau \right] e^{j2\pi ft} df \quad (2.198)$$

Since $x(\tau)e^{-j2\pi f\tau}e^{j2\pi ft}$ is absolutely integrable for $\tau \in [0, T]$ and $f \in [-\lambda, \lambda]$, it follows, according to Theorem 2.20, that the order of integration can be interchanged to yield

$$y(t) = \lim_{\lambda \to \infty} \int_0^T x(\tau) \left[\int_{-\lambda}^\lambda e^{-j2\pi f(\tau - t)} df \right] d\tau \quad (2.199)$$

Since sin is an odd function it follows that

$$\int_{-\lambda}^\lambda e^{-j2\pi f(\tau - t)} df = \int_{-\lambda}^\lambda \cos[2\pi f(\tau - t)] df = \frac{\sin[2\pi\lambda(\tau - t)]}{\pi(\tau - t)} \quad (2.200)$$

which is valid for all t, τ, including the case where $t = \tau$ whereupon

$\sin[2\pi\lambda(t-\tau)]/\pi(t-\tau) = 2\lambda$. Thus,

$$y(t) = \lim_{\lambda \to \infty} \int_0^T x(\tau) \frac{\sin(2\pi\lambda(\tau-t))}{\pi(\tau-t)} d\tau \qquad (2.201)$$

As x is piecewise smooth, it follows from Theorem 2.29 that

$$y(t) = \begin{cases} 0 & t \notin [0, T] \\ x(0^+)/2 & t = 0 \\ \dfrac{x(t^+) + x(t^-)}{2} & t \in (0, T) \\ x(T^-)/2 & t = T \end{cases} \qquad (2.202)$$

which is the required result.

APPENDIX 7: PROOF OF THEOREM 2.31

Assume $x \in L$. If $x \in L$, then the Fourier transform of x, denoted X, exists. Further, X is finite for all values of its argument. Thus, the integral I, defined according to

$$I(\lambda) = \int_{-\lambda}^{\lambda} |X(f)|^2 \, df \qquad (2.203)$$

is finite for all finite values of λ. Using the definition of the Fourier transform, I can be written as

$$I(\lambda) = \int_{-\lambda}^{\lambda} \left[\int_{-\infty}^{\infty} x(t) e^{-j2\pi ft} \, dt \right] \left[\int_{-\infty}^{\infty} x^*(\tau) e^{j2\pi f\tau} \, d\tau \right] df \qquad (2.204)$$

Since $x(t)e^{-j2\pi ft} x^*(\tau)e^{j2\pi f\tau}$ is absolutely integrable for $t, \tau \in (-\infty, \infty)$ and $f \in [-\lambda, \lambda]$, it follows, according to Theorem 2.20, that the order of integration can be interchanged to yield

$$I(\lambda) = \int_{-\infty}^{\infty} \int_{-\infty}^{\infty} x(t)x^*(\tau) \left[\int_{-\lambda}^{\lambda} e^{-j2\pi f(t-\tau)} \, df \right] dt \, d\tau \qquad (2.205)$$

Since sin is an odd function it follows that

$$\int_{-\lambda}^{\lambda} e^{-j2\pi f(t-\tau)} \, df = \int_{-\lambda}^{\lambda} \cos[2\pi f(t-\tau)] \, df = \frac{\sin[2\pi\lambda(t-\tau)]}{\pi(t-\tau)} \qquad (2.206)$$

which is valid for all t, τ, including the case where $t = \tau$, whereupon

$$\sin[2\pi\lambda(t-\tau)]/\pi(t-\tau) = 2\lambda$$

Thus,

$$I(\lambda) = \int_{-\infty}^{\infty} \int_{-\infty}^{\infty} x(t) x^*(\tau) \frac{\sin[2\pi\lambda(t-\tau)]}{\pi(t-\tau)} dt \, d\tau \qquad (2.207)$$

Consider the limit of I as $\lambda \to \infty$, that is,

$$\lim_{\lambda \to \infty} I(\lambda) = \lim_{\lambda \to \infty} \int_{-\infty}^{\infty} f(t, \lambda) \, dt \qquad (2.208)$$

where

$$f(t, \lambda) = x(t) \int_{-\infty}^{\infty} x^*(\tau) \frac{\sin[2\pi\lambda(t-\tau)]}{\pi(t-\tau)} d\tau \qquad (2.209)$$

To complete the proof, it is necessary to interchange the order of limit and integration in Eq. (2.208). According to the dominated convergence theorem (Theorem 2.24) a sufficient condition for the validity of this interchange is the existence of a function $g \in L$ such that $|f(t, \lambda)| < g(t)$ for $t \in \mathbf{R}$ and for all values of λ. The proof of the existence of such a function follows.

First, since $x \in L$, it follows from Theorems 2.17 and 2.18 that for all $\varepsilon, \delta > 0$ there exists an absolutely continuous function $\phi \in L$, such that

$$M\{\tau : |x^*(\tau) - \phi(\tau)| > \delta\} < \varepsilon \qquad \int_{-\infty}^{\infty} |x^*(\tau) - \phi(\tau)| \, d\tau < \varepsilon \qquad (2.210)$$

Now, for all finite values of λ, $\sin[2\pi\lambda(t-\tau)]/\pi(t-\tau)$ is continuous and bounded with a bound of 2λ. It then follows that there exists an $\varepsilon' > 0$, such that

$$\left| \int_{-\infty}^{\infty} x^*(\tau) \frac{\sin[2\pi\lambda(t-\tau)]}{\pi(t-\tau)} d\tau - \int_{-\infty}^{\infty} \phi(\tau) \frac{\sin[2\pi\lambda(t-\tau)]}{\pi(t-\tau)} d\tau \right|$$
$$\leq \int_{-\infty}^{\infty} \left| \frac{\sin[2\pi\lambda(t-\tau)]}{\pi(t-\tau)} \right| |x^*(\tau) - \phi(\tau)| \, d\tau < \varepsilon' \qquad (2.211)$$

Thus, there exists an absolutely continuous function ϕ, such that

$$\int_{-\infty}^{\infty} \phi(\tau) \frac{\sin[2\pi\lambda(t-\tau)]}{\pi(t-\tau)} d\tau$$

approximates

$$\int_{-\infty}^{\infty} x^*(\tau) \frac{\sin[2\pi\lambda(t-\tau)]}{\pi(t-\tau)} d\tau$$

arbitrarily closely.

Second, since ϕ is an absolutely continuous function, it follows from Theorem 2.10 that ϕ is differentiable at all points except on a set of zero measure. Thus, all points of $\mathbf{R}$, except for a set of zero measure, are Dirichlet points of ϕ. From Theorem 2.28 it follows that

$$\lim_{\lambda \to \infty} \int_{-\infty}^{\infty} \phi(\tau) \frac{\sin[2\pi\lambda(t-\tau)]}{\pi(t-\tau)} d\tau = \phi(t) \quad \text{a.e.} \tag{2.212}$$

which is independent of λ. Hence,

$$|f(t, \lambda)| < k|x(t)| |\phi(t)| \quad \text{a.e.} \tag{2.213}$$

for some fixed value of k. Since $x \in L$ and ϕ is bounded, it follows that $x\phi \in L$ and thus, from the dominated convergence theorem, the order of the limit and integration operation can be interchanged in Eq. (2.208) to yield

$$\lim_{\lambda \to \infty} I(\lambda) = \lim_{\lambda \to \infty} \int_{-\infty}^{\infty} f(t, \lambda) \, dt = \int_{-\infty}^{\infty} \lim_{\lambda \to \infty} f(t, \lambda) \, dt$$

$$= \int_{-\infty}^{\infty} x(t) \left[\lim_{\lambda \to \infty} \int_{-\infty}^{\infty} x^*(\tau) \frac{\sin[2\pi\lambda(t-\tau)]}{\pi(t-\tau)} d\tau \right] dt \tag{2.214}$$

Clearly, if t is a Dirichlet point of x^* with Dirichlet value $x^*(t)$, and this is the case for all $t \in \mathbf{R}$ except on a set of zero measure, then

$$\lim_{\lambda \to \infty} I(\lambda) = \int_{-\infty}^{\infty} |x(t)|^2 \, dt \tag{2.215}$$

as required.

In general, from Eq. (2.211) it follows that ϕ can be chosen, such that

$$\lim_{\lambda \to \infty} I(\lambda) = \int_{-\infty}^{\infty} x(t) \left[\lim_{\lambda \to \infty} \left(\int_{-\infty}^{\infty} \phi(\tau) \frac{\sin[2\pi\lambda(t-\tau)]}{\pi(t-\tau)} d\tau + \varepsilon_1 \right) \right] dt$$

$$= \int_{-\infty}^{\infty} x(t) [\phi(t) + \varepsilon_1] \, dt \tag{2.216}$$

where ε_1 is arbitrarily small. From Eq. (2.210) it follows that ϕ can be chosen such that $\phi(t) = x^*(t) + \delta(t)$ where $\delta(t)$ is arbitrarily small except on a set of arbitrarily small measure. According to Theorem 2.11, the integral of x over such a set can be made arbitrarily small. Thus, the conclusion again is

$$\lim_{\lambda \to \infty} I(\lambda) = \int_{-\infty}^{\infty} |x(t)|^2 \, dt \tag{2.217}$$

56 BACKGROUND: SIGNAL AND SYSTEM THEORY

Hence, $X \in L^2$ and the required integral relationship between x and X is proved.

APPENDIX 8: PROOF OF THEOREM 2.32

A.8.1 Proof of First Two Results

The summation is that of a geometric series with ratio, and first term, given by $e^{-j2\pi f/f_o}$. Using the result for geometric series

$$\sum_{i=1}^{N} aR^{i-1} = \begin{cases} \dfrac{a(1-R^N)}{1-R} & R \neq 1 \\ Na & R = 1 \end{cases} \quad (2.218)$$

it follows that

$$\sum_{i=1}^{N} e^{-j2\pi i f/f_o} = \begin{cases} \dfrac{e^{-j2\pi f/f_o}(1-e^{-j2\pi Nf/f_o})}{1-e^{-j2\pi f/f_o}} & e^{-j2\pi f/f_o} \neq 1 \\ N & e^{-j2\pi f/f_o} = 1 \end{cases} \quad (2.219)$$

The condition $e^{-j2\pi f/f_o} = 1$ implies $f/f_o \in Z$. Thus, after basic manipulation,

$$\left| \sum_{i=1}^{N} e^{-j2\pi i f/f_o} \right|^2 = \begin{cases} \dfrac{1 - \cos(2\pi Nf/f_o)}{1 - \cos(2\pi f/f_o)} = \dfrac{\sin^2(\pi Nf/f_o)}{\sin^2(\pi f/f_o)} & f/f_o \notin Z \\ N^2 & f/f_o \in Z \end{cases} \quad (2.220)$$

where the last result follows from the relationship $2\sin^2(A) = 1 - \cos(2A)$. In fact, when $f/f_o \in Z$, it is the case that $\sin^2(\pi Nf/f_o)/\sin^2(\pi f/f_o) = N^2$. To prove this, consider L'Hopital's rule (Spivak, 1994 p. 201): If both f and g have limits of zero as their arguments tend towards some number x_o, and if $f'(x_o)/g'(x_o) = k$, then $f(x_o)/g(x_o) = k$. Thus, using this rule twice, it follows that for any integer v

$$\begin{aligned}
\lim_{f/f_o \to v} \frac{\sin^2(\pi Nf/f_o)}{\sin^2(\pi f/f_o)} &= \lim_{f/f_o \to v} \frac{(2\pi N/f_o)\sin(\pi Nf/f_o)\cos(\pi Nf/f_o)}{(2\pi/f_o)\sin(\pi f/f_o)\cos(\pi f/f_o)} \\
&= \lim_{f/f_o \to v} \frac{N\sin(2\pi Nf/f_o)}{\sin(2\pi f/f_o)} \\
&= \lim_{f/f_o \to v} \frac{N(2\pi N/f_o)\cos(2\pi Nf/f_o)}{(2\pi/f_o)\cos(2\pi f/f_o)} = N^2
\end{aligned} \quad (2.221)$$

where the result $2\sin(A)\cos(A) = \sin(2A)$ has been used.

The second form in Eq. (2.157) arises from writing

$$\left| \sum_{i=1}^{N} e^{-j2\pi i f/f_o} \right|^2 = \sum_{i=1}^{N} \sum_{k=1}^{N} e^{-j2\pi(i-k)f/f_o} \quad (2.222)$$

and then separating the double summation into single summations where $k = i$, $k = i \pm 1, \ldots, k = i \pm (N-1)$.

A.8.2 Proof of Third Result

First, consider the case where $f \notin i f_o, i \in \mathbf{Z}$. For any $\varepsilon > 0$, there will exist a sufficiently large value of N, such that

$$\left| \frac{1}{N} \cdot \frac{\sin^2(\pi N f/f_o)}{\sin^2(\pi f/f_o)} \right| < \left| \frac{1}{N} \cdot \frac{1}{\sin^2(\pi f/f_o)} \right| < \varepsilon \quad (2.223)$$

Hence, for $f \notin i f_o, i \in \mathbf{Z}$

$$\lim_{N \to \infty} \frac{1}{N} \cdot \frac{\sin^2(\pi N f/f_o)}{\sin^2(\pi f/f_o)} = 0 \quad (2.224)$$

Second, consider a change of variable $\lambda = f - i f_o, i \in \mathbf{Z}$ in the following integral

$$I = \int_{if_o - \Delta f}^{if_o + \Delta f} \frac{1}{N} \cdot \frac{\sin^2(\pi N f/f_o)}{\sin^2(\pi f/f_o)} df \quad \begin{array}{l} i \in \mathbf{Z} \\ 0 < \Delta f < f_o \end{array} \quad (2.225)$$

which yields,

$$I = \int_{-\Delta f}^{\Delta f} \frac{1}{N} \cdot \frac{\sin^2(\pi N \lambda/f_o + iN\pi)}{\sin^2(\pi \lambda/f_o + i\pi)} d\lambda = \int_{-\Delta f}^{\Delta f} \frac{1}{N} \cdot \frac{\sin^2(\pi N \lambda/f_o)}{\sin^2(\pi \lambda/f_o)} d\lambda \quad i \in \mathbf{Z} \quad (2.226)$$

Consider a fixed $\varepsilon > 0$. It follows from Eq. (2.224) that there exists a sufficiently large value of N, such that the integrand in Eq. (2.226) will be less than $\varepsilon' = \varepsilon/\Delta f$ over the intervals $[-\Delta f, -\delta]$ and $[\delta, \Delta f]$, where $0 < \delta < \Delta f$ and δ is such that over the interval $-\delta \leqslant f < \delta$ the expression $\sin(\pi f/f_o)$ can be approximated by the linear term according to

$$\sin(\pi f/f_o) \approx \pi f/f_o \quad -\delta \leqslant f < \delta \quad (2.227)$$

With these definitions, there will exist an N, sufficiently large such that

$$\frac{1}{N}\int_{-\Delta f}^{-\delta} \frac{\sin^2(\pi N\lambda/f_o)}{\sin^2(\pi\lambda/f_o)}\, d\lambda < \varepsilon'\Delta f = \varepsilon \qquad \frac{1}{N}\int_{\delta}^{\Delta f} \frac{\sin^2(\pi N\lambda/f_o)}{\sin^2(\pi\lambda/f_o)}\, d\lambda < \varepsilon \tag{2.228}$$

Thus, within an error of 2ε, the integral I can be approximated by

$$I \approx \frac{1}{N}\int_{-\delta}^{\delta} \frac{\sin^2(\pi N\lambda/f_o)}{(\pi\lambda/f_o)^2}\, d\lambda = \int_{-N\delta}^{N\delta} \frac{\sin^2(\pi W/f_o)}{(\pi W/f_o)^2}\, dW \tag{2.229}$$

where the change of variable $W = N\lambda$ has been used. From the standard integral (Spiegel, 1968 p. 96)

$$\int_{0}^{\infty} \frac{\sin^2(px)}{(px)^2}\, dx = \frac{\pi}{2p} \tag{2.230}$$

it follows, as N becomes increasingly large, that $I \approx f_o$, and hence,

$$\lim_{N\to\infty} \frac{1}{N}\cdot\frac{\sin^2(\pi Nf/f_o)}{\sin^2(\pi f/f_o)} = f_o\delta(f - if_o) \qquad \begin{array}{l} f \in [if_o - \Delta f, if_o + \Delta f] \\ i \in \mathbb{Z} \end{array} \tag{2.231}$$

which is the required result.

A.8.3 Proof of Final Result

To prove the final result of the theorem, note that

$$\operatorname{sinc}(fT) = \begin{cases} \dfrac{\sin(\pi fT)}{\pi fT} & f \neq 0 \\ 1 & f = 0 \end{cases} \tag{2.232}$$

and it readily follows that

$$\lim_{T\to\infty} T\operatorname{sinc}^2(fT) = \begin{cases} 0 & f \neq 0 \\ \infty & f = 0 \end{cases} \tag{2.233}$$

Further, from the standard integral [see Eq. (2.230)], it directly follows that

$$\lim_{T\to\infty} \int_{-\infty}^{\infty} T\operatorname{sinc}^2(fT)\, df = \lim_{T\to\infty} T\int_{-\infty}^{\infty} \frac{\sin^2(\pi fT)\, df}{(\pi fT)^2} = 1 \tag{2.234}$$

Hence,

$$\lim_{T\to\infty} T\operatorname{sinc}^2(fT) = \delta(f) \tag{2.235}$$

3

The Power Spectral Density

3.1 INTRODUCTION

The power spectral density is widely used to characterize random processes in electronic and communication systems. One common application of the power spectral density is to characterize the noise in a system. From such a characterization the noise power, and hence, the system signal to noise ratio, can be evaluated. This chapter gives a detailed justification of the two distinct, but equivalent ways of defining the power spectral density. The first is via decomposition, as given by the Fourier transform, of signals comprising the random process; the second is through the Fourier transform of the time averaged autocorrelation function of waveforms comprising the random process. The first approach is used in later chapters and facilitates analysis to a greater degree than the second. Finally, the relationship between the power spectral density and autocorrelation function, as stated by the Wiener–Khintchine theorem, is justified. A brief historical account of the development of the theory underlying the power spectral density can be found in Gardner (1988 pp. 12f).

3.1.1 Relative Power Measures

In the following sections, the concepts of signal power and signal power spectral density are introduced and used. Strictly speaking, the concepts are that of relative signal power and relative signal power spectral density, as signals typically have units that lead to relative, not absolute, power measures. To simplify terminology, the word "relative" is dropped. The best justification for the use of relative power measures, is the signal to noise ratio which is defined as the signal power divided by the noise power. Provided both the signal and noise have the same units, for example, watts or volts squared, it does not matter whether relative or absolute power measures are used. Further, in many electronic circuit applications a relative power measure is appropriate as it is current and voltage levels, not power levels, that are of interest.

3.2 DEFINITION

The approach detailed in this section is consistent with that of Priestley (1981 ch. 4.3–4.8), Jenkins (1968 ch. 6), and Peebles (1993 ch. 7).

3.2.1 Characteristics of a Power Spectral Density

A power spectral density function, G, based on the standard sinusoidal or complex exponential basis set should have the following characteristics. First, to facilitate analysis it should be a continuous signal. Second, it should have the interpretation that $G(f_x)$ is directly proportional to the power in the sinusoidal components of the signal with a frequency of f_x Hz. Third, this proportionality should be such that the integral of the power spectral density over all possible frequencies equals the average signal power denoted $\bar{P}$, that is,

$$\bar{P} = \int_{-\infty}^{\infty} G(f)\, df \tag{3.1}$$

This last requirement is consistent with the sum of the power in the constituent waveforms equaling the total average power. In summary, a power spectral density function G, should be such that

(1) G is a continuous function.
(2) $G(f_x)$ is proportional to the power of the constituent sinusoidal signals with frequency f_x.
(3) $\bar{P} = \int_{-\infty}^{\infty} G(f)\, df.$

The following subsections give details of a power spectral density function that satisfies these three conditions or requirements.

3.2.2 Power Spectral Density of a Single Waveform

A natural basis for the power spectral density is the average power of a signal. For an interval $[0, T]$ the average power of a signal x, by definition, is

$$\bar{P}(T) = \frac{1}{T} \int_0^T |x(t)|^2\, dt \tag{3.2}$$

Assume that x is either piecewise smooth or of bounded variation. It then

DEFINITION   61

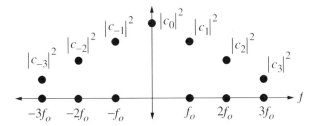

Figure 3.1 Display of power in sinusoidal components of a signal.

follows, from substitution of the Fourier series for the signal x [see Eq. (2.126)] into this equation, that

$$\bar{P}(T) = |a_0|^2 + 0.5 \sum_{i=1}^{\infty} |a_i|^2 + |b_i|^2 = \sum_{i=-\infty}^{\infty} |c_i|^2 = \sum_{i=-\infty}^{\infty} \frac{|X(T, if_o)|^2}{T^2} \quad (3.3)$$

where the last relationship follows from Eq. (2.137). As per Eq. (2.131), the power associated with signal components with a frequency of if_o Hz, namely, $a_i \cos(2\pi i f_o t)$ and $b_i \sin(2\pi i f_o t)$, is given by $|c_{-i}|^2 + |c_i|^2 = (|a_i|^2 + |b_i|^2)/2$. Consistent with this result, Figure 3.1 represents one way to display the power in the sinusoidal components of a single waveform, subject to the interpretation that the power in the sinusoidal components with a frequency of if_o is the sum of the values defined by the graph at frequencies of $-if_o$ and if_o Hz.

Note, for a real signal $|c_{-i}|^2 = |c_i|^2$ and the display is symmetric with respect to the vertical axis.

A problem with such a display is that the integral of the function defined by the graph is zero. To overcome this problem an alternative display, based on the relationship $c_i = X(T, if_o)/T$, can be constructed as shown in Figure 3.2.

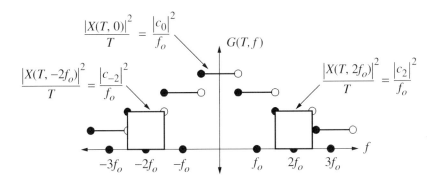

Figure 3.2 A power spectral density function. The shaded areas equal the power associated with sinusoidal components that have a frequency of $2f_o$ Hz.

With such a graph the area under the defined function, by construction, equals the average signal power.

The display in Figure 3.2 is consistent with writing the average power in the form,

$$\bar{P}(T) = \sum_{i=-\infty}^{\infty} \frac{|X(T, if_o)|^2}{T^2} \qquad (3.4)$$

The interpretation of the graph in Figure 3.2 is as follows: The area under each pair of levels of the graph associated with the frequencies $-if_o$ and if_o, equals the power in the sinusoidal waveforms with a frequency if_o. Consistent with this graph, the power spectral density function, G, can be defined as

$$G(T, f) = \frac{|X(T, if_o)|^2}{T} \qquad if_o - \frac{f_o}{2} \leqslant f < if_o + \frac{f_o}{2} \qquad (3.5)$$

or, more generally, according to

$$G(T, f) = \frac{1}{T} \left| X\left(T, \left\lfloor \frac{f + f_o/2}{f_o} \right\rfloor f_o \right) \right|^2 \qquad -\infty < f < \infty \qquad (3.6)$$

With such a definition it follows that

$$\bar{P}(T) = \int_{-\infty}^{\infty} G(T, f)\, df \qquad (3.7)$$

which is the third requirement of a power spectral density function.

Such a power spectral density function G, satisfies requirements (2) and (3) but is not a continuous function. Obtaining a continuous function for the power spectral density is discussed in the next subsection.

3.2.3 A Continuous Power Spectral Density Function

The basis for obtaining a continuous waveform for the power spectral density is Parseval's relationship (Theorem 2.31):

$$\int_0^T |x(t)|^2\, dt = \int_{-\infty}^{\infty} |X(T, f)|^2\, df \qquad (3.8)$$

Scaling both integrals by T yields

$$\bar{P}(T) = \frac{1}{T} \int_0^T |x(t)|^2\, dt = \int_{-\infty}^{\infty} \frac{|X(T, f)|^2}{T}\, df = \int_{-\infty}^{\infty} G(T, f)\, df \qquad (3.9)$$

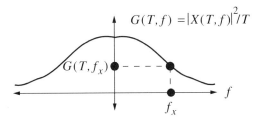

Figure 3.3 *Continuous power spectral density function based on Parseval's relationship.*

and a power spectral density function G, as per the following definition:

DEFINITION: POWER SPECTRAL DENSITY   The power spectral density of a signal x, evaluated on the interval $[0, T]$, is defined according to

$$G(T, f) = \frac{|X(T, f)|^2}{T} \qquad (3.10)$$

This power spectral density function is commonly called the periodogram (see Gardner, 1988 p. 13) or sample spectral density (Jenkins, 1968 p. 211; Parzen, 1962 p. 109).

The power spectral density function, as defined by Eq. (3.10), has the form shown in Figure 3.3 and it remains to show that it satisfies the three requirements of a power spectral density function. To this end note, first, that the integral of the power spectral density, by construction, equals the total power. Second, the power spectral density is a continuous function as stated by the following theorem (Champeney, 1987 p. 60).

THEOREM 3.1. CONTINUITY OF POWER SPECTRAL DENSITY   *If $x \in L[0, T]$ then the power spectral density function G, defined by Eq. (3.10), is continuous with respect to f for $f \in \mathbf{R}$.*

Proof. This result can be proved by first proving that $X(T, f)$ is continuous with respect to $f \in \mathbf{R}$ when $x \in L[0, T]$. The proof is straightforward and is omitted.

Third, the last requirement of a power spectral density function is that $G(T, f_x)$ should be proportional to the power in the constituent sinusoidal components that have a frequency of f_x Hz. This is not obviously the case, because a Fourier series decomposition on the interval $[0, T]$ only yields sinusoids with frequencies $f_o, 2f_o, \ldots$, where $f_o = 1/T$. It may well be the case that f_x is not an integer multiple of f_o. This issue is discussed in the following subsection.

3.2.3.1 Interpretation of Continuous Power Spectral Density Function

As a Fourier series decomposition of a signal on an interval $[0, T]$ yields sinusoidal components with frequencies $f_o, 2f_o, \ldots$ it is reasonable to conclude that $G(T, f_x)$ should only be interpreted for $f_x = if_o$, $i \in \mathbf{Z}^+$. The problem then is, how to interpret $G(T, if_o)$ for some integer value of i. The interpretation is given in the following theorem.

THEOREM 3.2. INTERPRETATION OF POWER SPECTRAL DENSITY *If $x \in L^2[0, T]$, and the power spectral density of x is defined according to*

$$G(T, f) = \frac{|X(T, f)|^2}{T} \qquad f_o = \frac{1}{T} \qquad (3.11)$$

then the average power in the sinusoidal components of x with a frequency if_o, and on the interval $[0, T]$, is given by

$$\begin{aligned} |c_{-i}|^2 + |c_i|^2 &= f_o[G(T, -if_o) + G(T, if_o)] \qquad i \in \mathbf{Z}^+ \\ |c_0|^2 &= f_o G(T, 0) \qquad i = 0 \end{aligned} \qquad (3.12)$$

Proof. Using the relationship $c_i = X(T, if_o)/T$ a step approximation to G can be defined, as shown in Figure 3.4 and consistent with that shown in Figure 3.2. With such a step approximation the area under each pair of levels with width f_o, centered at $\pm if_o$ and with respective heights $|X(T, -if_o)|^2/T$ and $|X(T, if_o)|^2/T$, equals $|c_{-i}|^2 + |c_i|^2$, and hence, the power in the sinusoidal components with a frequency of if_o Hz.

This theorem states that the power in the mean of a signal is given by $f_o G(T, 0)$. To confirm this, note that the mean μ_x of the signal x on $[0, T]$ is given by

$$\mu_x(T) = \frac{1}{T} \int_0^T x(t) \, dt = \frac{X(T, 0)}{T} \qquad (3.13)$$

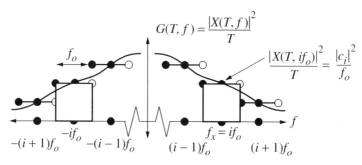

Figure 3.4 *Step approximation to power spectral density function. The area under the two levels associated with $f = -if_o$ and $f = if_o$ equals the power in the sinusoidal components of the signal with a frequency of if_o Hz.*

and this implies the following relationships:

$$\bar{P}_{\mu_x}(T) = \frac{1}{T}\int_0^T |\mu_x|^2 \, dt = |\mu_x|^2 = \frac{|X(T,0)|^2}{T^2} = |c_0|^2 = f_o G(T,0) \quad (3.14)$$

3.2.3.2 Power as Area Under the Power Spectral Density Graph
The power in the sinusoidal components with a frequency if_o can be approximated by the integral

$$|c_{-i}|^2 + |c_i|^2 \approx \int_{-if_o-f_o/2}^{-if_o+f_o/2} G(T,f) \, df + \int_{if_o-f_o/2}^{if_o+f_o/2} G(T,f) \, df$$

$$= 2 \int_{if_o-f_o/2}^{if_o+f_o/2} G(T,f) \, df \quad (3.15)$$

where the last equality in this equation only applies for real signals. How accurate this approximation is depends on the nature of the signal under consideration, and hence, G. The following example illustrates this point.

3.2.3.3 Example—Power Spectral Density of a Sinusoid
Consider a sinusoidal signal $A \sin(2\pi f_c t)$ on the interval $[0, T]$. From Eq. (3.10) it follows, after standard analysis, that the power spectral density can be written as

$$G(T,f) = \frac{A^2 T}{4} \left\{ \mathrm{sinc}^2\left[N\left(\frac{f}{f_c} - 1\right)\right] + \mathrm{sinc}^2\left[N\left(\frac{f}{f_c} + 1\right)\right] \right.$$

$$\left. - 2 \, \mathrm{sinc}\left[N\left(\frac{f}{f_c} - 1\right)\right] \mathrm{sinc}\left[N\left(\frac{f}{f_c} + 1\right)\right] \right\} \quad (3.16)$$

where $T = NT_c$ with $T_c = 1/f_c$. This power spectral density is shown in Figure 3.5 for the case where $A = 1$, $T = 1/f_o = 1$, and $f_c = 4$. Note that $G(T, if_o) = 0$ as expected, except for the case when $if_o = f_c$. However, it is clearly evident from this figure that

$$|c_{-i}|^2 + |c_i|^2 = 2|c_i|^2 = 0 \neq 2 \int_{if_o-f_o/2}^{if_o+f_o/2} G(T,f) \, df \quad \begin{cases} i = 1, 2, 3, 5, 6, \ldots \\ f_o = 1 \end{cases}$$

(3.17)

3.3 PROPERTIES

The following subsections detail basic properties of the defined power spectral density function.

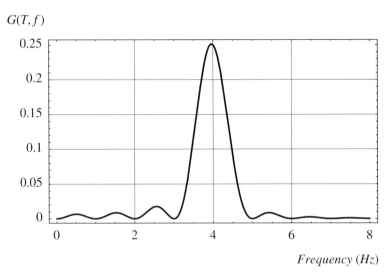

Figure 3.5 Power spectral density of a sinusoid with a frequency of 4 Hz, an amplitude of unity, and evaluated on a 1 sec interval ($f_o = 1$).

3.3.1 Symmetry in Power Spectral Density

For the case where x is real it follows that G is an even function with respect to f, that is, $G(T, -f) = G(T, f)$. This result follows from Eq. (2.136) which states:

$$X(T, -f) = X^*(T, f)$$

3.3.2 Resolution in Power Spectral Density

For a measurement interval of T seconds, the frequency resolution in the power spectral density is $f_o = 1/T$. Clearly, as T increases the resolution increases. In fact, for any resolution Δf in frequency, there exists an interval $[0, T]$, where $T = 1/\Delta f$, such that the rectangular areas of width Δf, centered at the frequencies $-f_x$ and f_x and with respective heights of $G(T, -f_x)$ and $G(T, f_x)$, equal the power in the sinusoidal components of the signal with a frequency f_x. The assumption here is that the frequency f_x is some integer multiple of the resolution Δf. This result is illustrated in Figure 3.4 provided f_o is interpreted as Δf. Note that, in general, $G(T, f)$ will vary with T.

3.3.3 Integrability of Power Spectral Density

An important property of the power spectral density function G, is that, in general, it is integrable.

THEOREM 3.3. INTEGRABILITY OF POWER SPECTRAL DENSITY   *If* $x \in L^2[0, T]$ *then* $G \in L$.

Proof. Given $x \in L^2[0, T]$ it follows from Parseval's relationship that

$$\int_{-\infty}^{\infty} |X(T, f)|^2 \, df$$

is finite which implies the integrability of G.

3.3.4 Power Spectral Density on Infinite Interval

Taking the limit as T tends toward infinity of the average power on the interval $[0, T]$ yields a definition for the average signal power on the interval $(0, \infty)$, denoted $\bar{P}_\infty$, that is,

$$\bar{P}_\infty = \lim_{T \to \infty} \frac{1}{T} \int_0^T |x(t)|^2 \, dt = \lim_{T \to \infty} \int_{-\infty}^{\infty} \frac{|X(T, f)|^2}{T} \, df = \lim_{T \to \infty} \int_{-\infty}^{\infty} G(T, f) \, df \tag{3.18}$$

If it is possible to interchange the order of integration and limit operations in the last equation, then $\bar{P}_\infty$ can be rewritten as

$$\bar{P}_\infty = \lim_{T \to \infty} \int_{-\infty}^{\infty} G(T, f) \, df = \int_{-\infty}^{\infty} \lim_{T \to \infty} G(T, f) \, df \tag{3.19}$$

and a power spectral density function G_∞, for the interval $[0, \infty]$ can be defined according to the following definition.

DEFINITION: POWER SPECTRAL DENSITY ON INFINITE INTERVAL

$$G_\infty(f) = \lim_{T \to \infty} G(T, f) \tag{3.20}$$

Note, the standard results that dictate whether it is possible to interchange the order of integration and limit operations are the Dominated and Monotone convergence theorems (Theorems 2.23 and 2.24).

3.4 RANDOM PROCESSES

Consider a random process X with ensemble

$$E_X = \{x : S_X \times [0, T] \to \mathbf{C}\} \tag{3.21}$$

68 THE POWER SPECTRAL DENSITY

and associated signal probabilities

$$P[x(i, t)] = P[x_i(t)] = p_i \qquad S_X \subseteq Z^+ \quad \text{countable case}$$
$$P[x(\lambda, t)|_{\lambda \in [\lambda_o, \lambda_o + d\lambda]}] = f_X(\lambda_o) \, d\lambda \qquad S_X \subseteq R \quad \text{uncountable case} \tag{3.22}$$

The average power in an individual waveform from the ensemble evaluated over the interval $[0, T]$ is

$$\bar{P}(\lambda, T) = \frac{1}{T} \int_0^T |x(\lambda, t)|^2 \, dt \tag{3.23}$$

For the countable case it is convenient to use a subscript rather than an argument according to $x(i, T) = x_i(T)$ and $\bar{P}(i, T) = \bar{P}_i(T)$.

The probabilities defined in Eq. (3.22) are the "natural" weighting factor to use in determining the average signal power according to

$$\bar{P}(T) = \sum_{i=1}^{\infty} p_i \bar{P}_i(T) = \sum_{i=1}^{\infty} p_i \frac{1}{T} \int_0^T |x_i(t)|^2 \, dt \quad \text{countable case} \tag{3.24}$$

$$\bar{P}(T) = \int_{-\infty}^{\infty} \bar{P}(\lambda, T) f_X(\lambda) \, d\lambda$$

$$= \int_{-\infty}^{\infty} \left[\frac{1}{T} \int_0^T |x(\lambda, t)|^2 \, dt \right] f_X(\lambda) \, d\lambda \quad \text{uncountable case} \tag{3.25}$$

Provided Parseval's relationship can be applied, and the order of summation/integration and integration can be interchanged, then the average signal power can be written as

$$\bar{P}(T) = \int_{-\infty}^{\infty} \frac{1}{T} \sum_{i=1}^{\infty} p_i |X_i(T, f)|^2 \, df = \int_{-\infty}^{\infty} G(T, f) \, df \quad \text{countable case} \tag{3.26}$$

$$\bar{P}(T) = \int_{-\infty}^{\infty} \left[\frac{1}{T} \int_{-\infty}^{\infty} |X(\lambda, T, f)|^2 f_X(\lambda) \, d\lambda \right] df$$

$$= \int_{-\infty}^{\infty} G(T, f) \, df \quad \text{uncountable case} \tag{3.27}$$

where

$$X_i(T, f) = \int_0^T x_i(t) e^{-j2\pi ft} \, dt \qquad X(\lambda, T, f) = \int_0^T x(\lambda, t) e^{-j2\pi ft} \, dt \tag{3.28}$$

and $G: R^2 \to R$ is the power spectral density function for the random process X, on the interval $[0, T]$, and defined according to:

DEFINITION: POWER SPECTRAL DENSITY ON FINITE INTERVAL

$$G(T, f) = \begin{cases} \dfrac{1}{T} \sum_{i=1}^{\infty} p_i |X_i(T, f)|^2 = \sum_{i=1}^{\infty} p_i G_i(T, f) & \text{countable case} \\ \dfrac{1}{T} \int_{-\infty}^{\infty} |X(\lambda, T, f)|^2 f_X(\lambda)\, d\lambda = \int_{-\infty}^{\infty} G(\lambda, T, f) f_X(\lambda)\, d\lambda & \text{uncountable case} \end{cases}$$

(3.29)

where $G_i(T, f)$ and $G(\lambda, T, f)$, respectively, are the power spectral densities for the ith and λth signals in the countable and uncountable ensembles.

Clearly, the power spectral density of a random process is the weighted average of the power spectral density of individual signals in the ensemble.

3.4.1 Power Spectral Density on Infinite Interval

The countable case is considered here. The analysis for the uncountable case follows in an analogous manner.

The average power of the random process on the interval $[0, \infty]$, denoted $\bar{P}_\infty$, by definition, is the weighted sum of the individual signal powers comprising the random process X as T tends towards infinity, that is,

$$\bar{P}_\infty = \sum_{i=1}^{\infty} p_i \lim_{T \to \infty} \bar{P}_i(T) = \sum_{i=1}^{\infty} p_i \lim_{T \to \infty} \frac{1}{T} \int_0^T |x_i(t)|^2\, dt \qquad (3.30)$$

By using Parseval's relationship, interchanging the order of the summation and limit operations, and then interchanging the order of the summation and integration, the average power can be written according to

$$\bar{P}_\infty = \lim_{T \to \infty} \int_{-\infty}^{\infty} \frac{1}{T} \sum_{i=1}^{\infty} p_i |X_i(T, f)|^2\, df = \lim_{T \to \infty} \int_{-\infty}^{\infty} G(T, f)\, df \qquad (3.31)$$

If it is possible to interchange the order of the limit and integral operations in this last equation, then the average power can be written as

$$\bar{P}_\infty = \int_{-\infty}^{\infty} \lim_{T \to \infty} G(T, f)\, df = \int_{-\infty}^{\infty} G_\infty(f)\, df \qquad (3.32)$$

where $G_\infty : \mathbf{R} \to \mathbf{R}$ is the power spectral density for the random process on the infinite interval $(0, \infty)$ and defined as:

DEFINITION: POWER SPECTRAL DENSITY ON INFINITE INTERVAL

$$G_\infty(f) = \lim_{T \to \infty} G(T, f) \qquad (3.33)$$

3.4.2 Example—PSD of Binary Digital Random Process

Consider a binary digital random process X, defined by either a pulse p or its negative $-p$ in each interval of D sec. On the interval $[0, ND]$ the ensemble for this random process is

$$E_X = \left\{ x(\gamma_1, \ldots, \gamma_N, t) = \sum_{k=1}^{N} \gamma_k p(t - (k-1)D), \gamma_k \in \{-1, 1\} \right\} \quad (3.34)$$

One of the 2^N possible waveforms in this ensemble is shown in Figure 3.6 for the case of a rectangular pulse function of duration $D/2$ sec. For the case where the pulses are independent from one interval of D sec to the next, the probability of a signal from the ensemble is

$$P[x(\gamma_1, \ldots, \gamma_N, t)] = P[\gamma_1, \ldots, \gamma_N] = P[\gamma_1]P[\gamma_2] \cdots P[\gamma_N] \quad (3.35)$$

For the case where the probability of a pulse p is equal to the probability of $-p$, that is, $P[\gamma_k = 1] = P[\gamma_k = -1] = 0.5$, then the probability of any signal from the ensemble is $1/2^N$. It follows from the definition of the power spectral density function, that the power spectral density of X, evaluated on the interval $[0, ND]$, is given by

$$G_X(ND, f) = \frac{1}{ND} \sum_{\gamma_1} \cdots \sum_{\gamma_N} \frac{1}{2^N} |X(\gamma_1, \ldots, \gamma_N, ND, f)|^2 \quad \gamma_i \in \{\pm 1\} \quad (3.36)$$

To evaluate the power spectral density, the assumption is made that the pulse function is zero outside the interval $[0, D]$. With this assumption, first note that

$$|X(\gamma_1, \ldots, \gamma_N, ND, f)|^2 = \sum_{i=1}^{N} \sum_{k=1}^{N} \gamma_i \gamma_k |P(f)|^2 e^{-j2\pi f(i-1)Df} e^{j2\pi f(k-1)Df}$$

$$= \sum_{i=1}^{N} |P(f)|^2 + \sum_{i=1}^{N} \sum_{\substack{k=1 \\ k \neq i}}^{N} \gamma_i \gamma_k |P(f)|^2 e^{-j2\pi f(i-k)Df} \quad (3.37)$$

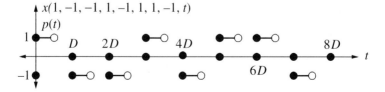

Figure 3.6 One waveform from a binary digital random process on the interval $[0, 8D]$.

where P is the Fourier transform of p. Next, note that substitution of Eq. (3.37) into Eq. (3.36) and interchanging the order of summation yields, for the second term:

$$\frac{1}{ND2^N} \sum_{i=1}^{N} \sum_{\substack{k=1 \\ k \neq i}}^{N} |P(f)|^2 e^{-j2\pi f(i-k)Df} \sum_{\gamma_1} \cdots \sum_{\gamma_N} \gamma_i \gamma_k \quad (3.38)$$

This summation is zero as $\gamma_i \in \{\pm 1\}$ and γ_i is independent of γ_k for $i \neq k$. Hence,

$$G_X(ND, f) = r|P(f)|^2 \quad (3.39)$$

where, $r = 1/D$. Note that the power spectral density in independent of the interval being considered. This is due to the fact that the pulse function is zero outside the interval $[0, D]$. The power spectral density is plotted in Figure 3.7 for the case where the pulse function is rectangular, that is,

$$p(t) = \begin{cases} 1 & 0 \leq t < W \\ 0 & \text{elsewhere} \end{cases} \quad P(f) = W\,\text{sinc}(fW)e^{-j\pi fW} \quad 0 < W \leq D \quad (3.40)$$

3.4.3 Miscellaneous Issues

3.4.3.1 Nonstationary Random Processes
The definition for the power spectral density [see Eq. (3.29)] is valid for single waveforms, stationary random processes, and nonstationary random processes, as it is based on

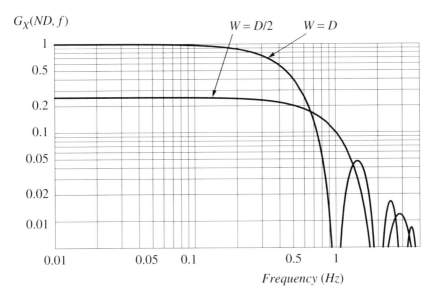

Figure 3.7 Power spectral density of binary digital random process for the case where $D = 1$.

average signal power over the interval $[0, T]$. However, care needs to be taken when interpreting the power spectral density when the random process is nonstationary.

3.4.3.2 Single-Sided Power Spectral Density
When x is real, it follows from Eq. (2.136) that G is an even function with respect to f, that is, $G(T, -f) = G(T, f)$. This leads some authors to define a single-sided power spectral density function according to

$$G_{ss}(T, f) = \begin{cases} 2G(T, f) & f \geq 0 \\ 0 & f < 0 \end{cases} \qquad (3.41)$$

Such a definition is not used in subsequent analysis.

3.4.3.3 Discrete Approximation for the Uncountable Case
In analysis, it is often convenient to replace the continuous random variable characterizing the waveforms in the ensemble by N outcomes of a discrete random variable. To this end, consider the random process X defined by the ensemble

$$E_X = \{x : \mathbf{R} \times [0, T] \to \mathbf{C}\} \qquad (3.42)$$

where

$$P[x(\lambda, t)|_{\lambda \in [\lambda_o, \lambda_o + d\lambda]}] = f_X(\lambda_o) \, d\lambda \qquad (3.43)$$

Next, consider a random process X_A with a finite number of outcomes defined by the ensemble

$$E_{X_A} = \left\{ x_{A_i} : [0, T] \to \mathbf{C} \quad \begin{matrix} i \in \{1, \ldots, N\} \\ x_{A_i}(t) = x(\lambda_i, t) \quad \lambda_i \in I_i \end{matrix} \right\} \qquad (3.44)$$

where $\{I_i\}_{i=1}^N$ is a partition of $\mathbf{R}$ and

$$P[x_{A_i}] = p_i = \int_{I_i} f_X(\lambda) \, d\lambda \qquad (3.45)$$

The power spectral density of X_A can then be defined according to

$$G_A(T, f) = \frac{1}{T} \sum_{i=1}^N p_i |X_{A_i}(T, f)|^2 \qquad (3.46)$$

The following theorem details when G_A approximates G.

THEOREM 3.4. DISCRETE APPROXIMATION TO UNCOUNTABLE RANDOM PROCESS
Assume T is fixed, $\sup\{|X(\lambda, T, f)|^2: \lambda \in R, f \in R\}$ is finite, and that all waveforms in the uncountable ensemble, E_X, are Lebesgue integrable on $[0, T]$. Further, assume that the signal defined by $y(\lambda) = x(\lambda, t)|_{t\,fixed}$ has bounded variation over all finite intervals of the form $[-\lambda_o, \lambda_o]$, that is, $\exists k_o(t) > 0$, such that for all finite N, for all partitions $\{I_i\}_{i=1}^N$ of $[-\lambda_o, \lambda_o]$, and with $\lambda_i \in I_i$, it is the case that

$$\sum_{i=1}^{N-1} |x(\lambda_{i+1}, t) - x(\lambda_i, t)| < k_o(t) \tag{3.47}$$

In addition, assume the bounded variation holds uniformly for $t \in [0, T]$, that is, $k_o(t)$ is independent of t. It then follows that $|G(T, f) - G_A(T, f)|$, for T and f fixed, can be made arbitrarily small by an appropriate partition $\{I_i\}_{i=1}^N$ of R.

Proof. The proof is given in Appendix 1.

3.5 EXISTENCE CRITERIA

Sufficient conditions for the validity of the definitions of G and G_∞, respectively, defined by Eqs. (3.29) and (3.33), are stated in the following theorem.

THEOREM 3.5. CONDITIONS FOR EXISTENCE OF POWER SPECTRAL DENSITY

(a) Finite Interval: If $x \in L^2[0, T]$ for each signal in the ensemble and the average power, defined by Eqs. (3.24) and (3.25) is finite, then the power spectral density function G, defined by Eq. (3.29), is valid.

The average power is guaranteed to be finite if there exists constants $\alpha, k_o, I_o > 0$, such that

$$\begin{aligned} \forall i > I_o \quad & p_i \bar{P}_i(T) < \frac{k_o}{i^{1+\alpha}} && \text{countable case} \\ \forall |\lambda| > I_o \quad & \bar{P}(\lambda, T) f_X(\lambda) < \frac{k_o}{|\lambda|^{1+\alpha}} && \text{uncountable case} \end{aligned} \tag{3.48}$$

If $\sup\{\bar{P}_i(T): i \in Z^+\}$ for the countable case and $\sup\{\bar{P}(\lambda, T): \lambda \in R\}$ for the uncountable case are finite, then the average power is guaranteed to be finite. Note, it can be the case that $\lim_{i \to \infty} \bar{P}_i(T)$ is infinite while the average power $\bar{P}(T)$ is finite.

(b) Infinite Interval. The power spectral density over the infinite interval G_∞, as defined by Eq. (3.33), is valid if first, $x \in L^2[0, T]$ for each signal in the ensemble and for all $T \in R^+$. Second, for the countable case, there exists a sequence $\{Q_i: i \in Z^+\}$ such that $p_i \bar{P}_i(T) < Q_i$ for every $i \in Z^+$, and $T \in R^+$ and $\Sigma_{i=1}^\infty Q_i$ is finite. For the uncountable case, there exists a function $Q \in L$, such

that $\bar{P}(\lambda, T) f_X(\lambda) < Q(\lambda)$ for all $T \in \mathbf{R}^+$. Third, there exists a function $H \in L$, such that $G(T, f) \leq H(f)$ for all $T \in \mathbf{R}^+$ and all f.

A less general, but simpler statement of conditions for the existence of G_∞ is: First, $x \in L^2[0, T]$ for each signal in the ensemble and for all $T \in \mathbf{R}^+$. Second, for the countable case $\sup\{\bar{P}_i(T): i \in \mathbf{Z}^+, T \in \mathbf{R}^+\}$ is finite and for the uncountable case $\sup\{\bar{P}(\lambda, T): \lambda \in \mathbf{R}, T \in \mathbf{R}^+\}$ is finite. Third, there exists a function $H \in L$, such that $G(T, f) \leq H(f)$ for all f and all $T \in \mathbf{R}^+$.

Proof. The proof of these conditions is given in Appendix 2.

Note, these are sufficient not necessary conditions.

3.5.1 Examples where Conditions are Violated

Three examples where it is not possible to find a function $H \in L$, such that $G(T, f) \leq H(f)$ for all f and all T are, (a) true $1/f$ noise that has a power spectral density function on the interval $[0, T]$ of the form

$$G(T, f) = \begin{cases} k_o T & f \ll 1/T \\ \dfrac{k_o}{|f|} & 1/T \ll f \ll T \\ \dfrac{k_1}{|f|^\beta} & f \gg T, \beta > 1 \end{cases} \quad (3.49)$$

and $\lim_{T \to \infty} G(T, f) = k_o/|f|$, (b) true white noise described by $G(T, f) = k$ independent of T. Both types of noise are abstractions and inconsistent with the nature of the physical universe. For example, a true white noise power spectral density is consistent with infinite power on all measurement intervals. The increasing higher level for the power spectral density of a true $1/f$ noise process is consistent with infinite power on an infinite time interval. (c) if $G(T, f)$ has an impulsive component at a frequency f_o then it is not possible to find a function $H \in L$, such that $G(T, f_o) \leq H(f_o)$ as $T \to \infty$. Such impulsive components arise when signals in the underlying random process have a periodic component. This is common in random processes encountered in engineering and is discussed in the following section.

3.6 IMPULSIVE CASE

Many random processes encountered in engineering are such that their power spectral density, evaluated on the interval $[0, T]$, contain a countable number of components that become impulsive on a set of zero measure as T approaches infinity. These components arise because of periodic components in signals comprising the random process. This case requires a distinct formula-

3.6.1 Conditions for Bounded Power Spectral Density

For the case of an infinite and countable ensemble, the power spectral density of the ith signal G_i, evaluated on a fixed interval $[0, T]$, may be bounded, but it might be the case that $\lim_{i \to \infty} G_i(T, f)$ becomes unbounded at specific values of f. The usual case with T fixed, however, is that the power spectral density of individual signals remains bounded for all signals in the ensemble, that is, $\sup\{G_i(T, f): i \in \mathbf{Z}^+, f \in \mathbf{R}\} < \infty$. This result is consistent with the average power in all signals in the ensemble being finite, as is stated in the following theorem.

THEOREM 3.6. BOUNDED POWER IMPLIES BOUNDED POWER SPECTRAL DENSITY
If, over a fixed interval $[0, T]$, the random process X is characterized by the ensemble

$$E_X = \left\{ x: S_X \times [0, T] \to \mathbf{C} \quad \begin{matrix} S_X \subseteq \mathbf{Z}^+ & \text{countable case} \\ S_X \subseteq \mathbf{R} & \text{uncountable case} \end{matrix} \right\} \quad (3.50)$$

and the average signal energy and power remain bounded on the ensemble, that is,

$$\sup \left\{ \int_0^T |x(\lambda, t)|^2 \, dt : \lambda \in S_X \right\} < \infty \quad (3.51)$$

then, the magnitude of individual signal Fourier transforms and individual signal power spectral densities remain bounded on the ensemble, that is,

$$\sup\{|X(\lambda, T, f)|: \lambda \in S_X, f \in \mathbf{R}\} < \infty \quad (3.52)$$

$$\sup\{G(\lambda, T, f): \lambda \in S_X, f \in \mathbf{R}\} < \infty \quad (3.53)$$

With individual power spectral densities being bounded the average power spectral density is bounded.

Proof. First, according to Theorem 2.13, it is the case that finite energy implies absolute integrability, that is,

$$\int_0^T |x(\lambda, t)|^2 \, dt < \infty \Rightarrow \int_0^T |x(\lambda, t)| \, dt < \infty \quad (3.54)$$

Second, absolute integrability implies the magnitude of the Fourier transform is bounded for all frequencies as

$$|X(\lambda, T, f)| = \left| \int_0^T x(\lambda, t) e^{-j2\pi ft} \, dt \right| \leq \int_0^T |x(\lambda, t)| \, dt < \infty \qquad f \in \mathbf{R} \quad (3.55)$$

Third, with the worst case of $|x(\lambda, t)| > 1$ it is the case that

$$\int_0^T |x(\lambda, t)|^2 \, dt > \int_0^T |x(\lambda, t)| \, dt \geq |X(\lambda, T, f)| \qquad (3.56)$$

For the case where $|x(\lambda, t)| \leq 1$, it is the case that $|X(\lambda, T, f)|$ is bounded by T. Hence, boundedness of energy implies boundedness of the Fourier transform, and hence, the power spectral density. The required result then follows.

3.6.2 Impulsive Power Spectral Density

Denoting the nonimpulsive or bounded component of $G(T, f)$ as $G_B(T, f)$ and the impulsive component as $G_I(T, f)$, the power spectral density can be written as

$$G(T, f) = G_B(T, f) + G_I(T, f) \qquad (3.57)$$

where

$$\lim_{T \to \infty} G_I(T, f) = \sum_i k_i \delta(f - f_i) \qquad (3.58)$$

$$G_B(f) = \lim_{T \to \infty} G_B(T, f) \qquad G_B \in L, G_B(f) \leq k_B$$

It then follows that the average power, assumed to be finite, is such that

$$\bar{P}_\infty = \lim_{T \to \infty} \int_{-\infty}^\infty G(T, f) \, df = \int_{-\infty}^\infty \lim_{T \to \infty} G_B(T, f) \, df + \lim_{T \to \infty} \int_{-\infty}^\infty G_I(T, f) \, df$$

$$(3.59)$$

but

$$\bar{P}_\infty \neq \int_{-\infty}^\infty \lim_{T \to \infty} [G_B(T, f) + G_I(T, f)] \, df$$

$$= \int_{-\infty}^\infty \lim_{T \to \infty} G_B(T, f) \, df = \int_{-\infty}^\infty G_B(f) \, df \qquad (3.60)$$

as per Eq. (2.148). Hence, it is not possible to define a power spectral density for the infinite interval according to

$$G_\infty(f) = \lim_{T \to \infty} G(T, f) = \lim_{T \to \infty} [G_B(T, f) + G_I(T, f)] \qquad (3.61)$$

which satisfies the requirement that the integral of the power spectral density equal the average power. However, for analytical convenience it is useful to define the power spectral density for the case where the power spectrum has impulsive components, according to Eq. (3.61). Because of the prevalence of impulsive components in the power spectral density, it is convenient to state this result through a formal definition.

DEFINITION: POWER SPECTRAL DENSITY — IMPULSIVE CASE AND INFINITE INTERVAL For the case where $G(T, f)$ is impulsive at a countable number of frequencies, but where the average power associated with all the impulsive components is finite, the power spectral density on the infinite interval $[0, \infty]$ is defined as:

$$G_\infty(f) = \lim_{T \to \infty} G(T, f) = \lim_{T \to \infty} [G_B(T, f) + G_I(T, f)] \qquad (3.62)$$

and the average power is given by

$$\bar{P}_\infty = \int_{-\infty}^{\infty} \lim_{T \to \infty} G_B(T, f)\, df + \lim_{T \to \infty} \int_{-\infty}^{\infty} G_I(T, f)\, df \qquad (3.63)$$

3.6.3 Integrated Spectrum

To overcome the inconsistencies inherent in the definition for the power spectral density, when $G(T, f)$ becomes impulsive as $T \to \infty$, a spectral distribution function F is defined according to (Champeney, 1987 ch. 11; Parzen, 1962 p. 110)

$$F(f) = \lim_{T \to \infty} \int_{-\infty}^{f} G(T, \lambda)\, d\lambda \qquad F(-\infty) = 0 \qquad (3.64)$$

Clearly, the average power over $(0, \infty)$ is

$$\bar{P}_\infty = F(\infty) \qquad (3.65)$$

The integrated spectrum of a power spectral density function $G(T, f)$ that contains an "impulsive" component is illustrated in Figure 3.8.

From the definition of the integrated spectrum, it follows that the average power due to signal components with frequencies between f_1 and f_2 is given by

$$F(f_2) - F(f_1) \qquad (3.66)$$

It is then possible to define the power spectral density G_∞, according to

$$G_\infty(f) = \begin{cases} \dfrac{d}{df} F(f) & \text{differentiable at } f \\ [F(f_o^+) - F(f_o^-)]\delta(f - f_o) & \text{not differentiable at } f_o \end{cases} \qquad (3.67)$$

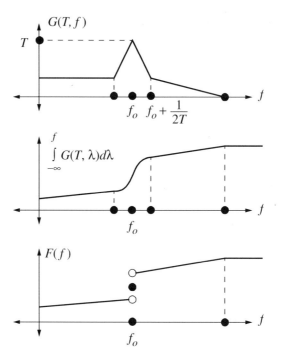

Figure 3.8 Illustration of the integrated spectral function for the case where G(T, f) has an "impulsive" component.

Again the average power is not equal to the integral of $G_\infty(f)$ but equal to $F(\infty)$.

For the case where there are no impulses in $G(T, f)$ as $T \to \infty$, and F is differentiable, such that

$$F(f) = \int_{-\infty}^{f} \lim_{T \to \infty} G(T, \lambda)\, d\lambda = \int_{-\infty}^{f} G_\infty(\lambda)\, d\lambda \qquad (3.68)$$

it follows that

$$G_\infty(f) = \frac{d}{df} F(f) \qquad (3.69)$$

3.7 POWER SPECTRAL DENSITY VIA AUTOCORRELATION

It is a common starting point to define the power spectral density as the Fourier transform of the autocorrelation function [see, for example, Champeney (1987 p. 102), Papoulis (2002 p. 408), and Thomas (1969 p. 97)]. In this section, the equivalence between this approach and the approach outlined so far, is given. The equivalence is summarized by the Wiener–Khintchine relations which are discussed. Appropriate references for this section are Jenkins (1968 pp. 213f), Peebles (1993 pp. 206f), Priestley (1981 pp. 210f), and Thomas (1969 pp. 96f).

3.7.1 Definition of the Autocorrelation Function

3.7.1.1 Definition of Autocorrelation Function—Single Waveform Case

With $x: \mathbf{R} \to \mathbf{C}$, the autocorrelation function of x on $[0, T]$ is defined according to

$$R(T, t, \tau) = \begin{cases} x(t)x^*(t - \tau) & t \in [0, T], t - \tau \in [0, T] \\ 0 & \text{elsewhere} \end{cases} \tag{3.70}$$

The autocorrelation function R is nonzero on the region of the t, τ plane as shown in Figure 3.9.

To interpret the autocorrelation function, consider a waveform x and its shifted counterpart, as shown in Figure 3.10. For a fixed time t, the shift τ required before there is a significant change between $x(t)$ and $x^*(t - \tau)$, and hence, $x(t)x^*(t - \tau)$, is an indication of the interval over which the signal is correlated at time t.

An average measure of the correlation time of the signal can be obtained through the time averaged autocorrelation function defined, according to

$$\bar{R}(T, \tau) = \begin{cases} \dfrac{1}{T} \displaystyle\int_0^{T+\tau} R(T, t, \tau)\, dt & \tau < 0 \\ \dfrac{1}{T} \displaystyle\int_\tau^{T} R(T, t, \tau)\, dt & \tau > 0 \end{cases} \tag{3.71}$$

where the integration limits are illustrated in Figure 3.9.

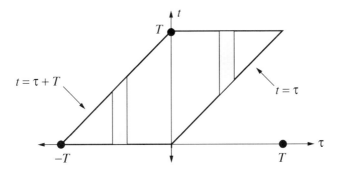

Figure 3.9 Region where autocorrelation function is nonzero.

80 THE POWER SPECTRAL DENSITY

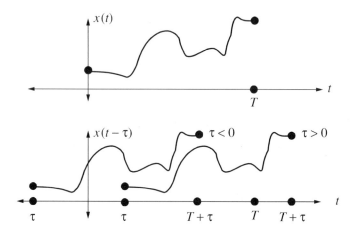

Figure 3.10 A waveform and its shifted counterpart (real case).

3.7.1.2 Notation The time averaged autocorrelation function, as defined by Eq. (3.71), is called the sample autocorrelation function (Parzen, 1962 p. 109) or the correlogram (Gardner, 1988 p. 13).

3.7.1.3 Definition of Autocorrelation Function — Random Process Case An autocorrelation function and a time averaged autocorrelation function can be defined for a random process via a weighted average. Consider a random process X defined by the countable ensemble E_X

$$E_X = \{x_i : \mathbf{R} \to \mathbf{C}, i \in \mathbf{Z}^+\} \qquad (3.72)$$

and a probability space, $P_X = \{p_i : p_i = P[x_i]\}$. Using the "natural" weighting of the waveform probabilities, the autocorrelation function and the time averaged autocorrelation function can be defined, respectively, according to

$$R(T, t, \tau) = \sum_{i=1}^{\infty} p_i R_i(T, t, \tau) = \sum_{i=1}^{\infty} p_i x_i(t) x_i^*(t - \tau) \qquad x_i \in E_X, p_i \in P_X \quad (3.73)$$

$$\bar{R}(T, \tau) = \sum_{i=1}^{\infty} p_i \bar{R}_i(T, \tau) \qquad p_i \in P_X \qquad (3.74)$$

where

$$\bar{R}_i(T, \tau) = \begin{cases} \dfrac{1}{T} \displaystyle\int_0^{T+\tau} R_i(T, t, \tau)\, dt & \tau < 0,\ R_i(T, t, \tau) = x_i(t) x_i^*(t - \tau) \\[2mm] \dfrac{1}{T} \displaystyle\int_\tau^{T} R_i(T, t, \tau)\, dt & \tau > 0,\ R_i(T, t, \tau) = x_i(t) x_i^*(t - \tau) \end{cases} \qquad (3.75)$$

The following theorem formally states conditions for the validity of the definition for $\bar{R}(T, \tau)$.

THEOREM 3.7. EXISTENCE OF AUTOCORRELATION FUNCTION *Assume T is fixed. If $x_i \in L^2[0, T]$ for all $x_i \in E_X$ and $\sup\{|\bar{R}_i(T, \tau)|: i \in Z^+, \tau \in [-T, T]\}$ is finite, then $\bar{R}(T, \tau) \in L[-T, T]$.*

Proof. Since $x_i \in L^2[0, T]$, it follows that for τ fixed, $x_i(t)x_i^*(t - \tau)$ is integrable over, for example, $[0, T + \tau]$ when $-T < \tau < 0$. This follows from the Schwarz inequality (Theorem 2.15), according to

$$\int_0^{T+\tau} x_i(t) x_i^*(t - \tau) \, dt \leqslant \sqrt{\int_0^{T+\tau} |x_i(t)|^2 \, dt} \sqrt{\int_0^{T+\tau} |x_i(t - \tau)|^2 \, dt} \qquad -T < \tau < 0 \tag{3.76}$$

Hence, $\bar{R}_i(T, \tau)$ is finite for $i \in Z^+$. If $\sup\{|\bar{R}_i(T, \tau)|: i \in Z^+, \tau \in [-T, T]\}$ is finite, the summation $\Sigma_{i=1}^{\infty} p_i \bar{R}_i(T, \tau) = \bar{R}(T, \tau)$ is also finite. Boundedness of $\bar{R}(T, \tau)$ for $\tau \in [-T, T]$ implies $\bar{R}(T, \tau) \in L[-T, T]$ as required.

3.7.2 Power Spectral Density—Autocorrelation Relationship

The following theorem states the relationship between the time averaged autocorrelation function and the power spectral density. This relationship yields an alternative definition for the power spectral density function.

THEOREM 3.8. POWER SPECTRAL DENSITY–AUTOCORRELATION RELATIONSHIP
If $x_i \in L^2[0, T]$, then the power spectral density of the ith waveform of a random process is given by

$$G_i(T, f) = \frac{|X_i(T, f)|^2}{T} = \int_{-T}^{T} \bar{R}_i(T, \tau) e^{-j2\pi f \tau} \, d\tau \tag{3.77}$$

If $x_i \in L^2[0, T]$ for all $x_i \in E_X$ and $\sup\{|\bar{R}_i(T, \tau)|: i \in Z^+, \tau \in [-T, T]\}$ is finite, then the power spectral density of the random process is given by the Fourier transform of the averaged autocorrelation function according to

$$G(T, f) = \int_{-T}^{T} \bar{R}(T, \tau) e^{-j2\pi f \tau} \, d\tau \tag{3.78}$$

This definition applies for both a single waveform and a random process, provided $\bar{R}$ is interpreted appropriately.

Proof. The proof of this result is given in Appendix 3.

From Eq. (3.78) it follows, according to Eq. (2.144), when $\bar{R}$ is continuous and $G \in L$, that $\bar{R}$ and G are a Fourier transform pair, that is,

$$\bar{R}(T, \tau) = \int_{-\infty}^{\infty} G(T, f) e^{j2\pi f \tau} df \tag{3.79}$$

Finite average power implies $G \in L$, and as $\bar{R}$ is defined in terms of an integral, it is continuous except when the underlying signals have impulsive components.

3.7.3 Autocorrelation Function on Infinite Interval

By definition the autocorrelation function on the infinite interval is

$$R_\infty(\tau) = \lim_{T \to \infty} \bar{R}(T, \tau) \tag{3.80}$$

This definition applies to both a single waveform and a random process provided $\bar{R}$ is interpreted appropriately.

For the nonimpulsive case, where $\lim_{T \to \infty} G(T, f)$ is bounded for all f, it follows that

$$G_\infty(f) = \lim_{T \to \infty} G(T, f) = \lim_{T \to \infty} \int_{-T}^{T} \bar{R}(T, \tau) e^{-j2\pi f \tau} d\tau \tag{3.81}$$

To relate G_∞ to R_∞, it is necessary to interchange the limit and integration operations in this equation. To achieve this, note, first, that the assumptions given in Theorem 3.7 imply $\bar{R} \in L[-T, T]$. Second, if there exists a function $Q \in L$ such that $\bar{R}(T, \tau) \leq Q(\tau)$ for all T, then from the dominated convergence theorem (Theorem 2.24) it follows that the order of limit and integral can be interchanged to yield

$$G_\infty(f) = \int_{-\infty}^{\infty} \lim_{T \to \infty} \bar{R}(T, \tau) e^{-j2\pi f \tau} d\tau = \int_{-\infty}^{\infty} R_\infty(\tau) e^{-j2\pi f \tau} d\tau \tag{3.82}$$

That is, the power spectral density function G_∞, is the Fourier transform of the autocorrelation function R_∞. Provided the average power is finite, consistent with $G_\infty \in L$, the integral of G_∞ equals the average power [i.e., there are no impulsive components in $G(T, f)$], and R_∞ is a continuous function, it follows that R_∞ is the inverse Fourier transform of G_∞ according to Eq. (2.144), that is,

$$R_\infty(f) = \int_{-\infty}^{\infty} G_\infty(\tau) e^{j2\pi f \tau} d\tau \qquad G_\infty(f) = \int_{-\infty}^{\infty} R_\infty(\tau) e^{-j2\pi f \tau} d\tau \tag{3.83}$$

In summary:

THEOREM 3.9. WIENER–KHINTCHINE RELATIONS—POPULAR DEFINITION *Assume for all $T \in \mathbf{R}^+$ it is the case that $x_i \in L^2[0, T]$ for all $x_i \in E_X$ and*

$$\sup\{|\bar{R}_i(T, \tau)|: i \in \mathbf{Z}^+, \tau \in [-T, T]\}$$

is finite. Further, assume that $G(T, f)$ is nonimpulsive, such that $\lim_{T \to \infty} G(T, f)$ is finite for all f, $G_\infty \in L$, there exists a function $Q \in L$ such that $\bar{R}(T, \tau) \leqslant Q(\tau)$ for all T, and R_∞ is a continuous function. With these assumptions, it follows that the Fourier transform relationships defined by Eq. (3.83) are valid.

Commonly, these Fourier transform relationships are called the Wiener–Khintchine relations (Champeney, 1987 p. 104; Parzen, 1962 p. 110; Peebles, 1993 p. 207).

3.7.3.1 Impulsive Case—Formal Definition of Wiener–Khintchine Relations

The general case is where G may contain impulsive components, so that it is not possible to interchange the order of limit and integration in the following equation:

$$R_\infty(\tau) = \lim_{T \to \infty} \bar{R}(T, \tau) = \lim_{T \to \infty} \int_{-\infty}^{\infty} G(T, f) e^{j2\pi f\tau} df \qquad (3.84)$$

For this case, a spectral distribution function, as defined in Eq. (3.64), is used, that is,

$$F(f) = \lim_{T \to \infty} \int_{-\infty}^{f} G(T, \lambda) d\lambda \qquad (3.85)$$

Note, $F(-\infty) = 0$, $\bar{P}_\infty = F(\infty)$, and F is a monotonically increasing function.

THEOREM 3.10. WIENER–KHINTCHINE THEOREM—FORMAL DEFINITION *The spectral theorem for the autocorrelation function is (Grimmett, 1992 p. 353; Parzen, 1962 p. 110; Champeney, 1987 p. 108)*

$$R_\infty(\tau) = \int_{-\infty}^{\infty} e^{j2\pi f\tau} dF(f) \qquad (3.86)$$

where the integral is a Riemann–Stieltjes integral (Apostol, 1974 ch 7; Parzen, 1960 p. 233). This result is a formal statement of the Wiener–Khintchine theorem (Priestley, 1981 p. 218).

Proof. The proof is given in Appendix 4.

3.7.3.2 Example
Consider the case where G is impulsive, such that

$$\lim_{T \to \infty} G(T, f) = \delta(f + f_o) + \delta(f - f_o) \qquad (3.87)$$

whereupon, it follows that

$$F(f) = \begin{cases} 0 & f < -f_o \\ 1 & -f_o \leq f < f_o \\ 2 & f_o \leq f \end{cases} \qquad (3.88)$$

and

$$R_\infty(\tau) = \int_{-\infty}^{\infty} e^{j2\pi f \tau} dF(f) = e^{-j2\pi f_o \tau} + e^{j2\pi f_o \tau} = 2\cos(2\pi f_o \tau) \qquad (3.89)$$

APPENDIX 1: PROOF OF THEOREM 3.4

Assume T is fixed. By definition

$$|G(T, f) - G_A(T, f)| = \frac{1}{T} \left| \int_{-\infty}^{\infty} f_X(\lambda) |X(\lambda, T, f)|^2 \, d\lambda - \sum_{i=1}^{N} |X_{A_i}(T, f)|^2 \int_{I_i} f_X(\lambda) \, d\lambda \right|$$

$$= \frac{1}{T} \left| \int_{-\infty}^{\infty} f_X(\lambda) [|X(\lambda, T, f)|^2 - |X_S(\lambda, T, f)|^2] \, d\lambda \right|$$

where $X_S(\lambda, T, f)$ is a step approximation, with respect to λ, to $X(\lambda, T, f)$ defined, when $\lambda \in I_i$, according to $X_S(\lambda, T, f) = X_{A_i}(T, f) = X(\lambda_i, T, f)$.

When $X_{max}^2(T) = \sup\{|X(\lambda, T, f)|^2 : \lambda \in R, f \in R\}$ is finite and $f_X \in L$, it follows, according to Theorem 2.11, that $\forall \varepsilon > 0$ there exists a $\lambda_o > 0$, such that

$$\int_{-\infty}^{-\lambda_o} f_X(\lambda) |X(\lambda, T, f)|^2 \, d\lambda < \varepsilon \qquad \int_{\lambda_o}^{\infty} f_X(\lambda) |X(\lambda, T, f)|^2 \, d\lambda < \varepsilon \qquad (3.90)$$

Hence,

$$|G(T, f) - G_A(T, f)| < \frac{1}{T} \left| \int_{-\lambda_o}^{\lambda_o} f_X(\lambda) [|X(\lambda, T, f)|^2 - |X_S(\lambda, T, f)|^2] \, d\lambda \right| + 2\varepsilon$$

$$\leq \frac{f_{max}}{T} \int_{-\lambda_o}^{\lambda_o} ||X(\lambda, T, f)|^2 - |X_S(\lambda, T, f)|^2| \, d\lambda + 2\varepsilon$$

$$(3.91)$$

where f_{max} is the maximum value of f_X. If $|X(\lambda, T, f)|^2$ is of bounded variation with respect to λ over the interval $[-\lambda_o, \lambda_o]$, then the required result follows directly from Theorem 2.19. To prove such bounded variation, consider the summation:

$$S = \sum_{i=1}^{N-1} ||X(\lambda_{i+1}, T, f)|^2 - |X(\lambda_i, T, f)|^2| \qquad \lambda_i, \lambda_{i+1} \in [-\lambda_o, \lambda_o]$$

$$= \sum_{i=1}^{N-1} |[|X(\lambda_{i+1}, T, f)| - |X(\lambda_i, T, f)|][|X(\lambda_{i+1}, T, f)| + |X(\lambda_i, T, f)|]| \quad (3.92)$$

$$\leqslant 2X_{max}(T) \sum_{i=1}^{N-1} ||X(\lambda_{i+1}, T, f)| - |X(\lambda_i, T, f)||$$

Using the triangle inequality and the definition for the Fourier transform yields:

$$S \leqslant 2X_{max}(T) \sum_{i=1}^{N-1} |X(\lambda_{i+1}, T, f) - X(\lambda_i, T, f)|$$

$$= 2X_{max}(T) \sum_{i=1}^{N-1} \left| \int_0^T [x(\lambda_{i+1}, t) - x(\lambda_i, t)] e^{-j2\pi ft} \, dt \right| \quad (3.93)$$

$$\leqslant 2X_{max}(T) \int_0^T \sum_{i=1}^{N-1} |x(\lambda_{i+1}, t) - x(\lambda_i, t)| \, dt$$

The assumption of bounded variation of $x(\lambda, t)$, with respect to λ, over all finite intervals implies there exists a constant k_o, that is independent of N, such that

$$\sum_{i=1}^{N-1} |x(\lambda_{i+1}, t) - x(\lambda_i, t)| \leqslant k_o(t) \quad (3.94)$$

Uniform bounded variation with respect to t implies that $k_o(t)$ is independent of t. Hence,

$$S \leqslant 2Tk_o X_{max}(T) \quad (3.95)$$

which proves bounded variation of $|X(\lambda, T, f)|^2$, with respect to λ, as required.

APPENDIX 2: PROOF OF THEOREM 3.5

The proof is given for the countable case; the proof for the uncountable case follows in an analogous manner.

A.2.1 Existence of PSD—Finite Case

The mathematical operations required for the definition of the power spectral density function, as per Eq. (3.29), are as follows:

$$\bar{P}(T) = \sum_{i=1}^{\infty} p_i \bar{P}_i(T) = \sum_{i=1}^{\infty} p_i \frac{1}{T} \int_0^T |x_i(t)|^2 \, dt = \sum_{i=1}^{\infty} p_i \frac{1}{T} \int_{-\infty}^{\infty} |X_i(T, f)|^2 \, df$$

$$= \int_{-\infty}^{\infty} \frac{1}{T} \sum_{i=1}^{\infty} p_i |X_i(T, f)|^2 \, df = \int_{-\infty}^{\infty} G(T, f) \, df \qquad (3.96)$$

There are three requirements that must be met before the mathematical operations implicit in this sequence of equations are valid.

(a) The average power $\bar{P}(T)$ must be finite, as is assumed by use of the equality signs in these equations; one cannot have equality between infinite numbers because an infinite number is larger than any other number. Finiteness of $\bar{P}(T)$ implies the individual signal powers, $\bar{P}_i(T)$ must be finite. This is guaranteed from Eq. (3.23), if $x_i \in L^2[0, T]$. Finiteness of $\bar{P}(T)$ also implies that $\sum_{i=1}^{N} p_i \bar{P}_i(T)$ must converge as N increases without bound. Standard results, such as the comparison test, [see, for example, Knopp (1956 p. 52f)] give sufficient conditions for such convergence, that is,

$$\exists \alpha > 0, \; \exists k_o > 0, \; \exists I_o > 0 \quad \text{s.t.} \quad \forall i > I_o \quad p_i \bar{P}_i(T) < \frac{k_o}{i^{1+\alpha}} \qquad (3.97)$$

If $\sup\{\bar{P}_i(T): i \in Z^+\}$ is finite, then convergence of the series is guaranteed. Note, it can be the case that $\lim_{i \to \infty} \bar{P}_i(T)$ is infinite, while the average power $\bar{P}(T)$ is finite.

(b) Parseval's relationship must hold. This is guaranteed, as per Theorem 2.31, if $x_i \in L^2[0, T]$.

(c) The order of summation and integration, as per the following equation, must be interchangeable:

$$\sum_{i=1}^{\infty} p_i \int_{-\infty}^{\infty} |X_i(T, f)|^2 \, df = \int_{-\infty}^{\infty} \sum_{i=1}^{\infty} p_i |X_i(T, f)|^2 \, df \qquad (3.98)$$

that is,

$$\lim_{N \to \infty} \int_{-\infty}^{\infty} \sum_{i=1}^{N} p_i |X_i(T, f)|^2 \, df = \int_{-\infty}^{\infty} \lim_{N \to \infty} \sum_{i=1}^{N} p_i |X_i(T, f)|^2 \, df \qquad (3.99)$$

This interchange is guaranteed if $x_i \in L^2[0, T]$, and the average power $\bar{P}(T)$ is finite, as the following argument demonstrates. First, with $x_i \in L^2[0, T]$ for all $i \in Z^+$ and with $\bar{P}(T)$ finite, it follows that $\sum_{i=1}^{\infty} p_i \int_{-\infty}^{\infty} |X_i(T, f)|^2 \, df$

is finite. Thus:

$$\sum_{i=1}^{\infty} p_i \int_{-\infty}^{\infty} |X_i(T, f)|^2 \, df > \sum_{i=1}^{N} p_i \int_{-\infty}^{\infty} |X_i(T, f)|^2 \, df = \int_{-\infty}^{\infty} \sum_{i=1}^{N} p_i |X_i(T, f)|^2 \, df$$
(3.100)

The interchange of summation and integration in this equation is possible because all the terms within the summation are positive, and a finite sum of finite integrals equals the integral of the finite sum. Second, define $F: \mathbf{R}^3 \to \mathbf{R}$ according to

$$F(N, T, f) = \sum_{i=1}^{N} p_i |X_i(T, f)|^2$$
(3.101)

Now, F is a monotonically increasing function with respect to N. Further, from Eq. (3.100) $\int_{-\infty}^{\infty} F(N, T, f) \, df$ is bounded above. Hence, according to the monotone convergence theorem (Theorem 2.23), it follows that

$$\lim_{N \to \infty} \int_{-\infty}^{\infty} F(N, T, f) \, df = \int_{-\infty}^{\infty} \lim_{N \to \infty} F(N, T, f) \, df$$
(3.102)

which is the required result.

In summary: If $x_i \in L^2[0, T]$ for all i, and the average power on the interval $[0, T]$ is finite, then the power spectral density function G, defined by Eq. (3.29), is valid.

A.2.2 Existence of PSD — Infinite Case

The power spectral density G_∞ depends on G being validly defined on all intervals of the form $[0, T]$, and that the sequence of mathematical operations, defined by the following equations, are valid:

$$\bar{P}_\infty = \sum_{i=1}^{\infty} \lim_{T \to \infty} p_i \bar{P}_i(T) = \lim_{T \to \infty} \sum_{i=1}^{\infty} p_i \bar{P}_i(T) = \lim_{T \to \infty} \sum_{i=1}^{\infty} p_i \frac{1}{T} \int_0^T |x_i(t)|^2 \, dt$$

$$= \lim_{T \to \infty} \sum_{i=1}^{\infty} p_i \frac{1}{T} \int_{-\infty}^{\infty} |X_i(T, f)|^2 \, df = \lim_{T \to \infty} \int_{-\infty}^{\infty} \frac{1}{T} \sum_{i=1}^{\infty} p_i |X_i(T, f)|^2 \, df$$

$$= \lim_{T \to \infty} \int_{-\infty}^{\infty} G(T, f) \, df = \int_{-\infty}^{\infty} \lim_{T \to \infty} G(T, f) \, df$$
(3.103)

By comparing Eq. (3.103) with Eq. (3.96), it follows, in addition to the requirements for the finite case, that, first, the limit and summation operations in the first line of this equation need to be interchangeable. Second, the limit and integral operations in the third line of this equation need to be interchangeable. These two additional requirements are discussed below.

Interchange of the limit and summation operations is valid if dominated convergence for the series with terms $p_i \bar{P}_i(T)$ can be established. Thus, if there exists a sequence $\{Q_i: i \in \mathbf{Z}^+\}$, such that $p_i \bar{P}_i(T) \leq Q_i$ for every i and for every finite value of T and $\Sigma_{i=1}^{\infty} Q_i$ converges, then the interchange is valid. For example, if independent of T, there exists constants $\alpha, k_o, I_o > 0$ such that

$$\forall i > I_o \quad p_i \bar{P}_i(T) < \frac{k_o}{i^{1+\alpha}} \tag{3.104}$$

then the interchange is valid. If $\sup\{\bar{P}_i(T): i \in \mathbf{Z}^+, T \in \mathbf{R}^+\}$ is finite, then again the interchange is valid.

Second, the limit and integral operations can be interchanged according to

$$\lim_{T \to \infty} \int_{-\infty}^{\infty} G(T, f) \, df = \int_{-\infty}^{\infty} \lim_{T \to \infty} G(T, f) \, df \tag{3.105}$$

if either the dominated or monotone convergence theorems (Theorems 2.23 and 2.24) can be applied. For example, if there exists a function $H \in L$, such that $G(T, f) \leq H(f)$ for all f and all T, then from the dominated convergence theorem the required interchange is valid.

In summary: If $x_i \in L^2[0, T]$ for $i \in \mathbf{Z}^+$ and $T \in \mathbf{R}^+$, there exists a sequence $\{Q_i: i \in \mathbf{Z}^+\}$ such that $p_i \bar{P}_i(T) \leq Q_i$ for $i \in \mathbf{Z}^+$ and $T \in \mathbf{R}^+$, $\Sigma_{i=1}^{\infty} Q_i$ is finite, and there exists a function $H \in L$, such that $G(T, f) \leq H(f)$ for all $T \in \mathbf{R}^+$ and all f, then G_∞ defined by Eq. (3.33) is valid.

APPENDIX 3: PROOF OF THEOREM 3.8

Consider the integral I, defined as

$$I = \int_{-T}^{T} \bar{R}_i(T, \tau) e^{-j2\pi f \tau} \, d\tau = \begin{cases} \int_{-T}^{0} \left[\frac{1}{T} \int_{0}^{T+\tau} x_i(t) x_i^*(t-\tau) \, dt \right] e^{-j2\pi f \tau} \, d\tau & \tau < 0 \\ \int_{0}^{T} \left[\frac{1}{T} \int_{\tau}^{T} x_i(t) x_i^*(t-\tau) \, dt \right] e^{-j2\pi f \tau} \, d\tau & \tau > 0 \end{cases} \tag{3.106}$$

where the second equivalence arises from Eq. (3.75). By assumption, $x_i \in L^2[0, T]$, which implies, consistent with Eq. (3.76), that

$$\int_{0}^{T+\tau} |x_i(t) x_i^*(t-\tau)| \, dt$$

is finite for $-T < \tau < 0$. A similar result holds for the case where $0 \leqslant \tau < T$. It then follows from Theorem 2.20 that the order of integration can be interchanged to yield (see Figure 3.9 for the region of integration)

$$I = \int_0^T \int_{t-T}^t \frac{1}{T} x_i(t) x_i^*(t-\tau) e^{-j2\pi f\tau} \, d\tau \, dt \qquad (3.107)$$

A change of variable $\xi = t - \tau$ for τ yields:

$$I = \int_0^T \int_0^T \frac{1}{T} x_i(t) x_i^*(\xi) e^{-j2\pi f(t-\xi)} \, d\xi \, dt$$

$$= \frac{1}{T} \left[\int_0^T x_i(t) e^{-j2\pi ft} \, dt \right] \left[\int_0^T x_i^*(\xi) e^{j2\pi f\xi} \, d\xi \right] = \frac{|X_i(T, f)|^2}{T} = G_i(T, f) \qquad (3.108)$$

which is the first required result.

To prove the second result, define I_2 according to

$$I_2 = \int_{-T}^T \bar{R}(T, \tau) e^{-j2\pi f\tau} \, d\tau = \int_{-T}^T \left[\lim_{N \to \infty} \sum_{i=1}^N p_i \bar{R}_i(T, \tau) \right] e^{-j2\pi f\tau} \, d\tau \qquad (3.109)$$

As $\sup\{|\bar{R}_i(T, \tau)|: i \in \mathbf{Z}^+, \tau \in [-T, T]\}$ is assumed to be finite, it follows from the dominated convergence theorem that the interchange of the limit and integral operations in this equation is valid. Thus,

$$I_2 = \lim_{N \to \infty} \int_{-T}^T \left[\sum_{i=1}^N p_i \bar{R}_i(T, \tau) \right] e^{-j2\pi f\tau} \, d\tau = \sum_{i=1}^\infty p_i \int_{-T}^T \bar{R}_i(T, \tau) e^{-j2\pi f\tau} \, d\tau$$

$$= \sum_{i=1}^\infty p_i G_i(T, f) = G(T, f) \qquad (3.110)$$

where the result that the integral of a finite sum equals the sum of corresponding finite integrals has been used.

APPENDIX 4: PROOF OF THEOREM 3.10

For the case where the random process underlying F has finite average power, it follows from Eq. (3.85) that F is a piecewise continuous function. It then follows that

$$\int_{-\infty}^\infty e^{j2\pi f\tau} \, dF(f) = \lim_{\Delta f \to 0} \sum_{i=-\infty}^\infty e^{j2\pi(i\Delta f)\tau} [F(i\Delta f) - F((i-1)\Delta f)] \qquad (3.111)$$

where the continuity of the exponential function has been used (Apostol, 1974 pp. 141, 148–9). Using the definition for F, as per Eq. (3.85), it follows that the right-hand side of this equation can be written as

$$\lim_{\Delta f \to 0} \sum_{i=-\infty}^{\infty} e^{j2\pi(i\Delta f)\tau} \left[\lim_{T \to \infty} \int_{-\infty}^{i\Delta f} G(T, f) \, df - \lim_{T \to \infty} \int_{-\infty}^{(i-1)\Delta f} G(T, f) \, df \right] \quad (3.112)$$

The finiteness of the average power guarantees that both limits defined by $T \to \infty$ are finite. It then follows that

$$\int_{-\infty}^{\infty} e^{j2\pi f \tau} \, dF(f) = \lim_{\Delta f \to 0} \sum_{i=-\infty}^{\infty} e^{j2\pi(i\Delta f)\tau} \left[\lim_{T \to \infty} \int_{(i-1)\Delta f}^{i\Delta f} G(T, f) \, df \right]$$

$$= \lim_{\Delta f \to 0} \sum_{i=-\infty}^{\infty} \left[\lim_{T \to \infty} \int_{(i-1)\Delta f}^{i\Delta f} G(T, f) e^{j2\pi(i\Delta f)\tau} \, df \right] \quad (3.113)$$

With the definitions

$$s_i(T) = \int_{(i-1)\Delta f}^{i\Delta f} G(T, f) e^{j2\pi(i\Delta f)\tau} \, df \qquad S_i(T) = \int_{(i-1)\Delta f}^{i\Delta f} G(T, f) \, df \quad (3.114)$$

$$S_i = \lim_{T \to \infty} S_i(T) = \lim_{T \to \infty} \int_{(i-1)\Delta f}^{i\Delta f} G(T, f) \, df \quad (3.115)$$

it follows that $|s_i(T)| \leq S_i(T)$ for $i \in Z$ and for all $\varepsilon_i > 0$ there exists a T_o, such that for all $T > T_o$ it is the case that $|S_i - S_i(T)| < \varepsilon_i$.

As

$$\sum_{i=-\infty}^{\infty} S_i = \lim_{T \to \infty} \int_{-\infty}^{\infty} G(T, f) \, df = \bar{P}_{\infty} \quad (3.116)$$

$$\sum_{i=-\infty}^{\infty} S_i(T) = \int_{-\infty}^{\infty} G(T, f) \, df = \bar{P}(T) \quad (3.117)$$

and $\lim_{T \to \infty} \bar{P}(T) = \bar{P}_{\infty}$, it follows that there exists a positive sequence $\{e_i\}$, such that $|s_i(T)| \leq S_i + e_i$ for $i \in Z$, and for all T greater than a constant T_o, and

$$\sum_{i=-\infty}^{\infty} e_i < \infty \quad (3.118)$$

With $\bar{P}_{\infty}$ being finite, it then follows from dominated convergence, that

$$\lim_{T \to \infty} \sum_{i=-\infty}^{\infty} s_i(T) = \sum_{i=-\infty}^{\infty} \lim_{T \to \infty} s_i(T) \quad (3.119)$$

Hence,

$$\int_{-\infty}^{\infty} e^{j2\pi f \tau} dF(f) = \lim_{T \to \infty} \int_{-\infty}^{\infty} G(T, f) e^{j2\pi i f \tau} df = \lim_{T \to \infty} \bar{R}(T, \tau) = R_{\infty}(\tau)$$

(3.120)

which is the required result.

4

Power Spectral Density Analysis

4.1 INTRODUCTION

In this chapter, general results for the power spectral density that facilitate evaluation of the power spectral density of specific random processes are given. First, the nature of the Fourier transform on the infinite interval is discussed and a criterion is given for the power spectral density to be bounded on this interval. Second, the use of an alternative power spectral density function that can be defined for the case where a signal consists of a sum of orthogonal or disjoint waveforms is discussed. Third, a theorem is proved that specifies when signal components outside of the interval $[0, T]$ can be included when evaluating the power spectral density. Including such signal components can greatly simplify analysis. Fourth, the cross power spectral density is defined and bounds on its level are established. Fifth, the power spectral density of the sum of an infinite number of random processes is derived. Sixth, the power spectral density of a periodic signal is derived and is shown to have the expected form, namely, impulsive component at integer multiples of the fundamental frequency. Finally, the power spectral density of a random process containing a periodic and a nonperiodic component is derived and it is shown, for the infinite interval, that the periodic and nonperiodic components can be treated separately.

4.2 BOUNDEDNESS OF POWER SPECTRAL DENSITY

To prove subsequent results, it is necessary to demarcate those random processes that have a bounded power spectral density on the interval $[0, \infty]$, from those that do not. Clearly, if for all $f \in \mathbf{R}$, there exists $k, T_o \in \mathbf{R}^+$, such that $\forall T > T_o$

$$|X(T, f)| < k\sqrt{T} \tag{4.1}$$

then

$$\lim_{T \to \infty} G_X(T, f) = \lim_{T \to \infty} \frac{|X(T, f)|^2}{T} < k^2 \quad f \in \mathbf{R} \tag{4.2}$$

Note, any signal that is periodic or contains a periodic component (including the degenerate case of a nonzero mean) will not satisfy this criterion. To see this, consider a periodic signal x, with period $T_p = 1/f_p$ that satisfies appropriate conditions (Theorem 2.26) such that it can be written, using an exponential Fourier series, as

$$x(t) = \sum_{i=-\infty}^{\infty} c_i e^{j2\pi i f_p t} \tag{4.3}$$

whereupon, for f equal to the qth harmonic, it follows that

$$X(T, qf_p) = c_q T + \int_0^T \sum_{\substack{i=-\infty \\ i \ne q}}^{\infty} c_i e^{j2\pi(i-q)f_p t} \, dt \tag{4.4}$$

Clearly, when $c_q \ne 0$, it is the case that both $|X(T, qf_p)|$ and $G(T, qf_p)$ increase in proportion to T. Thus, for a periodic signal, or the degenerate case of a signal with nonzero mean, the power spectral density is not bounded at specific frequencies. The unboundedness is restricted to a set of zero measure for a finite power random process.

4.2.1 Alternative Formulation for Boundedness

The following theorem gives an alternative criterion for the boundedness of $G_X(T, f)$ as $T \to \infty$.

THEOREM 4.1. BOUNDEDNESS OF POWER SPECTRAL DENSITY *Consider the sequence of functions* $X_1, \ldots, X_N$ *produced by a disjoint partition of the interval* $[0, T]$ *and defined according to*

$$X_i(f) = \int_{(i-1)\Delta T}^{i\Delta T} x(t) e^{-j2\pi f t} \, dt \quad i \in \{1, \ldots, N\} \tag{4.5}$$

such that

$$X(T, f) = \sum_{i=1}^{N} X_i(f) \qquad (4.6)$$

Here, $N = \lfloor T/\Delta T_o \rfloor$ for some fixed ΔT_o and ΔT is such that $N\Delta T = T$. If, for all $f \in \mathbf{R}$ there exists a $\Delta T_o > 0$ such that $\forall T \gg \Delta T_o$ the mean of $X_1(f), \ldots, X_N(f)$ decays according to $\sqrt{N}$ as T and N increase without bound, that is, there exists an integer $N_o > 0$ and a constant $k_x \in \mathbf{R}^+$, such that $\forall N > N_o$

$$\frac{1}{N}\left|\sum_{i=1}^{N} X_i(f)\right| < \frac{k_x}{\sqrt{N}} \qquad f \in \mathbf{R} \qquad (4.7)$$

then both

$$\sup\left\{\lim_{T \to \infty} \frac{|X(T, f)|}{\sqrt{T}} : f \in \mathbf{R}\right\} \quad \text{and} \quad \sup\left\{\lim_{T \to \infty} G(T, f) : f \in \mathbf{R}\right\}$$

are finite.

Proof. This result follows by simply noting that

$$\frac{1}{N}\left|\sum_{i=1}^{N} X_i(f)\right| < \frac{k_x}{\sqrt{N}} \Rightarrow \frac{\Delta T}{T}|X(T, f)| < \frac{k_x\sqrt{\Delta T}}{\sqrt{T}} \Rightarrow |X(T, f)| < \frac{k_x\sqrt{T}}{\sqrt{\Delta T}} \qquad (4.8)$$

and as $N \to \infty$ it is the case that $\Delta T \to \Delta T_o$.

4.2.1.1 Notes
This formulation is best understood by considering N outcomes $x_1, \ldots, x_N$ of N independent and identically distributed random variables with zero mean and variance σ^2. For N sufficiently large, and with a probability of 0.95, independence guarantees, as per the central limit theorem (Grimmett, 1992 p. 175; Larson, 1986 p. 322), that

$$-1.96\sqrt{N}\sigma < \sum_{i=1}^{N} x_i < 1.96\sqrt{N}\sigma \quad \Rightarrow \quad \frac{1}{N}\left|\sum_{i=1}^{N} x_i\right| < \frac{1.96\sigma}{\sqrt{N}} \qquad (4.9)$$

Hence, if there exists a time ΔT_o, such that for all longer time intervals it is the case that $X_1(f), \ldots, X_N(f)$ are independent samples, consistent with outcomes from N independent and identically distributed random variables, then the power spectral density of that process is guaranteed to be bounded. Such a result is consistent with the "correlation time" of the signal being less than ΔT_o.

4.2.2 Definition—Bounded Power Spectral Density

DEFINITION: BOUNDED POWER SPECTRAL DENSITY A random process X is said to have a bounded power spectral density if the above criteria hold, that is, if $\exists k_x \in \mathbf{R}^+$, such that

$$\sup\{G_X(T, f): T \in \mathbf{R}^+, f \in \mathbf{R}\} < k_x \quad (4.10)$$

In subsequent analysis, it will be assumed that random processes have a bounded power spectral density, or at most, have a bounded component plus an unbounded component due to periodic signal(s).

4.3 POWER SPECTRAL DENSITY VIA SIGNAL DECOMPOSITION

Consider an interval $[0, T]$ on which a signal x can be written as the sum of N disjoint, or orthogonal waveforms according to

$$x(t) = \sum_{i=1}^{N} x_i(t) \qquad t \in [0, T] \quad (4.11)$$

The average power on the interval $[0, T]$ is

$$\bar{P} = \frac{1}{T} \int_0^T |x(t)|^2 \, dt = \frac{1}{T} \int_0^T \sum_{i=1}^{N} |x_i(t)|^2 \, dt \quad (4.12)$$

From Parseval's relationship it follows that

$$\bar{P} = \int_{-\infty}^{\infty} \frac{|X(T, f)|^2}{T} \, df = \int_{-\infty}^{\infty} \sum_{i=1}^{N} \frac{|X_i(T, f)|^2}{T} \, df \quad (4.13)$$

which suggests, for the signal being considered, two alternative definitions for the power spectral density, respectively, denoted G_X and G_{XD}:

$$G_X(T, f) = \frac{|X(T, f)|^2}{T} \qquad G_{XD}(T, f) = \sum_{i=1}^{N} \frac{|X_i(T, f)|^2}{T} \quad (4.14)$$

By definition, G_X is the correct power spectral density. While G_{XD} is a valid power spectral density, as far as the average power is concerned, it may be the case that $G_X(T, f) \neq G_{XD}(T, f)$ almost everywhere, including the points if_o, $i \in \mathbf{Z}^+$ where $f_o = 1/T$. If this is the case, then $G_{XD}(T, f)$ does not have the interpretation required for a power spectral density, as per Theorem 3.2, namely, that the area of each pair of rectangles of width f_o, centered at $-if_o$ and if_o, and with respective heights $G_{XD}(T, -if_o)$ and $G_{XD}(T, if_o)$, is equal to the power in the sinusoidal components with a frequency if_o.

4.3.1 Example

Consider the waveform x and the pulse function p, as shown in Figure 4.1. The signal x can be written in terms of the pulse waveform p according to

$$x(t) = p\left(\frac{t}{2}\right) + p(t-3) \qquad (4.15)$$

and it follows that the power spectral density of this signal, evaluated on the interval $[0, 4]$, is

$$G_X(4, f) = \frac{|X(4, f)|^2}{T} = \frac{|2P(2f) + P(f)e^{-j2\pi 3f}|^2}{T} \qquad (4.16)$$

$$= \frac{4|P(2f)|^2 + |P(f)|^2 + 4\,\mathrm{Re}[P(2f)P^*(f)e^{j2\pi 3f}]}{T}$$

The first line in this equation follows from the relationships (McGillem, 1991 p. 146):

$$v(t - t_d) \leftrightarrow V(f)e^{-j2\pi f t_d} \qquad v(at) \leftrightarrow \frac{V(f/a)}{|a|} \qquad (4.17)$$

Now, x can be decomposed, in terms of disjoint signals defined by delayed versions of p, according to

$$x(t) = \sum_{i=1}^{3} x_i(t) = p(t) + p(t-1) + p(t-3) \qquad (4.18)$$

and the alternative power spectral density function G_{XD} can be defined as

$$G_{XD}(4, f) = \sum_{i=1}^{3} \frac{|X_i(4, f)|^2}{T} = \frac{3|P(f)|^2}{T} \qquad (4.19)$$

The power spectral densities G_X and G_{XD} are plotted in Figure 4.2 using the result that $P(f) = \mathrm{sinc}(f)e^{-j\pi f}$. Clearly, for this case $G_X(T, f) \neq G_{XD}(T, f)$ almost everywhere. In particular, when $T = 1/f_o = 4$ it is the case that

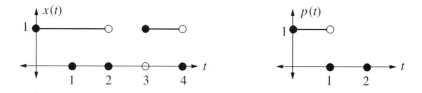

Figure 4.1 Graphs for the signal x and the pulse function p.

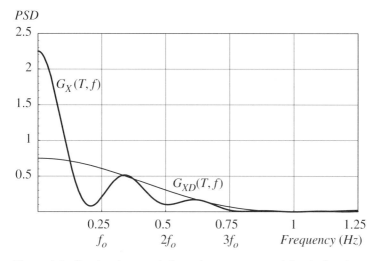

Figure 4.2 Graphs of true and alternative power spectral density functions.

$P(2f)|_{f=2f_o} = 0$, and

$$G_X(T, 2f_o) = \frac{|P(2f_o)|^2}{T} = \frac{\text{sinc}^2(0.5)}{4} = 0.1013$$

$$G_{XD}(T, 2f_o) = \frac{3|P(2f_o)|^2}{T} = \frac{3\,\text{sinc}^2(0.5)}{4} = 0.3039$$

(4.20)

Hence, for $f = 2f_o$, the alternative power spectral density, G_{XD}, does not predict the power in sinusoids with a frequency of $2f_o$.

4.3.2 Explanation

If the signal x can be written as a summation of N signals according to Eq. (4.11), then the power spectral density can be written as

$$\begin{aligned}G_X(T, f) &= \frac{|X(T, f)|^2}{T} = \frac{1}{T}\sum_{i=1}^{N}|X_i(T, f)|^2 + \frac{1}{T}\sum_{i=1}^{N}\sum_{\substack{j=1\\j\neq i}}^{N}X_i(T, f)X_j^*(T, f) \\ &= G_{XD}(T, f) + \frac{1}{T}\sum_{i=1}^{N}\sum_{\substack{j=1\\j\neq i}}^{N}X_i(T, f)X_j^*(T, f)\end{aligned}$$

(4.21)

Clearly, the cross product terms constitute the difference between the two power spectral densities. Since disjointness or orthogonality is a sufficient condition for the integral of each term in the double summation in this

98 POWER SPECTRAL DENSITY ANALYSIS

equation to be zero, it follows, in terms of the average power, that there is no difference between the two power spectral densities. However, when interpreted as the power in the sinusoidal components at a specific frequency, the alternative power spectral density G_{XD}, in general, is not correct.

4.4 SIMPLIFYING EVALUATION OF POWER SPECTRAL DENSITY

Consider the countable case and a random process X, defined by an ensemble $E_X = \{x_i : i \in \mathbf{Z}^+\}$ where $P[x_i] = p_i$. The power spectral density requires the evaluation of the Fourier transform of each waveform in the ensemble over the interval $[0, T]$, that is,

$$G_X(T, f) = \sum_{i=1}^{\infty} p_i \frac{|X_i(T, f)|^2}{T} \tag{4.22}$$

However, truncation of the signal through use of the interval $[0, T]$ often complicates analysis, while inclusion of some component of the signal outside of this interval can simplify analysis, as the following example illustrates.

4.4.1 Example

Consider the evaluation of the power spectral density on the interval $[0, T]$, of a signal that consists of a summation of pulse waveforms, that is,

$$x(t) = \sum_{i=0}^{\infty} p(t - iD) \tag{4.23}$$

where p has the form shown in Figure 4.3. The waveforms comprising the signal x are also shown in this figure, assuming, for illustrative purposes, that $T = 4D$. The Fourier transform of x on the interval $[0, T]$, for $T = 4D$, is given by

$$X(T, f) = P(f) + P(f)e^{-j2\pi fD} + P(f)e^{-j4\pi fD} + P_3(f) \tag{4.24}$$

where P is the Fourier transform of p, and P_3 is the Fourier transform of

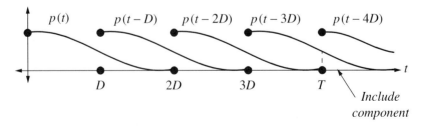

Figure 4.3 Pulse waveform p and waveforms comprising the signal x.

$p(t - 3D)$ evaluated on the interval $[0, T]$. Clearly, analysis can be simplified if the component of $p(t - 3D)$ outside of $[0, T]$, as shown in Figure 4.3, can be included, whereupon the approximation

$$X_A(T + D, f) = P(f) + P(f)e^{-j2\pi fD} + P(f)e^{-j4\pi fD} + P(f)e^{-j6\pi fD} \quad (4.25)$$

is obtained. Such a Fourier transform is consistent with the approximate signal x_A being defined as equal to the x on the interval $[0, T]$, but including the component of $p(t - 3D)$ outside this interval, that is,

$$x_A(t) = \begin{cases} \sum_{i=0}^{3} p(t - iD) & t \in [0, T + D) \\ 0 & \text{elsewhere} \end{cases} \quad (4.26)$$

The power spectral density of x_A over the interval $[0, T + D]$, but normalized by T rather than the interval length $T + D$, can be defined as

$$G_{XA}(T + D, f) = \frac{|X_A(T + D, f)|^2}{T} \quad (4.27)$$

and ideally, is such that $G_{XA}(T + D, f) \approx G_X(T, f)$ for all frequencies.

The inclusion of the contribution of a signal component outside the interval $[0, T]$, when evaluating the power spectral density, is justified if the contribution of the energy in this component to the average signal power is negligible. This result is formally stated by Theorem 4.2 in the following section.

4.4.2 Approximate Power Spectral Density

Define the interval, or in general the set of numbers, that simplifies analysis of the power spectral density as F. This set could, in the general case, consist of part of the interval $[0, T]$, and part of the remainder of the real line. It is convenient to partition F into two disjoint sets, that is,

$$F = F_I \cup F_O \quad \text{where} \quad F_I \subseteq [0, T] \quad F_O \subseteq [0, T]^C \quad (4.28)$$

In subsequent analysis, it is convenient to define a new set F_R according to

$$F_R = [0, T] \cap F_I^C \quad \text{s.t.} \quad [0, T] = F_I \cup F_R \quad (4.29)$$

The subscripts I, O, and R, respectively, stand for "inner," "outer," and "residual." The sets F_I, F_O, and F_R are graphically shown in Figure 4.4. The measure of the sets F_I, F_O, and F_R, respectively, are denoted M_I, M_O, and M_R, and the respective powers of the ith signal in these sets are denoted $P_i(F_I)$, $P_i(F_O)$, and $P_i(F_R)$.

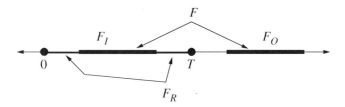

Figure 4.4 Definition of the sets F_I, F_O, and F_R.

Consider a random process X with ensemble $E_X = \{x_i : i \in \mathbf{Z}^+\}$. Define the random process X_A with ensemble E_{X_A}, consisting of waveforms that individually are identical on $[0, T]$ to a corresponding waveform from X, but may differ from the corresponding waveform outside this interval. Thus,

$$E_{X_A} = \left\{ x_{A_i} : x_{A_i}(t) = \begin{cases} x_i(t) & t \in [0, T], x_i \in E_X \\ \text{STFA} & t \notin [0, T] \end{cases} \quad i \in \mathbf{Z}^+ \right\} \quad (4.30)$$

Here, STFA means "specified to facilitate analysis," that is, outside $[0, T]$, x_{A_i} is specified in a manner that best facilitates analysis. The power spectral density of X_A, evaluated over the set F, is denoted G_{XA} and is defined according to

$$G_{XA}(F, f) = \sum_{i=1}^{\infty} \frac{p_i |X_{A_i}(F, f)|^2}{T} \qquad X_{A_i}(F, f) = \int_F x_{A_i}(t) e^{-j2\pi f t} \, dt \quad (4.31)$$

Note, the interval length used in the definition for G_{XA} is T and not the measure of the set F. This is because G_{XA} is an approximation to the true power spectral density G_X on the interval $[0, T]$. The following theorem quantifies how well the power spectral density of X_A approximates the power spectral density of X.

THEOREM 4.2. APPROXIMATION TO POWER SPECTRAL DENSITY *The integrated relative difference ε_R, between $G_{XA}(F, f)$ and $G_X(T, f)$ has the upper bound given by the following two equivalent expressions:*

$$\varepsilon_R = \frac{\int_{-\infty}^{\infty} |G_X(T, f) - G_{XA}(F, f)| \, df}{\int_{-\infty}^{\infty} G_X(T, f) \, df} \leq \frac{M_R}{T} \frac{\bar{P}(F_R)}{\bar{P}(T)} + \frac{M_O}{T} \frac{\bar{P}_A(F_O)}{\bar{P}(T)}$$

$$+ 2 \sqrt{\frac{M_I}{T}} \sqrt{\frac{\bar{P}(F_I)}{\bar{P}(T)}} \sqrt{\frac{M_R}{T}} \sqrt{\frac{\bar{P}(F_R)}{\bar{P}(T)}} + 2 \sqrt{\frac{M_I}{T}} \sqrt{\frac{\bar{P}(F_I)}{\bar{P}(T)}} \sqrt{\frac{M_O}{T}} \sqrt{\frac{\bar{P}_A(F_O)}{\bar{P}(T)}}$$

$$(4.32)$$

$$\varepsilon_R \leqslant \frac{\bar{E}(F_R)}{\bar{E}(T)} + \frac{\bar{E}_A(F_O)}{\bar{E}(T)} + 2\sqrt{\frac{\bar{E}(F_I)}{\bar{E}(T)}}\sqrt{\frac{\bar{E}(F_R)}{\bar{E}(T)}} + 2\sqrt{\frac{\bar{E}(F_I)}{\bar{E}(T)}}\sqrt{\frac{\bar{E}_A(F_O)}{\bar{E}(T)}} \quad (4.33)$$

where $\bar{P}(T)$ is the average power of X on $[0, T]$, $\bar{P}(F_R)$ and $\bar{P}(F_I)$, respectively, are the average power of X and X_A on the sets F_R and F_I, $\bar{P}_A(F_O)$ is the average power of X_A on the set F_O, and the symbol $\bar{E}$ denotes the average energy associated with the average power $\bar{P}$. These powers and energies are defined according to

$$\bar{P}(F_R) = \int_{-\infty}^{\infty} G_X(F_R, f)\, df = \int_{-\infty}^{\infty} G_{XA}(F_R, f)\, df = \frac{\bar{E}(F_R)}{M_R} \quad (4.34)$$

$$\bar{P}(F_I) = \int_{-\infty}^{\infty} G_X(F_I, f)\, df = \int_{-\infty}^{\infty} G_{XA}(F_I, f)\, df = \frac{\bar{E}(F_I)}{M_I} \quad (4.35)$$

$$\bar{P}_A(F_O) = \int_{-\infty}^{\infty} G_{XA}(F_O, f)\, df = \frac{\bar{E}_A(F_O)}{M_O} \quad (4.36)$$

Here, and for the countable case:

$$G_X(F_R, f) = \sum_{i=1}^{\infty} \frac{p_i |X_i(F_R, f)|^2}{M_R} = \sum_{i=1}^{\infty} \frac{p_i |X_{A_i}(F_R, f)|^2}{M_R} = G_{XA}(F_R, f) \quad (4.37)$$

$$G_X(F_I, f) = \sum_{i=1}^{\infty} \frac{p_i |X_i(F_I, f)|^2}{M_I} = \sum_{i=1}^{\infty} \frac{p_i |X_{A_i}(F_I, f)|^2}{M_I} = G_{XA}(F_I, f) \quad (4.38)$$

$$G_{XA}(F_O, f) = \sum_{i=1}^{\infty} \frac{p_i |X_{A_i}(F_O, f)|^2}{M_O} \quad (4.39)$$

Proof. The proof for the countable case is given in Appendix 1. The proof for the uncountable case follows in an analogous manner.

4.4.3 Specific Cases

As the measure of the set F_I approaches T, it follows that $\bar{P}(F_I)$ approaches $\bar{P}(T)$, and for this case, the upper bound on the integrated relative error can be approximated according to

$$\varepsilon_R \leqslant \frac{M_R}{T}\frac{\bar{P}(F_R)}{\bar{P}(T)} + \frac{M_O}{T}\frac{\bar{P}_A(F_O)}{\bar{P}(T)} + 2\sqrt{\frac{M_R}{T}}\sqrt{\frac{\bar{P}(F_R)}{\bar{P}(T)}} + 2\sqrt{\frac{M_O}{T}}\sqrt{\frac{\bar{P}_A(F_O)}{\bar{P}(T)}}$$

$$= \frac{\bar{E}(F_R)}{\bar{E}(T)} + \frac{\bar{E}_A(F_O)}{\bar{E}(T)} + 2\sqrt{\frac{\bar{E}(F_R)}{\bar{E}(T)}} + 2\sqrt{\frac{\bar{E}_A(F_O)}{\bar{E}(T)}} \quad (4.40)$$

Clearly, the measure of the sets F_O and F_R, relative to T, can be made sufficiently small so a required relative integrated error bound is achieved.

A common case encountered in analysis is where $F_O = [T, \infty]$, $F_I = [0, T]$ consistent with $F_R = \{\ \}$, and all signals of the random process X_A in the interval $F_O = [T, \infty]$ rapidly decay to zero. The following theorem details the bounds on the integrated relative difference for this case.

THEOREM 4.3. APPROXIMATION FOR POWER SPECTRAL DENSITY — INFINITE INTERVAL *If* $F = [0, NT]$, *with* $N \gg 1$, *which implies* $F_R = \{\ \}$, $F_I = [0, T]$, *and* $F_O = [T, NT]$ *then the integrated relative difference*, ε_R, *has the bound*

$$\varepsilon_R \leqslant \frac{(N-1)\bar{P}_A(F_O)}{\bar{P}(T)} + 2\sqrt{N-1}\sqrt{\frac{\bar{P}_A(F_O)}{\bar{P}(T)}} = \frac{\bar{E}_A(F_O)}{\bar{E}(T)} + 2\sqrt{\frac{\bar{E}_A(F_O)}{\bar{E}(T)}} \quad (4.41)$$

Proof. The proof of this theorem follows directly from Theorem 4.2.

As is clear from this theorem, when the ratio of the average energy in the interval F_O to the average energy in the interval $[0, T]$ approaches zero, the integrated relative error also approaches zero and the approximate power spectral density, G_{X_A}, approaches the true power spectral density G_X in a "mean" sense.

4.4.4 Example

Consider the case where

$$x(t) = \sum_{i=1}^{10} p(t - (i-1)D) \qquad 0 \leqslant t < 10D \quad (4.42)$$

with $p(t) = e^{-t/\tau}u(t)$ and $\tau = D = 1$. The true power spectral density of x evaluated on $[0, 10]$, as well as the approximate power spectral density obtained by including the tail of the exponential function in the interval $[10, \infty]$, are plotted in Figure 4.5. In Figure 4.6 the absolute difference between the true and approximate power spectral densities, that is, $|G_X(10, f) - G_{XA}(\infty, f)|$, is shown. The integrated relative error between the true and approximate power spectral densities, obtained by numerical integration, is 0.11, and is within the bound predicted by Eq. (4.41) of 0.27.

4.5 THE CROSS POWER SPECTRAL DENSITY

In subsequent analysis the cross power spectral density between two random processes is widely used. With the aim of defining the cross power spectral density, consider two random processes X and Y, defined on the interval $[0, T]$

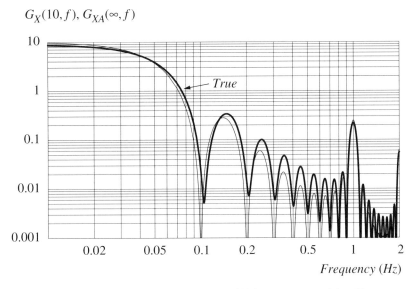

Figure 4.5 True (thick) and approximate (thin) power spectral densities.

by the respective ensembles E_X and E_Y:

$$E_X = \{x: S_X \times [0, T] \to \boldsymbol{C}\} \qquad (4.43)$$

$$E_Y = \{y: S_Y \times [0, T] \to \boldsymbol{C}\} \qquad (4.44)$$

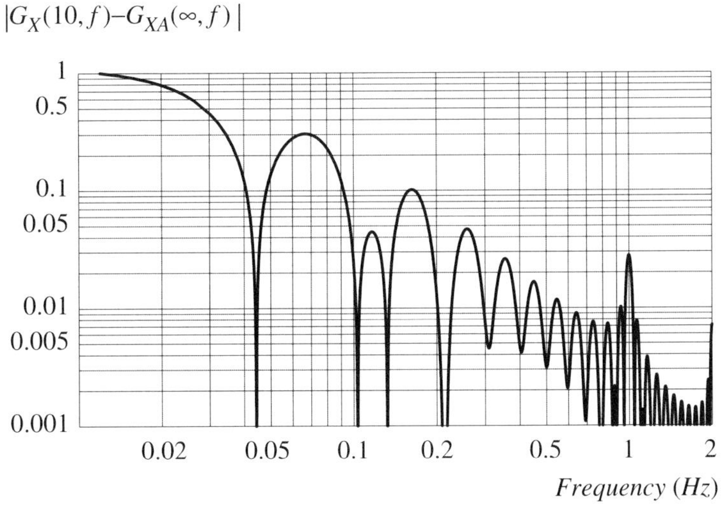

Figure 4.6 The absolute difference between the true and approximate power spectral densities shown in Figure 4.5.

where $S_X, S_Y \subseteq Z^+$ for the countable case and $S_X, S_Y \subseteq R$ for the uncountable case. For the countable case, the ith signal, $x(i, t)$ is written, for notational convenience as $x_i(t)$. The probabilities associated with the waveforms in the ensembles are as follows:

$$p_{xi} = P[x(i, t)] = P[x_i(t)] \qquad i \in Z^+ \quad \text{countable case}$$
$$P[x(\lambda, t)|_{\lambda \in I_x}] = \int_{I_x} f_X(\lambda) \, d\lambda \qquad I_x \subseteq R \quad \text{uncountable case}$$
(4.45)

$$p_{yj} = P[y(j, t)] = P[y_j(t)] \qquad j \in Z^+ \quad \text{countable case}$$
$$P[y(\lambda, t)|_{\lambda \in I_y}] = \int_{I_y} f_Y(\lambda) \, d\lambda \qquad I_y \subseteq R \quad \text{uncountable case}$$
(4.46)

where f_X and f_Y are probability density functions defining signal probabilities for the uncountable case.

DEFINITION: CROSS POWER SPECTRAL DENSITY The cross power spectral density between two random processes X and Y is, by definition (Peebles, 1993 p. 210)

$$G_{XY}(T, f)$$
$$= \begin{cases} \sum_{i=1}^{\infty} \sum_{j=1}^{\infty} p_{ij} \dfrac{X_i(T, f) Y_j^*(T, f)}{T} & \text{countable case} \\ \dfrac{1}{T} \int_{-\infty}^{\infty} \int_{-\infty}^{\infty} X(\lambda_x, T, f) Y^*(\lambda_y, T, f) f_{XY}(\lambda_x, \lambda_y) \, d\lambda_x \, d\lambda_y & \text{uncountable case} \end{cases}$$
(4.47)

where p_{ij} is the joint probability of x_i and y_j, that is, $p_{ij} = P[x_i, y_j]$, f_{XY} is the joint probability density function for x and y, that is,

$$P[x(\lambda_x, t), y(\lambda_y, t)|_{\lambda_x \in I_x, \lambda_y \in I_y}] = \int_{I_x} \int_{I_y} f_{XY}(\lambda_x, \lambda_y) \, d\lambda_x \, d\lambda_y \qquad (4.48)$$

and X and Y, respectively, are the Fourier transforms of x and y evaluated on the interval $[0, T]$.

From the definition it readily follows that

$$G_{YX}(T, f) = G_{XY}^*(T, f) \qquad (4.49)$$

4.5.1 Cross Power Spectral Density — Dependent Case

Consider the case where X and Y are fully correlated, such that the following dependencies hold for the countable and uncountable cases:

$$P[x_i(t)] = P[y_i(t)] = p_i \qquad P[x_i(t), y_j(t)] = 0 \quad i,j \in \mathbf{Z}^+, i \neq j$$

$$P[x(\lambda, t)|_{\lambda \in I}] = P[y(\lambda, t)|_{\lambda \in I}] = \int_I f(\lambda)\, d\lambda \qquad (4.50)$$

$$P[x(\lambda_x, t), y(\lambda_y, t)|_{\lambda_x \in I_x, \lambda_y \in I_y}] = 0 \qquad I_x \cap I_y = \emptyset$$

For this case, the cross power spectral density, given by Eq. (4.47), simplifies to

$$G_{XY}(T, f) = \begin{cases} \displaystyle\sum_{i=1}^{\infty} p_i \frac{X_i(T, f) Y_i^*(T, f)}{T} & \text{countable case} \\ \displaystyle\frac{1}{T} \int_{-\infty}^{\infty} X(\lambda, T, f) Y^*(\lambda, T, f) f(\lambda)\, d\lambda & \text{uncountable case} \end{cases} \qquad (4.51)$$

4.5.2 Cross Power Spectral Density — Independent Case

For the case where X is independent of Y, the cross power spectral density, given by Eq. (4.47), simplifies to

$$G_{XY}(T, f) = \frac{\bar{X}(T, f)\bar{Y}^*(T, f)}{T} \qquad (4.52)$$

where $\bar{X}$ is the mean Fourier transform of the random process X defined, for the countable case, according to

$$\bar{X}(T, f) = \sum_{i=1}^{\infty} p_{xi} X_i(T, f) = \sum_{i=1}^{\infty} p_{xi} \int_0^T x_i(t) e^{-j2\pi ft}\, dt = \int_0^T \mu_x(T, t) e^{-j2\pi ft}\, dt \qquad (4.53)$$

Here, μ_x is the mean signal of the random process X defined as

$$\mu_x(T, t) = \sum_{i=1}^{\infty} p_{xi} x_i(t) \qquad t \in [0, T] \qquad (4.54)$$

It has been assumed here that the order of summation and integration can be interchanged in Eq. (4.53). Similar definitions hold for $\bar{Y}^*$ and μ_y^* and for the uncountable case.

Consistent with stationarity, if the random processes have constant means on the interval $[0, T]$, that is, $\mu_x(T, t) = \mu_x$, the results:

$$2\sin^2(A) = 1 - \cos(2A) \qquad \int_0^T e^{-j2\pi ft}\, dt = \frac{1 - e^{-j2\pi fT}}{j2\pi f} \qquad (4.55)$$

$$\frac{\bar{X}(T, f)\bar{Y}^*(T, f)}{\mu_x \mu_y^*} = \frac{1 - e^{-j2\pi fT}}{j2\pi f} \frac{1 - e^{j2\pi fT}}{-j2\pi f} = \frac{2 - 2\cos(2\pi fT)}{(2\pi f)^2}$$

$$= \frac{1}{(\pi f)^2}\sin^2(\pi fT) = T^2 \operatorname{sinc}^2(fT) \qquad (4.56)$$

imply the cross power spectral density can be written as

$$G_{XY}(T, f) = \mu_x \mu_y^* T \operatorname{sinc}^2(fT) \qquad (4.57)$$

As T increases, Theorem 2.32 implies, for the infinite interval, that

$$G_{XY_\infty}(f) = \mu_x \mu_y^* \delta(f) \qquad (4.58)$$

4.5.3 Conditions for Cross Power Spectral Density to be Zero

A sufficient condition for the cross power spectral density $G_{XY}(T, f)$ to be zero for T fixed and for all $f \in \mathbf{R}$, is for X to be independent of Y, and for either or both of X and Y to have zero means on $[0, T]$.

4.5.4 Bounds on Cross Power Spectral Density

In analysis it is useful if the cross power spectral density between two random processes can be neglected. The following theorem states several bounds.

THEOREM 4.4. BOUNDS ON CROSS POWER SPECTRAL DENSITY *The cross power spectral density $G_{XY}(T, f)$ has the following bounds:*

$$|\operatorname{Re}[G_{XY}(T, f)]| \leq \sqrt{G_X(T, f)}\sqrt{G_Y(T, f)} \qquad (4.59)$$

$$|\operatorname{Re}[G_{XY}(T, f)]| \leq \frac{G_X(T, f) + G_Y(T, f)}{2} \qquad (4.60)$$

For the fully correlated case the bound is

$$|G_{XY}(T, f)| \leq \sqrt{G_X(T, f)}\sqrt{G_Y(T, f)} \qquad (4.61)$$

For the fully correlated case, and with $y(\lambda, t) = kx(\lambda, t)$ for all signals in the ensemble, where k is a real constant, it is the case that

$$|G_{XY}(T, f)| = \sqrt{G_X(T, f)}\sqrt{G_Y(T, f)}$$

If the random processes X and Y are independent then

$$|G_{XY}(T, f)| \leq \sqrt{G_X(T, f)}\sqrt{G_Y(T, f)} \quad (4.62)$$

For the case where $G_X(T, f) \gg G_Y(T, f)$ Eqs. (4.59), (4.61), and (4.62) imply that

$$|\text{Re}[G_{XY}(T, f)]| \ll G_X(T, f) \quad \text{general case} \quad (4.63)$$

$$|G_{XY}(T, f)| \ll G_X(T, f) \quad \begin{cases} \text{independent case} \\ \text{fully correlated case} \end{cases} \quad (4.64)$$

Proof. The proof for the countable case is given in Appendix 2. The proof for the uncountable case follows in analogous manner.

4.6 POWER SPECTRAL DENSITY OF A SUM OF RANDOM PROCESSES

Although the sum of two random processes is a subset of the general case of the sum of N random processes, it is instructive to consider this case separately.

4.6.1 Power Spectral Density—Sum of Two Random Processes

Define the random process Z according to

$$Z = X + Y \quad (4.65)$$

where X and Y are defined by Eqs. (4.43)–(4.46). This random process is defined for the countable case by the ensemble

$$E_Z = \{z_{ij} : [0, T] \to \mathbf{C}, z_{ij} = x_i + y_j, i, j \in \mathbf{Z}^+\} \quad (4.66)$$

and the probability space P_Z:

$$P_Z = \left\{ p_{ij} : p_{ij} = P[z_{ij}] = P[x_i, y_j], \sum_{i=1}^{\infty} \sum_{j=1}^{\infty} p_{ij} = 1 \right\} \quad (4.67)$$

Analogous definitions follow for the uncountable case. For the special case of X being independent of Y, it follows that $p_{ij} = p_{xi}p_{yj}$. The following theorem details the power spectral density of Z.

THEOREM 4.5. POWER SPECTRAL DENSITY OF THE SUM OF TWO RANDOM PROCESSES *The power spectral density of the random process* $Z = X \pm Y$ *is given by*

$$G_Z(T, f) = G_X(T, f) + G_Y(T, f) \pm 2\text{Re}[G_{XY}(T, f)] \qquad (4.68)$$

where G_X and G_Y, respectively, are the power spectral densities of X and Y, and G_{XY} is the cross power spectral density between X and Y.

For the case where X is independent of Y the power spectral density of Z is

$$G_Z(T, f) = G_X(T, f) + G_Y(T, f) \pm \frac{2\text{Re}[\bar{X}(T, f)\bar{Y}^*(T, f)]}{T} \qquad (4.69)$$

where $\bar{X}$ and $\bar{Y}^*$ are mean Fourier transforms defined in a manner consistent with Eq. (4.53).

For the case where X and Y are independent and stationary, such that their respective means, denoted μ_x and μ_y, are constant on $[0, T]$, the power spectral density is

$$G_Z(T, f) = G_X(T, f) + G_Y(T, f) \pm 2\text{Re}[\mu_x \mu_y^*] T \operatorname{sinc}^2(fT) \qquad (4.70)$$

As $T \to \infty$ the last term becomes impulsive to yield

$$G_{Z_\infty}(f) = G_{X_\infty}(f) + G_{Y_\infty}(f) \pm 2\text{Re}[\mu_x \mu_y^*]\delta(f) \qquad (4.71)$$

Proof. The proof of the first of these results is given in Appendix 3. The subsequent results follow from the first result using definitions of the cross power spectral density for the independent and stationary cases.

4.6.2 Power Spectral Density—Infinite Sum

Consider the infinite sequence of random processes $X_1, X_2, \ldots$ with respective ensembles $E_{X_1}, E_{X_2}, \ldots$ defined, on the interval $[0, T]$ and for the countable case, according to

$$E_{X_i} = \{x_i \colon \mathbf{Z}^+ \times [0, T] \to \mathbf{C}, i \in \mathbf{Z}^+\} \qquad (4.72)$$

where $P[x_i(\zeta_i, t)] = p_i(\zeta_i)$. The random processes Z and Z_N are defined by weighted summations according to

$$Z_N = \sum_{i=1}^{N} w_i X_i \qquad Z = \lim_{N \to \infty} Z_N = \sum_{i=1}^{\infty} w_i X_i \qquad (4.73)$$

where w_i is the weighting factor for the ith random process X_i. The ensemble associated with Z is

$$E_Z = \left\{ z \colon \mathbf{Z}^+ \times \cdots \times [0, T] \to \mathbf{C}, \; z(\zeta_1, \ldots, t) = \sum_{i=1}^{\infty} w_i x_i(\zeta_i, t) \right\} \quad (4.74)$$

The probability of a specific waveform $z(\zeta_1, \ldots, \zeta_N, \ldots, t)$ in the ensemble associated with the outcomes $\zeta_1, \ldots, \zeta_N, \ldots$ is likely to be zero. To avoid the complexities associated with an infinite number of signals, where each has a vanishingly small probability, it is convenient to partition this ensemble as follows:

$$E_Z = \bigcup_{\zeta_1 = 1}^{\infty} \cdots \bigcup_{\zeta_N = 1}^{\infty} E(\zeta_1, \ldots, \zeta_N) \quad (4.75)$$

where $E(\zeta_1, \ldots, \zeta_N)$ is the set of signals from E_Z whose first N waveforms in the summation for $z(\zeta_1, \ldots, \zeta_N, \ldots, t)$ are fixed and are $x_1(\zeta_1, t), \ldots, x_N(\zeta_N, t)$, that is,

$$E(\zeta_1, \ldots, \zeta_N) = \{ z(\zeta_1, \ldots, \zeta_N, \ldots, t) \colon \zeta_{N+1} \in \mathbf{Z}^+, \zeta_{N+2} \in \mathbf{Z}^+, \ldots \} \quad (4.76)$$

The probability of a waveform from $E(\zeta_1, \ldots, \zeta_N)$ equals the probability of the N outcomes $\zeta_1, \ldots, \zeta_N$, denoted $p(\zeta_1, \ldots, \zeta_N)$, and in general is nonzero. Further, this probability equals the probability of the corresponding signal from the ensemble of the random process Z_N, defined according to

$$E_{Z_N} = \left\{ z_N \colon \mathbf{Z}^+ \times \cdots \times \mathbf{Z}^+ \times [0, T] \to \mathbf{C}, \; z_N(\zeta_1, \ldots, \zeta_N, t) = \sum_{i=1}^{N} w_i x_i(\zeta_i, t) \right\} \quad (4.77)$$

Here,

$$P[z_N(\zeta_1, \ldots, \zeta_N, t)] = P[x_1(\zeta_1, t), \ldots, x_N(\zeta_N, t)] = p(\zeta_1, \ldots, \zeta_N) \quad (4.78)$$

Thus, associated with each waveform in the ensemble for Z_N is a set of waveforms in the ensemble for Z. The probability of the waveform in Z_N equals the probability of the corresponding set of waveforms in Z.

If it is the case that the power spectral density of a specific waveform in Z_N, $z_N(\zeta_1, \ldots, \zeta_N, t)$, closely approximates the power spectral density of all of the signals in the associated set $E(\zeta_1, \ldots, \zeta_N)$ of Z, that is, it is the case that

$$\frac{1}{T} \left| \sum_{i=1}^{N} w_i X_i(\zeta_i, T, f) \right|^2 \approx \frac{1}{T} \left| \sum_{i=1}^{\infty} w_i X_i(\zeta_i, T, f) \right|^2 \quad \zeta_{N+1} \in \mathbf{Z}^+ \quad \zeta_{N+2} \in \mathbf{Z}^+ \ldots$$

$$(4.79)$$

then the power spectral density of Z_N closely approximates the power spectral density of Z. To show that this can be the case, two assumptions are made: (a) The energy in all waveforms in the ensembles associated with the random processes, $X_1, \ldots, X_N, \ldots$, is bounded, that is,

$$\sup \left\{ \int_0^T |x_i(\zeta_i, t)|^2 \, dt : i \in \mathbf{Z}^+, \zeta_i \in \mathbf{Z}^+ \right\} < \infty \qquad (4.80)$$

This assumption implies, according to Theorem 3.6, that the magnitude of the Fourier transform of individual waveforms in the ensemble are bounded, that is,

$$X(T) = \sup\{|X_i(\zeta_i, T, f)| : i \in \mathbf{Z}^+, \zeta_i \in \mathbf{Z}^+, f \in \mathbf{R}\} < \infty \qquad (4.81)$$

(b) With absolute convergence of the weighting factor, that is, $\Sigma_{i=1}^{\infty} |w_i| < \infty$, it follows that, for all $\varepsilon > 0$, there exists a constant N_o, such that for all $N > N_o$ it is the case that $\Sigma_{i=N+1}^{\infty} |w_i| < \varepsilon$. These two assumptions allow the approximation given in Eq. (4.79) to be verified. To this end, consider the difference between the expressions given in Eq. (4.79):

$$B = \left| \left| \sum_{i=1}^{N} w_i X_i(\zeta_i, T, f) \right|^2 - \left| \sum_{i=1}^{\infty} w_i X_i(\zeta_i, T, f) \right|^2 \right|$$

$$= \left| \sum_{i=N+1}^{\infty} \sum_{j=N+1}^{\infty} W_i W_j^* + \sum_{i=1}^{N} \sum_{j=N+1}^{\infty} W_i W_j^* + \sum_{i=N+1}^{\infty} \sum_{j=1}^{N} W_i W_j^* \right| \qquad (4.82)$$

where $W_i = w_i X_i(\zeta_i, T, f)$. As $|X_i(\zeta_i, T, f)|$ is bounded for $i \in \mathbf{Z}^+$, $\zeta_i \in \mathbf{Z}^+$, and $f \in \mathbf{R}$ with a bound $X(T)$, it follows that

$$B < X^2(T) \left[\left| \sum_{i=N+1}^{\infty} |w_i| \right|^2 + 2 \sum_{i=1}^{N} |w_i| \sum_{j=N+1}^{\infty} |w_j| \right] < X^2(T) \left[\varepsilon^2 + \varepsilon \sum_{i=1}^{N} |w_i| \right] < \varepsilon_1 \qquad (4.83)$$

With the two given assumptions, the approximation between the power spectral density of Z_N and Z becomes closer as N increases, and it follows that

$$G_Z(T, f) = \lim_{N \to \infty} G_{Z_N}(T, f) \qquad (4.84)$$

The following theorem gives the power spectral density of Z.

THEOREM 4.6. POWER SPECTRAL DENSITY OF INFINITE SUM *If the energy in all waveforms in the ensembles associated with the random processes, $X_1, \ldots, X_N, \ldots$,* is bounded, that is,

$$\sup\left\{\int_0^T |x_i(\zeta_i, t)|^2 \, dt : i \in \mathbf{Z}^+, \begin{cases} \zeta_i \in \mathbf{Z}^+ & \text{countable case} \\ \zeta_i \in \mathbf{R} & \text{uncountable case} \end{cases}\right\} < \infty \quad (4.85)$$

and the weighting factor is such that $\sum_{i=1}^{\infty} |w_i| < \infty$, then the power spectral density of the random process $Z = \sum_{i=1}^{\infty} w_i X_i$ is given by

$$G_Z(T, f) = \sum_{i=1}^{\infty} |w_i|^2 G_i(T, f) + \sum_{i=1}^{\infty} \sum_{\substack{j=1 \\ j \neq i}}^{\infty} w_i w_j^* G_{ij}(T, f)$$

$$= \sum_{i=1}^{\infty} |w_i|^2 G_i(T, f) + \sum_{i=1}^{\infty} \sum_{\substack{j=1 \\ j > i}}^{\infty} 2Re[w_i w_j^* G_{ij}(T, f)] \quad (4.86)$$

Here, G_i is the power spectral density of X_i, and G_{ij} is the cross power spectral density between X_i and X_j defined for the countable case, consistent with Eq. (4.47), according to

$$G_{ij}(T, f) = \sum_{\zeta_i=1}^{\infty} \sum_{\zeta_j=1}^{\infty} p_{ij}(\zeta_i, \zeta_j) \frac{X_i(\zeta_i, T, f) X_j^*(\zeta_j, T, f)}{T} \quad (4.87)$$

An analogous definition holds for the uncountable case. The power spectral density of the random process Z, for the case where X_i is independent of X_j for $i \neq j$, is given by

$$G_Z(T, f) = \sum_{i=1}^{\infty} |w_i|^2 G_i(T, f) + \sum_{i=1}^{\infty} \sum_{\substack{j=1 \\ j > i}}^{\infty} 2Re\left[\frac{w_i w_j^* \bar{X}_i(T, f) \bar{X}_j^*(T, f)}{T}\right] \quad (4.88)$$

where $\bar{X}_i$ and $\bar{X}_j^*$ are mean Fourier transforms defined in a manner consistent with Eq. (4.53).

For the independent and stationary case it follows that

$$G_Z(T, f) = \sum_{i=1}^{\infty} |w_i|^2 G_i(T, f) + T \operatorname{sinc}^2(fT) \sum_{i=1}^{\infty} \sum_{\substack{j=1 \\ j > i}}^{\infty} 2Re[w_i w_j^* \mu_i \mu_j^*] \quad (4.89)$$

$$G_{Z_\infty}(f) = \sum_{i=1}^{\infty} |w_i|^2 G_{i_\infty}(f) + \delta(f)\left[\left|\sum_{i=1}^{\infty} w_i \mu_i\right|^2 - \sum_{i=1}^{\infty} |w_i|^2 |\mu_i|^2\right] \quad (4.90)$$

where μ_i is the mean of the ith random process defined consistent with Eq. (4.54).

For the independent and zero mean case, the power spectral density of the weighted sum of random processes is the weighted sum of individual power spectral densities, that is,

$$G_Z(T, f) = \sum_{i=1}^{\infty} |w_i|^2 G_i(T, f) \qquad G_{Z_\infty}(f) = \sum_{i=1}^{\infty} |w_i|^2 G_{i_\infty}(f) \qquad (4.91)$$

Proof. The proof of the first result is given, for the countable case, in Appendix 4. The proof for the uncountable case follows in an analogous manner. The subsequent results follow from the first result using the definitions of the cross power spectral for the independent and stationary cases. The summation associated with the Dirac delta function in Eq. (4.90) is proved in Appendix 4.

4.6.3 Power Spectral Density—Sum of N Random Processes

In many practical examples, a sum of N random processes is required. Accordingly, it is convenient to state the power spectral density for this case despite the fact that it is a subset of the more general case considered above.

THEOREM 4.7. POWER SPECTRAL DENSITY OF SUM OF N RANDOM PROCESSES
If the energy in all waveforms in the ensembles associated with the random processes $X_1, \ldots, X_N$ is bounded, that is,

$$\sup \left\{ \int_0^T |x_i(\zeta_i, t)|^2 \, dt : i \in \{1, \ldots, N\}, \begin{cases} \zeta_i \in \mathbf{Z}^+ & \text{countable case} \\ \zeta_i \in \mathbf{R} & \text{uncountable case} \end{cases} \right\} < \infty \qquad (4.92)$$

then the power spectral density of the random process $Z = \sum_{i=1}^{N} X_i$ is given by

$$\begin{aligned} G_Z(T, f) &= \sum_{i=1}^{N} G_i(T, f) + \sum_{\substack{i=1 \\ }}^{N} \sum_{\substack{j=1 \\ j \neq i}}^{N} G_{ij}(T, f) \\ &= \sum_{i=1}^{N} G_i(T, f) + \sum_{i=1}^{N} \sum_{\substack{j=1 \\ j > i}}^{N} 2\mathrm{Re}[G_{ij}(T, f)] \end{aligned} \qquad (4.93)$$

Results for the independent cases parallel those in Theorem 4.6 in a straightforward manner.

Proof. The proof follows from the previous theorem using weighting factors of unity for $1 \leq i \leq N$, and zero for $i > N$.

4.7 POWER SPECTRAL DENSITY OF A PERIODIC SIGNAL

Many random processes have periodic components, and accordingly, a precise statement of the power spectral density of a periodic signal is necessary.

4.7.1 Power Spectral Density—Arbitrary Interval Case

Consider a periodic signal x, with period T_p, that is, $x(t + kT_p) = x(t)$ for all $k \in Z$. Consider the general case where the power spectral density is to be evaluated on the interval $[0, T]$, where T is not an integer multiple of the signal period T_p. It is convenient to define an integer $N = \lfloor T/T_p \rfloor$ whereupon $NT_p \leq T < (N + 1)T_p$. Denote the power spectral density on the interval $[0, NT_p]$ at a specific frequency f, as $G_{XA}(NT_p, f)$. Then, with the definitions $F_I = [0, NT_p]$, $F_R = [NT_p, T]$, and $F_O = \{\}$ it follows, from Theorem 4.2, that

$$\varepsilon_R = \frac{\int_{-\infty}^{\infty} |G_X(T, f) - G_{XA}(NT_p, f)| \, df}{\int_{-\infty}^{\infty} G_X(T, f) \, df} \leq \frac{\bar{E}(F_R)}{\bar{E}(T)} + \frac{2\sqrt{\bar{E}(F_I)}}{\sqrt{\bar{E}(T)}} \frac{\sqrt{\bar{E}(F_R)}}{\sqrt{\bar{E}(T)}} \quad (4.94)$$

Since the measure of F_R is less than T_p, the periodicity of the signal implies the ratio $\bar{E}(F_R)/\bar{E}(T)$ has an upper bound of $1/N$. It then follows, as T is increased such that, $NT_p/T \approx 1$ and $\bar{E}(F_I)/\bar{E}(T) \approx 1$, that an approximate upper bound for ε_R is

$$\varepsilon_R \leq \frac{2}{\sqrt{N}} \quad (4.95)$$

Hence, by choosing T sufficiently large, the integrated error involved in computing the power spectral density over an interval $[0, NT_p]$, rather than $[0, T]$, can be made arbitrarily small. Accordingly, when ascertaining the power spectral density of a periodic signal over an interval, whose measure is large relative to the period T_p, it is appropriate to use an interval that is an integer multiple of the signal period.

THEOREM 4.8. POWER SPECTRAL DENSITY OF A PERIODIC SIGNAL Consider a periodic signal $x \in L[0, NT_p]$, with a period of $T_p = 1/f_p$ sec, that can be represented by an exponential Fourier series according to

$$x(t) = \sum_{i=-\infty}^{\infty} c_i e^{j2\pi i f_p t} \quad c_i = \frac{1}{T_p} \int_0^{T_p} x(t) e^{-j2\pi i f_p t} \, dt = \frac{X(T_p, i f_p)}{T_p} \quad (4.96)$$

On the interval $[0, NT_p]$, with a fundamental frequency $f_o = 1/NT_p = f_p/N$, the Fourier transform and the power spectral density of the signal x are

$$X(NT_p, f) = X(T_p, f) \sum_{i=0}^{N-1} e^{-j2\pi f i T_p} = X(T_p, f) \frac{1 - e^{-j2\pi f NT_p}}{1 - e^{-j2\pi f T_p}} \quad (4.97)$$

$$G_X(NT_p, f) = G_X(T_p, f) \frac{1}{N} \frac{\sin^2(\pi N f/f_p)}{\sin^2(\pi f/f_p)} \quad (4.98)$$

where $X(T_p, f)$ and $G_X(T_p, f)$, respectively, are the Fourier transform and power spectral density at a frequency f and evaluated over one period, that is,

$$G_X(T_p, f) = \frac{|X(T_p, f)|^2}{T_p} \qquad G_X(T_p, if_p) = \frac{|c_i|^2}{f_p} \qquad i \in Z \qquad (4.99)$$

The power spectral density level at the ith harmonic frequency if_p, is given by

$$G_X(NT_p, if_p) = \frac{|c_i|^2}{f_o} \qquad f_o = \frac{1}{NT_p} \qquad (4.100)$$

For f fixed, there exists constants k_1, k_2, $T_o > 0$, and $0 < \alpha < 1$, such that the power spectral density varies with T, for all $T > T_o$, according to

$$G_X(T, f) < k_1/T \qquad f \notin \{\ldots, -f_p, 0, f_p, \ldots\}$$
$$\alpha k_2 T < G_X(T, f) < k_2 T \qquad f \in \{\ldots, -f_p, 0, f_p, \ldots\}, G_X(T, f) \neq 0$$
$$(4.101)$$

Consistent with these bounds the power spectral density on the infinite interval is

$$G_{X_\infty}(f) = \sum_{i=-\infty}^{\infty} |c_i|^2 \delta(f - if_p) \qquad (4.102)$$

Proof. The proof is given in Appendix 5.

4.7.2 Notes

According to Theorem 3.2, the average power in the sinusoidal components of the signal on the interval $[0, NT_p]$, with a frequency kf_o where $k \in Z^+$, is given by

$$f_o G_X(NT_p, -kf_o) + f_o G_X(NT_p, kf_o) \qquad f_o = 1/NT_p \qquad (4.103)$$

This result has to be reconciled with two facts related to periodic signals that can be inferred from Eq. (4.96). First, the only sinusoidal components that have nonzero power, are those with a frequency which is some multiple of the period frequency $f_p = Nf_o$. This implies that $G_X(NT_p, -kf_o) + G_X(NT_p, kf_o)$ must be zero for $k \notin \{0, N, 2N, \ldots\}$. That this is the case, follows from Eq. (4.98) because $\sin\{\pi Nf/f_p\} = \sin(\pi f/f_o) = 0$ for $f \in \{0, f_o, 2f_o, \ldots\}$.

The second fact is that the power in the sinusoidal components of the signal with a frequency if_p is given by $|c_{-i}|^2 + |c_i|^2$. This is the case, as a comparison of Eqs. (4.100) and (4.103) indicates.

4.7.3 Example

Consider the power spectral density of a periodic pulse train shown in Figure 4.7. The Fourier series of such a pulse train is

$$x(t) = \sum_{i=-\infty}^{\infty} c_i e^{j2\pi i f_p t} \qquad c_i = \frac{A}{T_p} \int_0^W e^{-j2\pi i f_p t} dt = \frac{AW \operatorname{sinc}(if_p W) e^{-j\pi i f_p W}}{T_p}$$

(4.104)

where the Fourier transform result in Theorem 2.33 has been used in the evaluation of c_i. This same result yields the Fourier transform

$$X(T_p, f) = AW \operatorname{sinc}(fW) e^{-j\pi fW}$$
(4.105)

and it follows from Eqs. (4.98) and (4.99) that

$$G_X(NT_p, f) = \frac{A^2 W^2}{T_p} \operatorname{sinc}^2(fW) \frac{1}{N} \frac{\sin^2(\pi f/f_o)}{\sin^2(\pi f/f_p)} \qquad f_o = \frac{f_p}{N} = \frac{1}{NT_p}$$
(4.106)

The power spectral density, evaluated on the interval $[0, NT_p]$, has the form shown in Figures 4.8 and 4.9 for the respective cases of $N = 4$ and $N = 8$.

From Figures 4.8 and 4.9, note first, that the impulsive components in the spectrum only occur at multiples of the period frequency f_p, and second, as the interval length is increased, the impulsive components increase in height according to $G_X(NT_p, if_p) = |c_i|^2/f_o$, and decrease in width in proportion to f_o. Finally, the power spectral density is zero at all integer multiples of the fundamental frequency $f_o = f_p/N$ that do not coincide with a multiple of the period frequency f_p. It is easy to conclude from these figures the result, stated in Eq. (4.102), that the power spectral density G_{X_c} is zero for all frequency except where it is undefined at integer multiples of the fundamental frequency f_p, that is, it has the form shown in Figure 4.10. Consistent with this figure and Eq. (4.102), the area under each pair of impulses at frequencies of $-if_p$ and if_p is equal to the power in the sinusoids with the frequency if_p, namely $|c_{-i}|^2 + |c_i|^2$. From Eq. (4.104) it follows that $|c_{-i}| = |c_i|$, and for the parameter

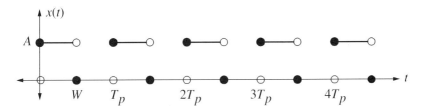

Figure 4.7 Periodic pulse train.

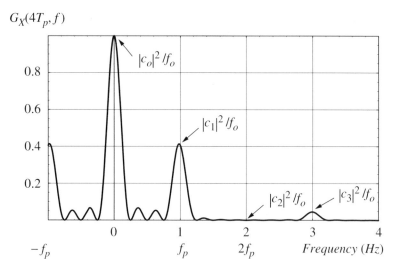

Figure 4.8 Power spectral density of pulse train on an interval of [0, $4T_p$] for the case where $A = 1$, $T_p = 1$, $f_p = 1$, $W = 0.5$, and $f_o = 0.25$.

values used in Figures 4.8 and 4.9, it is the case that $|c_0|^2 = 0.25$, $|c_1|^2 = 0.101$, $|c_2|^2 = |c_4|^2 = 0$, and $|c_3|^2 = 0.011$.

In conclusion, the power spectral density of a periodic signal over a semi-infinite interval $[0, \infty]$ has the expected form, namely, it consists of impulses at multiples of the period frequency f_p and the area under each pair

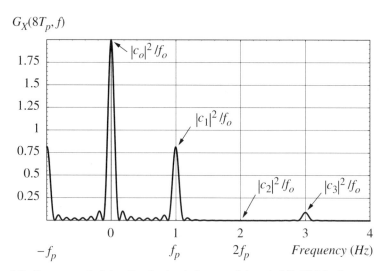

Figure 4.9 Power spectral density of pulse train on an interval of [0, $8T_p$] for the case where $A = 1$, $T_p = 1$, $f_p = 1$, $W = 0.5$, and $f_o = 0.125$.

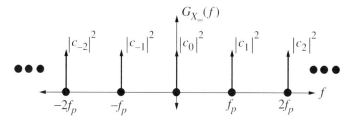

Figure 4.10 Power spectral density of a periodic signal evaluated on an infinite interval.

of impulses at $-if_p$ and if_p is equal to the power in the constituent sinusoidal component with a frequency if_p.

4.7.4 Generalizations

The following theorem describes the power spectral density of a random process X, that is characterized by an ensemble E_X and where each waveform in the ensemble is periodic, that is,

$$E_X = \{x_i : x_i(t + kT_i) = x_i(t),\ k \in Z,\ i \in Z^+\} \qquad (4.107)$$

Further, $P[x_i(t)] = p_i$ and

$$x_i(t) = \sum_{k=-\infty}^{\infty} c_{ik} e^{j2\pi k f_i t} \qquad c_{ik} = \frac{1}{T_i}\int_0^{T_i} x_i(t) e^{-j2\pi k f_i t}\, dt \qquad f_i = \frac{1}{T_i} \qquad (4.108)$$

THEOREM 4.9. POWER SPECTRAL DENSITY OF PERIODIC RANDOM PROCESS
Assuming that $T \gg T_i$, $i \in Z^+$ and the average power in the periodic signals is bounded, that is, $\sup\{\int_0^T |x_i(t)|^2\, dt : i \in Z^+\} < \infty$, then the power spectral density of the random process X is

$$G_X(T, f) \approx \sum_{i=1}^{\infty} p_i G_i(T_i, f) \frac{1}{N_i} \frac{\sin^2(\pi N_i f/f_i)}{\sin^2(\pi f/f_i)} \qquad N_i = \left\lfloor \frac{T}{T_i} \right\rfloor \qquad (4.109)$$

where $G_i(T_i, f)$ is the power spectral density of the ith periodic signal evaluated over its period of $T_i = 1/f_i$ sec, that is,

$$G_i(T_i, f) = \frac{|X_i(T_i, f)|^2}{T_i} \qquad G_i(T_i, kf_i) = \frac{|c_{ik}|^2}{f_i} \qquad k \in Z \qquad (4.110)$$

On the infinite interval $[0, \infty]$ the power spectral density is

$$G_{X_\infty}(f) = \sum_{i=1}^{\infty} p_i \sum_{k=-\infty}^{\infty} |c_{ik}|^2 \delta(f - kf_i) \qquad (4.111)$$

Proof. This theorem is a straightforward generalization of the previous theorem.

This random process is not ergodic as a single outcome of the random process is not representative of other waveforms in the ensemble (Papoulis, 2002 pp. 523f).

4.7.4.1 Power Spectral Density of an Infinite Sum of Periodic Signals
The following theorem details the power spectral density of a signal x, defined as an infinite sum of periodic signals, where each has a distinct period, that is,

$$x(t) = \sum_{i=1}^{\infty} w_i x_i(t) \quad \begin{cases} x_i(t + kT_i) = x_i(t) & k \in \mathbf{Z}, i \in \mathbf{Z}^+, T_i = 1/f_i \\ pT_i \neq qT_j; pT_i < T, qT_j < T & p, q \in \mathbf{Z}^+, i \neq j \end{cases} \qquad (4.112)$$

THEOREM 4.10. POWER SPECTRAL DENSITY OF AN INFINITE SUM OF PERIODIC SIGNALS *If the signal energy of all periodic signals in the ensemble is bounded, that is,*

$$\sup\left\{\int_0^T |x_i(t)|^2 \, dt : i \in \mathbf{Z}^+\right\} < \infty \quad \text{and} \quad \sum_{i=1}^{\infty} |w_i| < \infty$$

then on the interval $[0, T]$ the power spectral density of x is given by

$$G_X(T, f) = \sum_{i=1}^{\infty} |w_i|^2 G_i(T, f) + \frac{1}{T} \sum_{i=1}^{\infty} \sum_{\substack{j=1 \\ j \neq i}}^{\infty} w_i w_j^* X_i(T, f) X_j^*(T, f) \qquad (4.113)$$

where

$$X_i(T, f) = T \sum_{p=-\infty}^{\infty} c_{ip} e^{-j\pi(f - pf_i)T} \operatorname{sinc}[(f - pf_i)T] \qquad (4.114)$$

For the case of $T \gg T_i, i \in \mathbf{Z}^+$

$$G_i(T, f) \approx G_i(T_i, f) \frac{1}{N_i} \frac{\sin^2(\pi N_i f / f_i)}{\sin^2(\pi f / f_i)} \qquad N_i = \left\lfloor \frac{T}{T_i} \right\rfloor \qquad (4.115)$$

On the infinite interval $(0, \infty)$, the power spectral density is

$$G_{X_x}(f) = \sum_{i=1}^{\infty} |w_i|^2 \sum_{\substack{k=-\infty \\ k \neq 0}}^{\infty} |c_{ik}|^2 \delta(f - kf_i) + \delta(f) \left|\sum_{i=1}^{\infty} w_i c_{i0}\right|^2 \quad (4.116)$$

Proof. The proof of this theorem is given in Appendix 6.

4.8 POWER SPECTRAL DENSITY — PERIODIC COMPONENT CASE

It is often the case that some waveforms in the ensemble of a random process contain a periodic component. It is useful to be able to treat the periodic component separately from the random component when evaluating the power spectral density. To this end, consider the case where each signal in the ensemble of a random process X can be written as

$$x_i(t) = x_{iB}(t) + x_{iP}(t) \qquad i \in Z^+ \quad (4.117)$$

where x_{iB} is the component of x_i, which has a bounded power spectral density, and x_{iP} is the periodic component (including the mean) of x_i. If the period of x_{iP} is $T_i = 1/f_i$, then this component can be written as an exponential Fourier series according to

$$x_{iP}(t) = \begin{cases} \sum_{k=-\infty}^{\infty} c_{ik} e^{j2\pi k f_i t} & 0 \leq t < T \\ 0 & \text{elsewhere} \end{cases} \quad (4.118)$$

$$c_{ik} = \frac{1}{T_i} \int_0^{T_i} x_{iP}(t) e^{-j2\pi k f_i t} dt$$

Consistent with such definitions, the ensemble for X on $[0, T]$ can be defined as

$$E_X = \{x_i : x_i(t) = x_{iB}(t) + x_{iP}(t), \ i \in Z^+\} \quad (4.119)$$

where $P[x_i(t)] = p_i$. Two new random processes, X_B and X_P can be defined consistent, respectively, with the periodic and nonperiodic components of X and with respective ensembles E_{XB} and E_{XP}:

$$E_{XB} = \{x_{iB} : i \in Z^+\} \qquad E_{XP} = \{x_{iP} : i \in Z^+\} \quad (4.120)$$

Here, $P[x_{iB}(t)] = P[x_{iP}(t)] = p_i$. The power spectral densities of these random processes, respectively, are denoted G_{XP} and G_{XB}. The following theorem states the relationship between the power spectral density of X and G_{XP} and G_{XB}.

THEOREM 4.11. POWER SPECTRAL DENSITY — PERIODIC COMPONENT CASE *If, for T fixed, it is the case that the energy in both the periodic and nonperiodic components of the ensemble for X is finite, that is,*

$$\sup\left\{\int_0^T |x_{iB}(t)|^2 \, dt : i \in \mathbf{Z}^+\right\} < \infty \qquad \sup\left\{\int_0^T |x_{iP}(t)|^2 \, dt : i \in \mathbf{Z}^+\right\} < \infty \quad (4.121)$$

then the power spectral density of the random process X, defined by the ensemble as per Eq. (4.119), is given by

$$G_X(T, f) = G_{XB}(T, f) + G_{XP}(T, f) + \frac{2}{T} \sum_{i=1}^{\infty} p_i \mathrm{Re}[X_{iB}(T, f) X_{iP}^*(T, f)]$$

$$= G_{XB}(T, f) + G_{XP}(T, f) \qquad (4.122)$$

$$+ 2 \sum_{i=1}^{\infty} p_i \sum_{k=-\infty}^{\infty} \mathrm{sinc}[(f - kf_i)T] \mathrm{Re}[c_{ik}^* e^{j\pi(f - kf_i)T} X_{iB}(T, f)]$$

where

$$X_{iP}(T, f) = T \sum_{k=-\infty}^{\infty} c_{ik} \mathrm{sinc}[(f - kf_i)T] e^{-j\pi(f - kf_i)T}$$

$$\approx X_{iP}(T_i, f) \left[\frac{1 - e^{-j2\pi N_i f T_i}}{1 - e^{-j2\pi f T_i}}\right] \qquad T \gg T_i \qquad (4.123)$$

Assuming $T \gg T_i$, $i \in \mathbf{Z}^+$, *the power spectral density can be written in the form:*

$$G_X(T, f) \approx G_{XB}(T, f) + \sum_{i=1}^{\infty} p_i G_{iP}(T_i, f) \frac{1}{N_i} \frac{\sin^2[\pi N_i f/f_i]}{\sin^2[\pi f/f_i]}$$

$$+ \frac{2}{T} \sum_{i=1}^{\infty} p_i \mathrm{Re}\left[X_{iP}^*(T_i, f) \left(\frac{1 - e^{j2\pi N_i f T_i}}{1 - e^{j2\pi f T_i}}\right) X_{iB}(T, f)\right] \qquad (4.124)$$

where G_{iP} *is the power spectral density of the ith periodic signal whose period is* $T_i = 1/f_i$, *and where* $N_i = \lfloor T/T_i \rfloor$.

For the infinite interval, the periodic and bounded components can be treated separately, and

$$G_{X_\infty}(f) = G_{XB_\infty}(f) + G_{XP_\infty}(f)$$

$$= G_{XB_\infty}(f) + \sum_{i=1}^{\infty} p_i \sum_{k=-\infty}^{\infty} |c_{ik}|^2 \delta(f - kf_i) \qquad (4.125)$$

Proof. The proof of this theorem is given in Appendix 7.

4.8.1 Power Spectral Density — Nonzero Mean Case

Consider the countable case where each waveform in the ensemble of a random process X can be written as

$$x_i(t) = v_i(t) + \bar{x}_i \tag{4.126}$$

where $\bar{x}_i$ is the mean on $[0, T]$, and v_i has zero mean and no periodic component on the same interval, that is,

$$\bar{x}_i = \frac{1}{T} \int_0^T x_i(t)\, dt \qquad \frac{1}{T} \int_0^T v_i(t)\, dt = 0 \tag{4.127}$$

Consistent with such definitions, the ensemble for X on $[0, T]$ can be defined as

$$E_X = \{x_i : x_i(t) = v_i(t) + \bar{x}_i,\ t \in [0, T],\ i \in \mathbf{Z}^+\} \tag{4.128}$$

where $P[x_i(t)] = p_i$. The zero mean random process V is defined by the ensemble

$$E_V = \{v_i : v_i(t) = x_i(t) - \bar{x}_i,\ t \in [0, T],\ i \in \mathbf{Z}^+\} \tag{4.129}$$

where $P[v_i(t)] = p_i$. The following theorem states the relationship between the power spectral densities of X and V.

THEOREM 4.12. POWER SPECTRAL DENSITY OF NONZERO MEAN RANDOM PROCESS *For the case where*

$$\sum_{i=1}^{\infty} p_i |\bar{x}_i| < \infty \qquad \sum_{i=1}^{\infty} p_i |\bar{x}_i|^2 < \infty \qquad \sup\left\{\int_0^T |v_i(t)|^2\, dt : i \in \mathbf{Z}^+\right\} < \infty \tag{4.130}$$

and $G_V(T, f)$ is bounded as T becomes unbounded, the power spectral density of the random process X, defined by the ensemble as per Eq. (4.128), is given by

$$G_X(T, f) = G_V(T, f) + T \operatorname{sinc}^2\left(\frac{f}{f_o}\right) \sum_{i=1}^{\infty} p_i |\bar{x}_i|^2 \tag{4.131}$$

$$+ 2 \sum_{i=1}^{\infty} p_i \operatorname{sinc}\left(\frac{f}{f_o}\right) \operatorname{Re}[\bar{x}_i^* e^{j\pi f/f_o} V_i(T, f)]$$

$$G_{X_\infty}(f) = G_{V_\infty}(f) + \delta(f) \sum_{i=1}^{\infty} p_i |\bar{x}_i|^2 \tag{4.132}$$

where $f_o = 1/T$. Analogous results hold for the uncountable case.

122 POWER SPECTRAL DENSITY ANALYSIS

Proof. The proof of this theorem for the countable case is given in Appendix 8.

4.8.1.1 Notes First, if there exists a signal in the ensemble that has nonzero probability of occurring and a nonzero mean, then the power spectral density has an impulsive component at $f = 0$. Second, when the power spectral density close to zero frequency is not of interest and the power spectral density over the semi-infinite interval $[0, \infty]$ is being evaluated, then the mean of each waveform in the ensemble can be ignored.

4.9 GRAPHING IMPULSIVE POWER SPECTRAL DENSITIES

A common case is where many waveforms, comprising the random process, have both a periodic and an aperiodic component. The information usually required for the periodic component is its power and not its spectral form, which is impulsive at set frequencies. Accordingly, for the infinite interval or an interval that is long in comparison with the period of the periodic component, it is appropriate to plot the power in the impulsive component, along with the continuous power spectral density associated with the bounded and aperiodic component of the random process. The following example illustrates this.

4.9.1 Example

Consider the random process X defined by binary digital signaling, as per Example 3.4.2, where there is an additive sinusoidal component to each signal, such that the ensemble is defined according to

$$E_X = \left\{ x(\gamma_1, \ldots, \gamma_N, t) = A \sin(2\pi f_c t) + \sum_{k=1}^{N} \gamma_k p(t - (k-1)D) \quad \gamma_k \in \{-1, 1\} \right\} \tag{4.133}$$

From Eq. (3.39) and Theorem 4.11 the power spectral density on the infinite interval is given by

$$G_{X_\infty}(f) = r|P(f)|^2 + \frac{A^2}{4}[\delta(f - f_c) + \delta(f + f_c)] \tag{4.134}$$

This power spectral density is plotted in Figure 4.11 for the case where the pulse function is rectangular, that is,

$$p(t) = \begin{cases} 1 & 0 \leq t < W \\ 0 & \text{elsewhere} \end{cases} \quad \begin{array}{l} P(f) = W \operatorname{sinc}(fW) e^{-j\pi fW} \\ 0 < W \leq D \end{array} \tag{4.135}$$

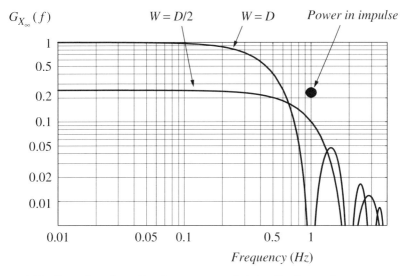

Figure 4.11 Illustration of graphing the power spectral density of a random process whose power spectral density contains both a bounded and an impulsive component.

$D = 1$, $A = 1$, $f_c = 1$, and the power associated with the impulse, rather than the impulse, is displayed.

APPENDIX 1: PROOF OF THEOREM 4.2

The proof relies on the following lemma:

LEMMA If $F = F_1 \cup F_2$ and F_1 and F_2 are disjoint sets, then the Fourier transform of a signal z, denoted Z, and evaluated over F, can be written as the sum of the Fourier transforms of z over F_1 and over F_2, that is, $Z(F, f) = Z(F_1, f) + Z(F_2, f)$. Further,

$$|Z(F, f)|^2 = |Z(F_1, f)|^2 + |Z(F_2, f)|^2 + 2\text{Re}[Z(F_1, f)Z^*(F_2, f)] \quad (4.136)$$

Proof. The proof of this result follows directly from writing out the definition of the Fourier transform for z, and using the fact that F_1 and F_2 are disjoint sets.

Proof (continued). Consider the ith waveform x_i, in the ensemble of X with a Fourier transform X_i. Since $x_{A_i}(t) = x(t)$ over $[0, T]$, $[0, T] = F_I \cup F_R$, $F = F_I \cup F_O$, and F_I, F_O, and F_R are disjoint sets, it follows from the lemma that

$$|X_i(T, f)|^2 - |X_{A_i}(F, f)|^2 = |X_i(F_R, f)|^2 + 2\text{Re}[X_i(F_I, f)X_i^*(F_R, f)]$$
$$- |X_{A_i}(F_O, f)|^2 - 2\text{Re}[X_i(F_I, f)X_{A_i}^*(F_O, f)]$$

$$(4.137)$$

Then

$$\sum_{i=1}^{\infty} \frac{p_i |X_i(T, f)|^2}{T} - \sum_{i=1}^{\infty} \frac{p_i |X_{A_i}(F, f)|^2}{T}$$

$$= \sum_{i=1}^{\infty} \frac{p_i |X_i(F_R, f)|^2}{T} - \sum_{i=1}^{\infty} \frac{p_i |X_{A_i}(F_O, f)|^2}{T}$$

$$+ \frac{2}{T} \sum_{i=1}^{\infty} p_i \mathrm{Re}[X_i(F_I, f) X_i^*(F_R, f)]$$

$$- \frac{2}{T} \sum_{i=1}^{\infty} p_i \mathrm{Re}[X_i(F_I, f) X_{A_i}^*(F_O, f)]$$

(4.138)

and

$$G_X(T, f) - G_{XA}(F, f) = \frac{M_R}{T} G_X(F_R, f) - \frac{M_O}{T} G_{XA}(F_O, f)$$

$$+ \frac{2}{T} \sum_{i=1}^{\infty} p_i \mathrm{Re}[X_i(F_I, f) X_i^*(F_R, f)] \quad (4.139)$$

$$- \frac{2}{T} \sum_{i=1}^{\infty} p_i \mathrm{Re}[X_i(F_I, f) X_{A_i}^*(F_O, f)]$$

Using the result $|A \pm B| \leq |A| + |B|$, it follows that

$$|G_X(T, f) - G_{XA}(F, f)| \leq \frac{M_R}{T} G_X(F_R, f) + \frac{M_O}{T} G_{XA}(F_O, f)$$

$$+ \frac{2}{T} \sum_{i=1}^{\infty} p_i |X_i(F_I, f) X_i^*(F_R, f)| \quad (4.140)$$

$$+ \frac{2}{T} \sum_{i=1}^{\infty} p_i |X_i(F_I, f) X_{A_i}^*(F_O, f)|$$

To obtain bounds on the last two terms in this equation, note that

$$|X_i(F_I, f)| |X_i(F_R, f)| = \sqrt{M_I} \sqrt{M_R} \sqrt{G_i(F_I, f)} \sqrt{G_i(F_R, f)} \quad (4.141)$$

Hence,

$$|G_X(T, f) - G_{XA}(F, f)| \leq \frac{M_R}{T} G_X(F_R, f) + \frac{M_O}{T} G_{XA}(F_O, f)$$

$$+ \frac{2\sqrt{M_I}\sqrt{M_R}}{T} \sum_{i=1}^{\infty} \sqrt{p_i G_i(F_I, f)} \sqrt{p_i G_i(F_R, f)}$$

$$+ \frac{2\sqrt{M_I}\sqrt{M_O}}{T} \sum_{i=1}^{\infty} \sqrt{p_i G_i(F_I, f)} \sqrt{p_i G_{A_i}(F_O, f)}$$

(4.142)

Using the Schwarz inequality for summations (Theorem 2.15), it follows that

$$\sum_{i=1}^{\infty} \sqrt{p_i G_i(F_I, f)} \sqrt{p_i G_i(F_R, f)} = \sqrt{\sum_{i=1}^{\infty} p_i G_i(F_I, f)} \sqrt{\sum_{i=1}^{\infty} p_i G_i(F_R, f)} \quad (4.143)$$

Hence,

$$|G_X(T, f) - G_{XA}(F, f)| \leqslant \frac{M_R}{T} G_X(F_R, f) + \frac{M_O}{T} G_{XA}(F_O, f)$$

$$+ \frac{2\sqrt{M_I}\sqrt{M_R}}{T} \sqrt{G_X(F_I, f)} \sqrt{G_X(F_R, f)}$$

$$+ \frac{2\sqrt{M_I}\sqrt{M_O}}{T} \sqrt{G_X(F_I, f)} \sqrt{G_{XA}(F_O, f)} \quad (4.144)$$

and thus,

$$\varepsilon_R = \frac{\int_{-\infty}^{\infty} |G_X(T, f) - G_{XA}(F, f)| \, df}{\int_{-\infty}^{\infty} G_X(T, f) \, df} \leqslant \frac{M_R}{T} \frac{\bar{P}(F_R)}{\bar{P}(T)} + \frac{M_O}{T} \frac{\bar{P}_A(F_O)}{\bar{P}(T)}$$

$$+ \frac{2\sqrt{M_I}\sqrt{M_R}}{T} \int_{-\infty}^{\infty} \frac{\sqrt{G_X(F_I, f)} \sqrt{G_X(F_R, f)}}{\bar{P}(T)} \, df \quad (4.145)$$

$$+ \frac{2\sqrt{M_I}\sqrt{M_O}}{T} \int_{-\infty}^{\infty} \frac{\sqrt{G_X(F_I, f)} \sqrt{G_{XA}(F_O, f)}}{\bar{P}(T)} \, df$$

The Schwarz inequality for integrals (Theorem 2.15) implies

$$\int_{-\infty}^{\infty} \sqrt{G_X(F_I, f)} \sqrt{G_X(F_R, f)} \, df = \sqrt{\int_{-\infty}^{\infty} G_X(F_I, f) \, df} \sqrt{\int_{-\infty}^{\infty} G_X(F_R, f) \, df}$$

$$(4.146)$$

and hence, the required result follows:

$$\varepsilon_R \leqslant \frac{M_R}{T} \frac{\bar{P}(F_R)}{\bar{P}(T)} + \frac{M_O}{T} \frac{\bar{P}_A(F_O)}{\bar{P}(T)} + \frac{2\sqrt{M_I}\sqrt{M_R}}{T} \frac{\sqrt{\bar{P}(F_I)} \sqrt{\bar{P}(F_R)}}{\bar{P}(T)}$$

$$+ \frac{2\sqrt{M_I}\sqrt{M_O}}{T} \frac{\sqrt{\bar{P}(F_I)} \sqrt{\bar{P}_A(F_O)}}{\bar{P}(T)} \quad (4.147)$$

APPENDIX 2: PROOF OF THEOREM 4.4

The proof of the first result follows the proof given by Papoulis (2002 p. 211) to a related problem. Consider a real variable α and the result:

$$0 \leqslant \frac{1}{T} \sum_i \sum_j p_{ij} |\alpha X_i(T, f) \pm Y_j(T, f)|^2$$

$$= \frac{\alpha^2}{T} \sum_i \sum_j p_{ij} |X_i(T, f)|^2 + \frac{1}{T} \sum_i \sum_j p_{ij} |Y_j(T, f)|^2 \qquad (4.148)$$

$$\pm 2\alpha \text{Re} \left[\frac{1}{T} \sum_i \sum_j p_{ij} X_i(T, f) Y_j^*(T, f) \right]$$

Using results from conditional probability, $p_{ij} = p_{i/j} p_j$ and $\Sigma_i p_{i/j} = 1$, as well as definitions for the power spectral density and the cross power spectral density, it follows that

$$0 \leqslant \alpha^2 G_X(T, f) \pm 2\alpha \text{Re}[G_{XY}(T, f)] + G_Y(T, f) \qquad (4.149)$$

The right-hand side of the inequality is a quadratic equation with respect to α with roots given by

$$r_1, r_2 = \frac{\mp \text{Re}[G_{XY}(T, f)]}{G_X(T, f)} \left[1 \pm \sqrt{1 - \frac{G_X(T, f) G_Y(T, f)}{\{\text{Re}[G_{XY}(T, f)]\}^2}} \right] \qquad (4.150)$$

Thus,

$$0 \leqslant (\alpha - r_1)(\alpha - r_2) \qquad (4.151)$$

This inequality must hold for all $\alpha \in \mathbf{R}$ which implies that the graph of the quadratic, with respect to α, must not cross below the horizontal axis. This is consistent with either imaginary roots or equal real roots. Thus, the argument of the square root operator must be $\leqslant 0$, that is,

$$|\text{Re}[G_{XY}(T, f)]| \leqslant \sqrt{G_X(T, f)} \sqrt{G_Y(T, f)} \qquad (4.152)$$

which is the required result.

To prove the second result, consider the inequality defined by Eq. (4.149) for the case where $\alpha = 1$ (Larson, 1986 p. 435):

$$0 \leqslant G_X(T, f) + G_Y(T, f) \pm 2\text{Re}[G_{XY}(T, f)] \qquad (4.153)$$

APPENDIX 2: PROOF OF THEOREM 4.4

It then directly follows that

$$2|\text{Re}[G_{XY}(T, f)]| \leq G_X(T, f) + G_Y(T, f) \qquad (4.154)$$

which is the required result.

For the fully correlated case the cross power spectral density is

$$G_{XY}(T, f) = \sum_i \frac{p_i X_i(T, f) Y_i^*(T, f)}{T} = \sum_i \left[\frac{\sqrt{p_i} X_i(T, f)}{\sqrt{T}} \right] \left[\frac{\sqrt{p_i} Y_i^*(T, f)}{\sqrt{T}} \right] \qquad (4.155)$$

Direct use of the Schwarz inequality (Theorem 2.15) yields the required result, namely,

$$G_{XY}(T, f) \leq \sqrt{\sum_i \frac{p_i |X_i(T, f)|^2}{T}} \sqrt{\sum_i \frac{p_i |Y_i(T, f)|^2}{T}} = \sqrt{G_X(T, f)} \sqrt{G_Y(T, f)} \qquad (4.156)$$

To establish when equality exists between $|G_{XY}(T, f)|$ and a function of $G_X(T, f)$ and $G_Y(T, f)$, consider the fully correlated case given by Eq. (4.155). If $x_i(t) = k y_i(t)$ for $i \in \mathbb{Z}^+$, then $G_X(T, f) = k^2 G_Y(T, f)$, and

$$|G_{XY}(T, f)| = \sum_i \frac{p_i |X_i(T, f)|^2}{|k|T} = \frac{G_X(T, f)}{|k|} = |k| \sum_i \frac{p_i |Y_i(T, f)|^2}{T} = |k| G_Y(T, f)$$

$$= \sqrt{G_X(T, f)} \sqrt{G_Y(T, f)} \qquad (4.157)$$

which establishes the required equality.

To establish the bounds for the independent case, consider the summation defining G_{XY} when $p_{ij} = p_{xi} p_{yj}$, whereupon

$$\left| \sum_i \sum_j \frac{p_{ij} X_i(T, f) Y_j^*(T, f)}{T} \right| \leq \sum_i \sqrt{p_{xi}} \left[\frac{\sqrt{p_{xi}} |X_i(T, f)|}{\sqrt{T}} \right] \sum_j \sqrt{p_{yj}} \left[\frac{\sqrt{p_{yj}} |X_j(T, f)|}{\sqrt{T}} \right] \qquad (4.158)$$

From the Schwarz inequality (Theorem 2.15), it follows that

$$\sum_i \sqrt{p_{xi}} \left[\frac{\sqrt{p_{xi}} |X_i(T, f)|}{\sqrt{T}} \right] \leq \sqrt{\sum_i p_{xi}} \sqrt{\sum_i \frac{p_{xi} |X_i(T, f)|^2}{T}} = \sqrt{G_X(T, f)} \qquad (4.159)$$

and hence, the inequality for the independent case follows, namely,

$$\left| \frac{1}{T} \sum_i \sum_j p_{ij} X_i(T, f) Y_j^*(T, f) \right| \leq \sqrt{G_X(T, f)} \sqrt{G_Y(T, f)} \qquad (4.160)$$

APPENDIX 3: PROOF OF THEOREM 4.5

From the definition of the power spectral density, it follows that the power spectral density of $Z = X \pm Y$, evaluated on the interval $[0, T]$, is given by

$$G_Z(T, f) = \sum_{i=1}^{\infty} \sum_{j=1}^{\infty} \frac{p_{ij}}{T} |Z_{ij}(T, f)|^2 = \sum_{i=1}^{\infty} \sum_{j=1}^{\infty} \frac{p_{ij}}{T} |X_i(T, f) \pm Y_j(T, f)|^2$$

$$= \sum_{i=1}^{\infty} \sum_{j=1}^{\infty} \frac{p_{ij}}{T} \left[|X_i(T, f)|^2 \pm 2\mathrm{Re}[X_i(T, f) Y_j^*(T, f)] + |Y_j(T, f)|^2 \right]$$

(4.161)

Using the results from conditional probability,

$$p_{kl} = p_k p_{l/k} = p_l p_{k/l} \qquad \sum_{k=1}^{\infty} p_{k/l} = 1 \qquad \sum_{l=1}^{\infty} p_{l/k} = 1 \qquad (4.162)$$

and, assuming that the order of the Re operator and summation can be interchanged, the required result follows, that is,

$$G_Z(T, f) = G_X(T, f) + G_Y(T, f) \pm 2\mathrm{Re}[G_{XY}(T, f)] \qquad (4.163)$$

APPENDIX 4: PROOF OF THEOREM 4.6

As $G_Z(T, f) = \lim_{N \to \infty} G_{Z_N}(T, f)$, it suffices to determine the power spectral density of Z_N. The power spectral density of Z_N, evaluated on the interval $[0, T]$, is

$$G_{Z_N}(T, f) = \frac{1}{T} \sum_{\zeta_1 = 1}^{\infty} \cdots \sum_{\zeta_N = 1}^{\infty} p(\zeta_1, \ldots, \zeta_N) |Z_N(\zeta_1, \ldots, \zeta_N, T, f)|^2 \qquad (4.164)$$

where

$$|Z_N(\zeta_1, \ldots, \zeta_N, T, f)|^2 = \left| \sum_{i=1}^{N} w_i X_i(\zeta_i, T, f) \right|^2$$

$$= \sum_{i=1}^{N} \sum_{j=1}^{N} w_i w_j^* X_i(\zeta_i, T, f) X_j^*(\zeta_j, T, f)$$

(4.165)

Z_N as given by this expression is finite since, according to Theorem 3.6, bounded signal energy [see Eq. (4.85)] guarantees that

$$\sup\{|X_i(\zeta_i, T, f)|: i \in \{1, \ldots, N\}, \zeta_i \in \mathbf{Z}^+, f \in \mathbf{R}\} < \infty \quad (4.166)$$

Thus, $G_{Z_N}(T, f)$ equals

$$\frac{1}{T} \sum_{\zeta_1=1}^{\infty} \cdots \sum_{\zeta_N=1}^{\infty} \sum_{i=1}^{N} \sum_{j=1}^{N} p(\zeta_1, \ldots, \zeta_N) w_i w_j^* X_i(\zeta_i, T, f) X_j^*(\zeta_j, T, f) \quad (4.167)$$

Equation (4.166) results in the summation in Eq. (4.167) being absolutely convergent and hence, according to Theorem 2.21, the order of summation can be interchanged to yield

$$G_{Z_N}(T, f) = \sum_{i=1}^{N} \sum_{\zeta_1=1}^{\infty} \cdots \sum_{\zeta_N=1}^{\infty} p(\zeta_1, \ldots, \zeta_N) \frac{|w_i|^2 |X_i(\zeta_i, T, f)|^2}{T}$$

$$+ \sum_{i=1}^{N} \sum_{\substack{j=1 \\ j \neq i}}^{N} \sum_{\zeta_1=1}^{\infty} \cdots \sum_{\zeta_N=1}^{\infty} p(\zeta_1, \ldots, \zeta_N) \frac{w_i w_j^*}{T} X_i(\zeta_i, T, f) X_j^*(\zeta_j, T, f)$$

$$(4.168)$$

Using the results from conditional probability,

$$p(\zeta_1, \ldots, \zeta_N) = p_i(\zeta_i) p(\zeta_1, \ldots, \zeta_{i-1}, \zeta_{i+1}, \ldots, \zeta_N / \zeta_i) \quad (4.169)$$

$$\sum_{\zeta_1=1}^{\infty} \cdots \sum_{\zeta_{i-1}=1}^{\infty} \sum_{\zeta_{i+1}=1}^{\infty} \cdots \sum_{\zeta_N}^{\infty} p(\zeta_1, \ldots, \zeta_{i-1}, \zeta_{i+1}, \ldots, \zeta_N / \zeta_i) = 1 \quad (4.170)$$

it follows that the first summation in Eq. (4.168) can be written as

$$\sum_{i=1}^{N} |w_i|^2 \sum_{\zeta_1=1}^{\infty} \cdots \sum_{\zeta_{i-1}=1}^{\infty} \sum_{\zeta_{i+1}=1}^{\infty} \cdots \sum_{\zeta_N=1}^{\infty}$$

$$p(\zeta_1, \ldots, \zeta_{i-1}, \zeta_{i+1}, \ldots, \zeta_N / \zeta_i) \sum_{\zeta_i=1}^{\infty} p_i(\zeta_i) \frac{|X_i(\zeta_i, T, f)|^2}{T} = \sum_{i=1}^{N} |w_i|^2 G_i(T, f)$$

$$(4.171)$$

Similarly, the summation with respect to $\zeta_1, \ldots, \zeta_N$ in the second term of Eq. (4.168), can be written as

$$w_i w_j^* \sum_{\zeta_1=1}^{\infty} \cdots \sum_{\zeta_{i-1}=1}^{\infty} \sum_{\zeta_{i+1}=1}^{\infty} \cdots \sum_{\zeta_{j-1}=1}^{\infty} \sum_{\zeta_{j+1}=1}^{\infty} \cdots \sum_{\zeta_N=1}^{\infty} p(\zeta/\zeta_i, \zeta_j)$$

$$\times \left[\sum_{\zeta_i=1}^{\infty} \sum_{\zeta_j=1}^{\infty} p_{ij}(\zeta_i, \zeta_j) \frac{X_i(\zeta_i, T, f) X_j^*(\zeta_j, T, f)}{T} \right] = w_i w_j^* G_{ij}(T, f) \quad (4.172)$$

where $\zeta = \zeta_1, \ldots, \zeta_{i-1}, \zeta_{i+1}, \ldots, \zeta_{j-1}, \zeta_{j+1}, \ldots, \zeta_N$. Substitution of these two results into Eq. (4.168) yields the required result, namely,

$$G_{Z_N}(T, f) = \sum_{i=1}^{N} |w_i|^2 G_i(T, f) + \sum_{i=1}^{N} \sum_{\substack{j=1 \\ j \neq i}}^{N} w_i w_j^* G_{ij}(T, f) \quad (4.173)$$

To prove the final result, as in Eq. (4.90), note that $\lim_{T \to \infty} T \operatorname{sinc}^2(fT) = \delta(f)$ and

$$\sum_{i=1}^{N} \sum_{\substack{j=1 \\ j \neq i}}^{N} w_i w_j^* \mu_i \mu_j^* = \sum_{i=1}^{N} \sum_{j=1}^{N} w_i w_j^* \mu_i \mu_j^* - \sum_{i=1}^{N} |w_i|^2 |\mu_i|^2 \quad (4.174)$$

$$= \left| \sum_{i=1}^{N} w_i \mu_i \right|^2 - \sum_{i=1}^{N} |w_i|^2 |\mu_i|^2$$

APPENDIX 5: PROOF OF THEOREM 4.8

By definition, the power spectral density on the interval $[0, NT_p]$ is

$$G_X(NT_p, f) = \frac{|X(NT_p, f)|^2}{NT_p} \qquad X(NT_p, f) = \int_0^{NT_p} x(t) e^{-j2\pi ft} \, dt \quad (4.175)$$

On the interval $[0, NT_p]$ the Fourier transform X can be written as

$$X(NT_p, f) = \int_0^{T_p} x(t) e^{-j2\pi ft} \, dt + \int_{T_p}^{2T_p} x(t) e^{-j2\pi ft} \, dt + \cdots + \int_{(N-1)T_p}^{NT_p} x(t) e^{-j2\pi ft} \, dt \quad (4.176)$$

and, since x is periodic with period T_p, the changes of variables, $\zeta = t - iT_p$, $i \in \{1, \ldots, N-1\}$, results in

$$X(NT_p, f) = \left[\int_0^{T_p} x(t) e^{-j2\pi ft} \, dt \right] [1 + e^{-j2\pi fT_p} + \cdots + e^{-j2\pi f(N-1)T_p}]$$

$$= X(T_p, f) \sum_{i=0}^{N-1} e^{-j2\pi f iT_p} = X(T_p, f) \frac{1 - e^{-j2\pi fNT_p}}{1 - e^{-j2\pi fT_p}} \quad (4.177)$$

where the last result comes from Theorem 2.32. Thus, with $N = \lfloor T/T_p \rfloor$ it follows that

$$X(T, f) = X(T_p, f) \frac{1 - e^{-j2\pi fNT_p}}{1 - e^{-j2\pi fT_p}} + \int_{NT_p}^{T} x(t) e^{-j2\pi ft} \, dt \quad (4.178)$$

When $f \in \{\ldots, -f_p, 0, f_p, \ldots\}$, it follows that $1 - e^{-j2\pi f T_p} = 0$, and from Theorem 2.32 that $|(1 - e^{-j2\pi f N T_p})/(1 - e^{-j2\pi f T_p})| = N$. Thus,

$$|X(T, f)| = \left\| \left[\frac{T}{T_p}\right] X(T_p, f) + \int_{NT_p}^{T} x(t) e^{-j2\pi f t} \, dt \right\| \quad (4.179)$$

Hence, when $f \in \{\ldots, -f_p, 0, f_p, \ldots\}$ and $G(T, f) \neq 0$, it is the case that $G(T, f)$ approaches infinity in a manner proportional to T and thus, there exists constants k_2, $T_o > 0$, and $0 < \alpha < 1$, such that for all $T > T_o$ it is the case that

$$\alpha k_2 T < G(T, f) = \frac{|X(T, f)|^2}{T} < k_2 T \quad (4.180)$$

When $f \notin \{\ldots, -f_p, 0, f_p, \ldots\}$ it follows that $1 - e^{-j2\pi f T_p} \neq 0$. Consequently, from Eq. (4.178), it follows that

$$|X(T, f)| < \frac{2|X(T_p, f)|}{|1 - e^{-j2\pi f T_p}|} + \left| \int_{NT_p}^{(N+1)T_p} |x(t)| \, dt \right| \quad (4.181)$$

which is independent of T. Hence, when $f \notin \{\ldots, -f_p, 0, f_p, \ldots\}$ there exists constants k_1, $T_o > 0$, such that for all $T > T_o$, it is the case that

$$G(T, f) = \frac{|X(T, f)|^2}{T} < \frac{k_1}{T} \quad (4.182)$$

which completes the proof of the required bounds on $G(T, f)$ as T increases.

To determine expressions for $G_X(NT_p, f)$ note, from Eq. (4.177), that

$$G_X(NT_p, f) = \frac{|X(T_p, f)|^2}{NT_p} \left| \sum_{i=0}^{N-1} e^{-j2\pi f i T_p} \right|^2 = G_X(T_p, f) \frac{1}{N} \left| \sum_{i=0}^{N-1} e^{-j2\pi i f / f_p} \right|^2$$

(4.183)

Using a result from Theorem 2.32, the required form for the power spectral density on the interval $[0, NT_p]$, that is,

$$G_X(NT_p, f) = G_X(T_p, f) \frac{1}{N} \frac{\sin^2(\pi N f / f_p)}{\sin^2(\pi f / f_p)} \quad (4.184)$$

follows. As $\sin^2(\pi N f / f_p)/\sin^2(\pi f / f_p) = N^2$ when $f = i f_p$ (Theorem 2.32), the result

$$G_X(NT_p, i f_p) = N G_X(T_p, i f_p) = \frac{N |X(T_p, i f_p)|^2}{T_p} = NT_p |c_i|^2 = \frac{|c_i|^2}{f_o} \quad (4.185)$$

follows for $i \in Z$. The third equality in this equation uses Eq. (4.96).

On the infinite interval $(0, \infty)$, another result from Theorem 2.32 substituted into Eq. (4.184), yields

$$G_{X_\infty}(f) = \lim_{N \to \infty} G_X(NT_p, f) = G_X(T_p, f) \sum_{i=-\infty}^{\infty} f_p \delta(f - if_p) \qquad (4.186)$$

Since

$$c_i = \frac{X(T_p, if_p)}{T_p} \qquad G_X(T_p, if_p) = \frac{|X(T_p, if_p)|^2}{T_p} = \frac{|c_i|^2}{f_p} \qquad i \in Z \qquad (4.187)$$

it follows that

$$G_{X_\infty}(f) = \sum_{i=-\infty}^{\infty} |c_i|^2 \delta(f - if_p) \qquad (4.188)$$

which is the last required result.

APPENDIX 6: PROOF OF THEOREM 4.10

It directly follows from Theorem 4.6 that

$$G_X(T, f) = \sum_{i=1}^{\infty} |w_i|^2 G_i(T, f) + \frac{1}{T} \sum_{i=1}^{\infty} \sum_{\substack{j=1 \\ j \ne i}}^{\infty} w_i w_j^* X_i(T, f) X_j^*(T, f) \qquad (4.189)$$

which is the first required result. It follows from Eq. (4.108) and Theorem 2.33 that

$$X_i(T, f) = T \sum_{k=-\infty}^{\infty} c_{ik} e^{-j\pi(f - kf_i)T} \operatorname{sinc}[(f - kf_i)T] \qquad (4.190)$$

For the infinite interval, it is required to show that the double summation, DS, in Eq. (4.189) is insignificant compared with the first summation as $T \to \infty$, except when $f = 0$. To this end, note that

$$DS = \frac{1}{T} \sum_{i=1}^{\infty} \sum_{\substack{j=1 \\ j \ne i}}^{\infty} w_i w_j^* X_i(T, f) X_j^*(T, f) \le \sum_{i=1}^{\infty} \sum_{\substack{j=1 \\ j \ne i}}^{\infty} |w_i||w_j^*| \frac{|X_i(T, f)|}{\sqrt{T}} \frac{|X_j(T, f)|}{\sqrt{T}}$$

$$= \sum_{i=1}^{\infty} \sum_{\substack{j=1 \\ j \ne i}}^{\infty} |w_i||w_j^*| \sqrt{G_i(T, f)} \sqrt{G_j(T, f)} \qquad (4.191)$$

Finite signal energy as per the assumption in the theorem guarantees, from Theorem 3.6, that

$$\sup\{G_i(T, f): i \in Z^+, f \in R\}$$

is finite. Absolute convergence of the weightings then guarantees that the bound in this equation is finite. From Theorem 4.8 it follows that there exists constants $k_i, r_i, T_o > 0, 0 < \alpha < 1$, such that for all $T > T_o$ it is the case that

$$\begin{array}{ll} G_i(T, f) < k_i/T & f \notin \{\ldots, -f_i, 0, f_i, \ldots\} \\ \alpha r_i T < G_i(T, f) < r_i T & f \in \{\ldots, -f_i, 0, f_i, \ldots\} \end{array} \quad (4.192)$$

Thus, when f is not a multiple of one of the fundamental frequencies of the periodic waveforms, that is, $f \notin \{qf_i: q \in Z, i \in Z^+\}$, it is the case that

$$DS = \frac{1}{T} \sum_{i=1}^{\infty} \sum_{\substack{j=1 \\ j \neq i}}^{\infty} w_i w_j^* X_i(T, f) X_j^*(T, f) \leq \sum_{i=1}^{\infty} \sum_{\substack{j=1 \\ j \neq i}}^{\infty} |w_i||w_j^*| \sqrt{\frac{k_i}{T}} \sqrt{\frac{k_j}{T}} \quad (4.193)$$

which clearly converges to zero as $T \to \infty$.

The second case is where f is a multiple of one of the fundamental frequencies of the periodic waveforms, but not equal to zero, that is,

$$f \in \{qf_i: q \in Z, q \neq 0, i \in Z^+\}$$

For the case where $f = q_n f_n$ and $q_n \neq 0$, it follows from Eq. (4.192) that

$$DS = \sum_{\substack{i=1 \\ i \neq n}}^{\infty} \sum_{\substack{j=1 \\ j \neq i,n}}^{\infty} \frac{w_i w_j^* X_i(T, f) X_j^*(T, f)}{T} + 2\text{Re} \sum_{\substack{j=1 \\ j \neq n}}^{\infty} \frac{w_n w_j^* X_n(T, f) X_j^*(T, f)}{T}$$

$$\leq \sum_{\substack{i=1 \\ i \neq n}}^{\infty} \sum_{\substack{j=1 \\ j \neq i,n}}^{\infty} |w_i||w_j^*| \sqrt{\frac{k_i}{T}} \sqrt{\frac{k_j}{T}} + 2 \sum_{\substack{j=1 \\ j \neq n}}^{\infty} |w_n||w_j^*|\sqrt{r_n T} \sqrt{\frac{k_j}{T}} \quad (4.194)$$

which clearly is bounded as $T \to \infty$. However, for $f = q_n f_n$, it is the case that $G_n(T, f)$, and hence the first summation in Eq. (4.189), becomes unbounded as $T \to \infty$. Thus, the double summation can be neglected as required, except when $f = 0$.

For the case where f is in the neighborhood of zero, it follows from Eq. (4.190) and T sufficiently large, that

$$X_i(T, f) \approx T c_{i0} \, \text{sinc}(fT) \quad (4.195)$$

Hence,

$$\frac{1}{T} \sum_{i=1}^{\infty} \sum_{j=1}^{\infty} w_i w_j^* X_i(T, f) X_j^*(T, f) \approx T \operatorname{sinc}^2(fT) \sum_{i=1}^{\infty} \sum_{j=1}^{\infty} w_i w_j^* c_{i0} c_{j0}^* \quad (4.196)$$

Thus, as $T \to \infty$, it follows from Theorem 2.32 and for $f \simeq 0$, that

$$\frac{1}{T} \sum_{i=1}^{\infty} \sum_{j=1}^{\infty} w_i w_j^* X_i(T, f) X_j^*(T, f) = \delta(f) \sum_{i=1}^{\infty} \sum_{j=1}^{\infty} w_i w_j^* c_{i0} c_{j0}^* = \delta(f) \left| \sum_{i=1}^{\infty} w_i c_{i0} \right|^2$$

(4.197)

which is the final result required for the infinite interval.

APPENDIX 7: PROOF OF THEOREM 4.11

By definition, the power spectral density of X is

$$G_X(T, f) = \sum_{i=1}^{\infty} p_i \frac{|X_i(T, f)|^2}{T} \qquad X_i(T, f) = X_{iB}(T, f) + X_{iP}(T, f), \quad i \in \mathbf{Z}^+$$

(4.198)

Direct evaluation yields

$$G_X(T, f) = G_{XB}(T, f) + G_{XP}(T, f) + \frac{2}{T} \sum_{i=1}^{\infty} p_i \operatorname{Re}[X_{iB}(T, f) X_{iP}^*(T, f)] \quad (4.199)$$

The summations in this equation are finite because of the assumptions stated in Eq. (4.121), which, according to Theorem 3.6, imply that

$$\sup\{|X_{iB}(T, f)| : i \in \mathbf{Z}^+\} < \infty \qquad \sup\{G_{iB}(T, f) : i \in \mathbf{Z}^+\} < \infty$$

and similarly for X_{iP} and G_{iP}. This is the first required result.

From Eq. (4.118) and Theorem 2.33, it follows for $i \in \mathbf{Z}^+$ that

$$X_{iP}(T, f) = T \sum_{k=-\infty}^{\infty} c_{ik} e^{-j\pi(f-kf_i)T} \operatorname{sinc}[(f - kf_i)T] \quad (4.200)$$

Also, when $T \gg T_i$, that is, $N_i = \lfloor T/T_i \rfloor \gg 1$, it follows from Theorem 4.8 that

$$X_{iP}(T, f) \approx X_{iP}(T_i, f) \left[\frac{1 - e^{-j2\pi N_i f T_i}}{1 - e^{-j2\pi f T_i}} \right] \quad (4.201)$$

$$G_{XP}(T, f) \approx \sum_{i=1}^{\infty} p_i G_{iP}(T_i, f) \frac{1}{N_i} \frac{\sin^2(\pi N_i f / f_i)}{\sin^2(\pi f / f_i)} \quad (4.202)$$

and hence,

$$G_X(T, f) \approx G_{XB}(T, f) + \sum_{i=1}^{\infty} p_i G_{iP}(T_i, f) \frac{1}{N_i} \frac{\sin^2(\pi N_i f/f_i)}{\sin^2(\pi f/f_i)}$$
$$+ \frac{2}{T} \sum_{i=1}^{\infty} p_i \text{Re}\left[X_{iP}^*(T_i, f) \left(\frac{1 - e^{j2\pi N_i f T_i}}{1 - e^{j2\pi f T_i}} \right) X_{iB}(T, f) \right]$$

(4.203)

Direct substitution of Eq. (4.200) into Eq. (4.199) yields the alternative form for G_X,

$$G_X(T, f) = G_{XB}(T, f) + G_{XP}(T, f)$$
$$+ 2 \sum_{i=1}^{\infty} p_i \sum_{k=-\infty}^{\infty} \text{sinc}[(f - kf_i)T] \text{Re}[c_{ik}^* e^{j\pi(f - kf_i)T} X_{iB}(T, f)]$$

(4.204)

To obtain an expression for $G_{X_\infty}(f) = \lim_{T \to \infty} G_X(T, f)$, there are two cases to consider. The first case is where f is not a multiple of one of the fundamental frequencies of the periodic waveforms, that is, $f \notin \{pf_i: p \in \mathbf{Z}, i \in \mathbf{Z}^+\}$. For this case, note that the summation in Eq. (4.199), denoted T_3, has the bound,

$$T_3 \leq 2 \sum_{i=1}^{\infty} p_i \frac{|X_{iB}(T, f)| \, |X_{iP}(T, f)|}{\sqrt{T} \, \sqrt{T}} = 2 \sum_{i=1}^{\infty} p_i \sqrt{G_{iB}(T, f)} \sqrt{G_{iP}(T, f)} \quad (4.205)$$

For the case where f is not equal to a multiple of one of the fundamental frequencies of the periodic signals, it follows from Theorem 4.8 that $G_{iP}(T, f)$ converges to zero as $T \to \infty$. As noted above $\sup\{G_{iB}(T, f): i \in \mathbf{Z}^+\} < \infty$ and hence, $\lim_{T \to \infty} T_3 = 0$.

The second case is where f is a multiple of one of the fundamental frequencies of the periodic waveforms, that is, $f \in \{pf_i: p \in \mathbf{Z}, i \in \mathbf{Z}^+\}$. According to Theorem 4.8, it is the case that $\lim_{T \to \infty} G_{iP}(T, f)$ is infinite, and accordingly, T_3 also becomes unbounded as $T \to \infty$ assuming $G_{iB}(T, f) \neq 0$. Hence, potentially, T_3 becomes impulsive. However, according to Eq. (4.101), $G_{iP}(T, f)$ increases in proportion to T for T sufficiently large, and as $G_{iB}(T, f)$ is bounded it follows that T_3 increases in proportion to $\sqrt{T}$. Thus, there will always exist a sufficiently large value of T, such that

$$G_{XP}(T, f) = \sum_{i=1}^{\infty} p_i G_{iP}(T, f) \gg 2 \sum_{i=1}^{\infty} p_i \sqrt{G_{iB}(T, f)} \sqrt{G_{iP}(T, f)} > T_3 \quad (4.206)$$

Hence,

$$G_{X_\infty}(f) = G_{XB_\infty}(f) + \sum_{i=1}^{\infty} p_i \sum_{k=-\infty}^{\infty} |c_{ik}|^2 \delta(f - kf_i) \quad (4.207)$$

where the last term follows from Theorem 4.9. This is the required result.

APPENDIX 8: PROOF OF THEOREM 4.12

By definition the power spectral density of X is

$$G_X(T, f) = \sum_{i=1}^{\infty} p_i \frac{|X_i(T, f)|^2}{T} \qquad X_i(T, f) = V_i(T, f) + \bar{x}_i P(T, f) \quad (4.208)$$

where P is the Fourier transform of the pulse function with unit height on the interval $[0, T]$, and zero elsewhere. From Theorem 2.33

$$P(T, f) = T \operatorname{sinc}(f/f_o) e^{-j\pi f/f_o} \qquad f_o = 1/T \quad (4.209)$$

and it then follows that

$$G_X(T, f) = G_V(T, f) + T \operatorname{sinc}^2\left(\frac{f}{f_o}\right) \sum_{i=1}^{\infty} p_i |\bar{x}_i|^2$$
$$+ 2 \sum_{i=1}^{\infty} p_i \operatorname{sinc}\left(\frac{f}{f_o}\right) \operatorname{Re}[\bar{x}_i^* e^{j\pi f/f_o} V_i(T, f)] \quad (4.210)$$

The assumptions made in Eq. (4.130) ensure, according to Theorem 3.6, that

$$\sup\{|V_i(T, f)|: i \in \mathbf{Z}^+\} < \infty \qquad \text{and} \qquad \sup\{G_{V_i}(T, f): i \in \mathbf{Z}^+\} < \infty$$

These results, along with the absolute convergence of the summation of $p_i|\bar{x}_i|^2$ and $p_i|\bar{x}_i|$, ensure that G_X, as given by this equation, is finite. This is the first required result.

To obtain an expression for $G_{X_\infty}(f) = \lim_{T \to \infty} G_X(T, f)$, first, note from Theorem 2.32 that $\lim_{T \to \infty} T \operatorname{sinc}^2(Tf) = \delta(f)$. Second, for f fixed and $f \neq 0$ it is the case that $|\operatorname{sinc}(fT)|$ decays according to $1/T$ as T increases. As

$$\sup\{G_{V_i}(T, f): i \in \mathbf{Z}^+, T \in \mathbf{R}^+\} < \infty$$

it then follows that the last term in Eq. (4.210) has the bound,

$$\left| 2 \sum_{i=1}^{\infty} p_i \operatorname{sinc}\left(\frac{f}{f_o}\right) \operatorname{Re}[\bar{x}_i^* e^{j\pi f/f_o} V_i(T, f)] \right|$$

$$\leq \frac{2k}{T} \sum_{i=1}^{\infty} p_i |\bar{x}_i| |V_i(T, f)| \qquad (4.211)$$

$$< \frac{2k}{T} \sqrt{T} \sup\{\sqrt{G_{V_i}(T, f)} : i \in \mathbf{Z}^+, T \in \mathbf{R}^+\} \sum_{i=1}^{\infty} p_i |\bar{x}_i|$$

for some appropriate constant k and when $f \neq 0$. Clearly, this summation approaches zero for T sufficiently large. When $f = 0$, the second term in Eq. (4.210) increases in proportion to T, while the last term in this equation potentially increases in proportion to $\sqrt{T}$. Hence, when $f = 0$ the last term can be neglected. Thus,

$$G_{X_x}(f) = G_{V_x}(f) + \delta(f) \sum_{i=1}^{\infty} p_i |\bar{x}_i|^2 \qquad (4.212)$$

which is the required result.

5

Power Spectral Density of Standard Random Processes — Part 1

5.1 INTRODUCTION

In Chapters 5 and 6 the power spectral density of commonly encountered random processes are given in detail. Specifically, the power spectral density of random processes associated with signaling, quantization, jitter, and shot noise are discussed in this chapter, while the power spectral density associated with sampling, quadrature amplitude modulation, random walks, and $1/f$ noise, are discussed in Chapter 6.

In this chapter, the random processes discussed have a general form that is associated with signaling, and the terminology of a signaling random process is introduced. The results associated with signaling random processes are used in Chapter 7, to detail an approach for determining the power spectral density of a random process after a nonlinear memoryless transformation.

5.2 SIGNALING RANDOM PROCESSES

As defined below, the signal form associated with signaling is such that signaling random processes are found in models for a diverse range of physical processes. For example, signaling random processes include baseband and certain bandpass communication processes. The signal form of interest is that of an information signal.

DEFINITION: INFORMATION SIGNAL   An information signal is one generated by a sum of signals from a "signaling set," where one signal is associated with each signaling interval of D sec. With a signaling set E_Φ, an information signal has the form

$$\sum_{i=1}^{\infty} \phi(\gamma_i, t - (i-1)D) \tag{5.1}$$

where $\phi \in E_\Phi$ and γ_i is an index variable which defines the signal from E_Φ that is associated with the ith signaling interval $[(i-1)D, iD]$.

DEFINITION: SIGNALING RANDOM PROCESS A signaling random process X, is one whose ensemble consists of information signals. The ensemble E_X characterizing such a random process for the interval $[0, ND]$ is

$$E_X = \left\{ x(\gamma_1, \ldots, \gamma_N, t) = \sum_{i=1}^{N} \phi(\gamma_i, t - (i-1)D), \; \gamma_i \in S_\Gamma, \phi \in E_\Phi \right\} \tag{5.2}$$

where $\gamma_i \in S_\Gamma$ and the vector $(\gamma_1, \ldots, \gamma_N)$ is an element of $S_X = S_\Gamma \times \cdots \times S_\Gamma$, which is an index set to distinguish between waveforms in the ensemble. $S_\Gamma \subseteq \mathbf{Z}^+$ for the countable case and $S_\Gamma \subseteq \mathbf{R}$ for the uncountable case. Equivalently, $\gamma_1, \ldots, \gamma_N$ are the respective outcomes of N identically distributed random variables $\Gamma_1, \ldots, \Gamma_N$, and Γ is used to denote any one of these. The sample space associated with Γ is S_Γ.

Associated with each element of the set S_Γ, or equivalently with each outcome of the random variable Γ, is a signal, and this association defines the set or ensemble of signaling waveforms, E_Φ:

$$E_\Phi = \{\phi(\gamma, t): \gamma \in S_\Gamma\} \tag{5.3}$$

The probability of any given signal from the signaling set is given by the probability of the associated outcome from Γ, that is,

$$P[\phi(\gamma, t)] = P[\gamma] = p_\gamma \qquad \text{countable case}$$
$$P[\phi(\gamma, t)|_{\gamma \in [\gamma_o, \gamma_o + d\gamma]}] = P[\gamma \in [\gamma_o, \gamma_o + d\gamma]] = f_\Gamma(\gamma_o) \, d\gamma \qquad \text{uncountable case}$$
$$\tag{5.4}$$

where f_Γ is the probability density function of the random variable Γ for the uncountable case. The probability associated with waveforms in E_X are

$$P[x(\gamma_1, \ldots, \gamma_N, t)] = P[\gamma_1, \ldots, \gamma_N] = p_{\gamma_1 \ldots \gamma_N} \qquad \text{countable case}$$
$$P[x(\gamma_1, \ldots, \gamma_N, t)|_{\gamma_i \in I_i}] = \int_{I_1} \cdots \int_{I_N} f_{\Gamma_1 \ldots \Gamma_N}(\gamma_1, \ldots, \gamma_N) \, d\gamma_1 \ldots d\gamma_N \qquad \text{uncountable case}$$
$$\tag{5.5}$$

where $p_{\gamma_1\ldots\gamma_N}$ and $f_{\Gamma_1\ldots\Gamma_N}$, respectively, are the joint probability of $\gamma_1,\ldots,\gamma_N$ for the countable case and the joint probability density function of $\Gamma_1\ldots\Gamma_N$ for the uncountable case. For the independent case

$$p_{\gamma_1\ldots\gamma_N} = p_{\gamma_1}\cdots p_{\gamma_N} \qquad f_{\Gamma_1\ldots\Gamma_N}(\gamma_1,\ldots,\gamma_N) = f_{\Gamma_1}(\gamma_1)\cdots f_{\Gamma_N}(\gamma_N) \qquad (5.6)$$

Finally, the Fourier transform of a signaling waveform $\phi \in E_\Phi$, evaluated over the interval $(-\infty, \infty)$, is

$$\Phi(\gamma, f) = \int_{-\infty}^{\infty} \phi(\gamma, t) e^{-j2\pi f t}\, dt \qquad \gamma \in S_\Gamma \qquad (5.7)$$

5.2.0.1 Example—Standard Communication Signals
Each outcome of a signaling random process X is an individual signal, and with an appropriate signaling set, is suitable for use in a communication system. For the case of signaling at a constant rate $r = 1/D$, there are two standard information signals defined on the interval $[0, ND]$ according to

$$y(\gamma_1,\ldots,\gamma_N, t) = \sum_{i=1}^{N} A_{\gamma_i}\phi(t-(i-1)D) \qquad \begin{cases} \gamma_i \in S_\Gamma = \{1,\ldots,M\} \\ A_{\gamma_i} \in \{A_1,\ldots,A_M\} \\ P[A_i] = P[\gamma_i] = p_{\gamma_i} \end{cases} \qquad (5.8)$$

$$x(\gamma_1,\ldots,\gamma_N, t) = \sum_{i=1}^{N} \phi(\gamma_i, t-(i-1)D) \qquad \begin{cases} \gamma_i \in S_\Gamma = \{1,\ldots,M\} \\ \phi \in E_\Phi \\ P[\phi(\gamma_i, t)] = P[\gamma_i] = p_{\gamma_i} \end{cases} \qquad (5.9)$$

where $E_\Phi = \{\phi(\gamma_i, t): \gamma_i \in S_\Gamma\}$. The signaling waveforms are of a pulse form for baseband communication and $A_1,\ldots,A_M$ are signaling amplitudes.

The first signal defined above is one where a constant pulse shape is used, and the information is encoded through use of different amplitudes. The second is where different signaling waveforms are used to convey information. Clearly, the second form is more general and includes the first as a subcase. An example of the second signaling form is shown in Figure 5.1, where $E_\Phi = \{\phi(1, t), \phi(2, t)\}$, and

$$\phi(1, t) = \begin{cases} A\cos(2\pi f_c t)\sin(\pi t/D) & t \in [0, D] \\ 0 & \text{elsewhere} \end{cases} \qquad (5.10)$$

$$\phi(2, t) = \begin{cases} A\sin(2\pi f_c t)\sin(\pi t/D) & t \in [0, D] \\ 0 & \text{elsewhere} \end{cases} \qquad (5.11)$$

The plotted signal is $x(1, 2, 2, 1, 1, t)$, for the case where $A = 1$, $D = 1$, and $f_c = 4$ Hz.

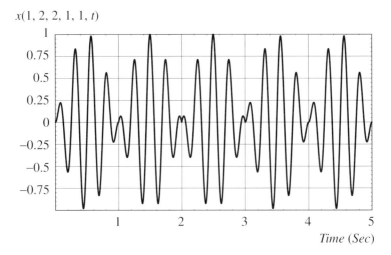

Figure 5.1 Baseband signaling waveform, $A = 1$, $D = 1$, and $f_c = 4$.

Note, if the duration of all signaling waveforms in the signaling set is kD sec, then at any time t after the first k transient signaling intervals, there are potentially k nonzero waveforms comprising the signal x.

5.2.0.2 Generality of Information Signal Form

A broad class of signals can be written in an information signal form. To illustrate this, consider first, the fact that any bandlimited signal x can be written, on the interval $(-\infty, \infty)$, in the information signal form (Gabel, 1987),

$$x(t) = \sum_{i=-\infty}^{\infty} x(iD) \operatorname{sinc}\left(\frac{t - iD}{D}\right) \qquad (5.12)$$

Second, consider that any signal x with bounded variation can be written, on an interval $[0, ND]$, in the information signal form according to

$$x(\gamma_1, \ldots, \gamma_N, t) = \sum_{\gamma_1 = 1}^{N} \phi_{\gamma_i}(t - (i - 1)D) \qquad \phi_{\gamma_i} \in S_\Phi, \gamma_i \in R \qquad (5.13)$$

where S_Φ is the set of signals with bounded variation on the interval $[0, D]$ and which are zero outside this interval.

5.2.1 Power Spectral Density of a Signaling Random Process

The following theorem details the power spectral density of a signaling random process.

THEOREM 5.1. POWER SPECTRAL DENSITY OF A SIGNALING RANDOM PROCESS
Assuming the effect of including components of the signaling waveform outside of the interval $[0, ND]$ *is negligible, the dependency between signaling waveforms depends on the difference between the location of the signaling intervals and not on their absolute location, and the ith signaling waveform is independent of the jth signaling waveform for* $|i - j| > m$, *then the power spectral density of the random process* X, *defined by the ensemble as per Eq.* (5.2), *is*

$$G_X(ND, f) = r\overline{|\Phi(f)|^2} - r|\mu_\Phi(f)|^2 + r|\mu_\Phi(f)|^2 \left[\frac{1}{N} \frac{\sin^2(\pi N f/r)}{\sin^2(\pi f/r)}\right]$$

$$+ 2r \sum_{i=1}^{m} \left[1 - \frac{i}{N}\right] \text{Re}[e^{j2\pi i D f}(R_{\Phi_1 \Phi_{1+i}}(f) - |\mu_\Phi(f)|^2)]$$

(5.14)

$$G_{X_\infty}(f) = r\overline{|\Phi(f)|^2} - r|\mu_\Phi(f)|^2 + r^2|\mu_\Phi(f)|^2 \sum_{n=-\infty}^{\infty} \delta(f - nr)$$

$$+ 2r \sum_{i=1}^{m} \text{Re}[e^{j2\pi i D f}(R_{\Phi_1 \Phi_{1+i}}(f) - |\mu_\Phi(f)|^2)]$$

(5.15)

where $r = 1/D$ and

$$\mu_\Phi(f) = \begin{cases} \sum_{\gamma=1}^{\infty} p_\gamma \Phi(\gamma, f) & \text{countable case} \\ \int_{-\infty}^{\infty} \Phi(\gamma, f) f_\Gamma(\gamma) \, d\gamma & \text{uncountable case} \end{cases}$$

(5.16)

$$\overline{|\Phi(f)|^2} = \begin{cases} \sum_{\gamma=1}^{\infty} p_\gamma |\Phi(\gamma, f)|^2 & \text{countable case} \\ \int_{-\infty}^{\infty} |\Phi(\gamma, f)|^2 f_\Gamma(\gamma) \, d\gamma & \text{uncountable case} \end{cases}$$

(5.17)

$$R_{\Phi_1 \Phi_{1+i}}(f)$$

$$= \begin{cases} \sum_{\gamma_1=1}^{\infty} \sum_{\gamma_{1+i}=1}^{\infty} p_{\gamma_1 \gamma_{1+i}} \Phi(\gamma_1, f) \Phi^*(\gamma_{1+i}, f) & \text{countable case} \\ \int_{-\infty}^{\infty} \int_{-\infty}^{\infty} \Phi(\gamma_1, f) \Phi^*(\gamma_{1+i}, f) f_{\Gamma_1 \Gamma_{1+i}}(\gamma_1, \gamma_{1+i}) \, d\gamma_1 \, d\gamma_{1+i} & \text{uncountable case} \end{cases}$$

(5.18)

For the independent case $R_{\Phi_i \Phi_{1+i}}(f) = |\mu_\Phi(f)|^2$ and

$$G_X(ND, f) = r\overline{|\Phi(f)|^2} - r|\mu_\Phi(f)|^2 + r|\mu_\Phi(f)|^2 \left[\frac{1}{N}\frac{\sin^2(\pi Nf/r)}{\sin^2(\pi f/r)}\right] \quad (5.19)$$

$$G_{X_\infty}(f) = r\overline{|\Phi(f)|^2} - r|\mu_\Phi(f)|^2 + r^2|\mu_\Phi(f)|^2 \sum_{n=-\infty}^{\infty} \delta(f - nr) \quad (5.20)$$

Proof. The proof is given in Appendix 1.

5.2.1.1 Notes
The results stated in the above theorem for the independent, countable, and infinite interval case are consistent with those of van den Elzen (1970).

The given expressions for $G_X(ND, f)$ and $G_{X_\infty}(f)$ can be written in a simpler form with the variance definition

$$\sigma_\Phi^2(f) = \overline{|\Phi(f)|^2} - |\mu_\Phi(f)|^2 \quad (5.21)$$

however, the given forms best facilitate evaluation of the power spectral density.

As the discrete and independent case, where there are M possible signaling waveforms, commonly occurs the following explicit expressions are useful:

$$G_X(ND, f) = r \sum_{\gamma=1}^{M} p_\gamma |\Phi(\gamma, f)|^2 + r \left|\sum_{\gamma=1}^{M} p_\gamma \Phi(\gamma, f)\right|^2 \left[\frac{1}{N}\frac{\sin^2(\pi Nf/r)}{\sin^2(\pi f/r)} - 1\right] \quad (5.22)$$

$$G_{X_\infty}(f) = r \sum_{\gamma=1}^{M} p_\gamma |\Phi(\gamma, f)|^2 - r \left|\sum_{\gamma=1}^{M} p_\gamma \Phi(\gamma, f)\right|^2$$

$$+ r^2 \left|\sum_{\gamma=1}^{M} p_\gamma \Phi(\gamma, f)\right|^2 \sum_{n=-\infty}^{\infty} \delta(f - nr) \quad (5.23)$$

The equations specified in Theorem 5.1 can be considerably simplified if the mean of the Fourier transform of the signaling waveforms μ_Φ, is zero. A sufficient condition for this is for the mean of the signaling waveforms to be zero, that is, $\mu_\phi(t) = 0$ for $t \in (-\infty, \infty)$, where μ_ϕ is defined, respectively, for the countable and uncountable cases according to

$$\mu_\phi(t) = \sum_{\gamma=1}^{\infty} p_\gamma \phi(\gamma, t) \qquad \mu_\phi(t) = \int_{-\infty}^{\infty} \phi(\gamma, t) f_\Gamma(\gamma) \, d\gamma \quad (5.24)$$

When $\mu_\phi(t) = 0$ for $t \in (-\infty, \infty)$, it follows for the countable case that

$$\mu_\Phi(f) = \sum_{\gamma=1}^{\infty} p_\gamma \Phi(\gamma, f) = \sum_{\gamma=1}^{\infty} p_\gamma \int_{-\infty}^{\infty} \phi(\gamma, t) e^{-j2\pi f t} \, dt$$

$$= \int_{-\infty}^{\infty} \left[\sum_{\gamma=1}^{\infty} p_\gamma \phi(\gamma, t)\right] e^{-j2\pi f t} \, dt = 0 \quad (5.25)$$

The interchange of summation and integration in this equation is valid, according to the dominated convergence theorem, when there exists a function $g \in L$, such that $\sum_{\gamma=1}^{M} p_\gamma \phi(\gamma, t) < g(t)$ for all values of $M \in Z^+$. A typical case is where $\sup\{|\phi(\gamma, t)|: \gamma \in Z^+\}$ is bounded and integrable on the infinite interval, and for this case the interchange is valid. A similar argument can be used for the uncountable case.

5.2.1.2 Case 1: Mean of Signaling Waveforms is Zero

For the case where the mean of the signaling waveforms is such that $\mu_\Phi(f) = 0$ for $f \in R$, the results given in Theorem 5.1, for the power spectral density of a signaling random process, simplify to

$$G_X(ND, f) = r\overline{|\Phi(f)|^2} + 2r \sum_{i=1}^{m} \left[1 - \frac{i}{N}\right] \text{Re}[e^{j2\pi i Df} R_{\Phi_1 \Phi_{1+i}}(f)]$$

$$G_{X_x}(f) = r\overline{|\Phi(f)|^2} + 2r \sum_{i=1}^{m} \text{Re}[e^{j2\pi i Df} R_{\Phi_1 \Phi_{1+i}}(f)]$$

(5.26)

When $\mu_\Phi = 0$ and the signaling waveforms in different signaling intervals are independent, the simple result

$$G_X(ND, f) = G_{X_x}(f) = r\overline{|\Phi(f)|^2} = G_\Phi(D, f) \quad (5.27)$$

holds, where G_Φ is the power spectral density of the random process defined by the ensemble E_Φ, as per Eq. (5.3), that is,

$$G_\Phi(D, f) = \begin{cases} \dfrac{1}{D} \sum_{\gamma=1}^{\infty} p_\gamma |\Phi(\gamma, f)|^2 & \text{countable case} \\ \dfrac{1}{D} \int_{-\infty}^{\infty} |\Phi(\gamma, f)|^2 f_\Gamma(\gamma) \, d\gamma & \text{uncountable case} \end{cases} \quad (5.28)$$

Consistent with Eq. (5.7), the contribution of the signaling waveform components outside of the interval $[0, D]$ are included in this power spectral density definition.

5.2.1.3 Case 2: Information Encoded in Pulse Amplitudes

Consider the case where information is encoded in the pulse amplitudes, such that $\phi(\gamma, t) = A(\gamma)\phi(t)$, and

$$E_\Phi = \left\{ A(\gamma)\phi(t): \ A(\gamma) = \begin{cases} A_\gamma & \gamma \in Z^+ \quad \text{countable case} \\ A(\gamma) & \gamma \in R \quad \text{uncountable case} \end{cases} \right\} \quad (5.29)$$

where

$$P[A_\gamma] = P[\gamma] = p_\gamma \qquad P[A(\gamma)|_{\gamma \in [\gamma_o, \gamma_o + d\gamma]}] = f_\Gamma(\gamma_o) \, d\gamma \quad (5.30)$$

SIGNALING RANDOM PROCESSES   145

It then follows that

$$\mu_\Phi(f) = \begin{cases} \Phi(f) \sum_{\gamma=1}^{\infty} p_\gamma A_\gamma = \mu_A \Phi(f) & \text{countable case} \\ \Phi(f) \int_{-\infty}^{\infty} A(\gamma) f_\Gamma(\gamma)\, d\gamma = \mu_A \Phi(f) & \text{uncountable case} \end{cases} \quad (5.31)$$

$$\overline{|\Phi(f)|^2} = \begin{cases} |\Phi(f)|^2 \sum_{\gamma=1}^{\infty} p_\gamma |A_\gamma|^2 = |\Phi(f)|^2 \overline{|A|^2} & \text{countable case} \\ |\Phi(f)|^2 \int_{-\infty}^{\infty} |A(\gamma)|^2 f_\Gamma(\gamma)\, d\gamma = |\Phi(f)|^2 \overline{|A|^2} & \text{uncountable case} \end{cases}$$

(5.32)

where, μ_A and $\overline{|A|^2}$, respectively, are the mean and mean square value of the signaling amplitudes. Further,

$$R_{\Phi_1 \Phi_{1+i}}(f) =$$

$$\begin{cases} |\Phi(f)|^2 \sum_{\gamma_1=1}^{\infty} \sum_{\gamma_{1+i}=1}^{\infty} p_{\gamma_1 \gamma_{1+i}} A_{\gamma_1} A^*_{\gamma_{1+i}} & \text{countable case} \\ |\Phi(f)|^2 \int_{-\infty}^{\infty} \int_{-\infty}^{\infty} A(\gamma_1) A^*(\gamma_{1+i}) f_{\Gamma_1 \Gamma_{1+i}}(\gamma_1, \gamma_{1+i})\, d\gamma_1\, d\gamma_{1+i} & \text{uncountable case} \end{cases}$$

$$= |\Phi(f)|^2 R_{A_1 A_{1+i}} \quad (5.33)$$

where, $p_{\gamma_1 \gamma_{1+i}}$ is the joint probability of γ_1 in the first signaling interval and γ_{1+i} in the $1+i$th signaling interval, or equivalently, the joint probability of the amplitude A_{γ_1} in the first signaling interval, and the amplitude $A_{\gamma_{1+i}}$ in the $1+i$th signaling interval. Similarly, $f_{\Gamma_1 \Gamma_{1+i}}$ is the joint probability density function for amplitudes in the first and $1+i$th signaling intervals. The definition for $R_{A_1 A_{1+i}}$ is obvious from this equation. With these definitions, it follows that

$$G_X(ND, f) = r|\Phi(f)|^2 \left\{ \overline{|A|^2} - |\mu_A|^2 + |\mu_A|^2 \left[\frac{1}{N} \frac{\sin^2(\pi Nf/r)}{\sin^2(\pi f/r)} \right] \right.$$

$$\left. + 2 \sum_{i=1}^{m} \left[1 - \frac{i}{N} \right] \text{Re}[e^{j2\pi i Df}(R_{A_1 A_{1+i}} - |\mu_A|^2)] \right\} \quad (5.34)$$

$$G_{X_\infty}(f) = r|\Phi(f)|^2 \left\{ \overline{|A|^2} - |\mu_A|^2 + r|\mu_A|^2 \sum_{n=-\infty}^{\infty} \delta(f - nr) \right.$$

$$\left. + 2 \sum_{i=1}^{m} \text{Re}[e^{j2\pi i Df}(R_{A_1 A_{1+i}} - |\mu_A|^2)] \right\} \quad (5.35)$$

The variance definition

$$\sigma_A^2 = \overline{|A|^2} - |\mu_A|^2 \qquad (5.36)$$

can simplify the form of these equations. Carlson (1986 pp. 388–389) gives equivalent results.

For the independent case, where $R_{A_1 A_{1+i}} = |\mu_A|^2$, it follows that

$$G_X(ND, f) = r|\Phi(f)|^2 \left[\overline{|A|^2} - |\mu_A|^2 + |\mu_A|^2 \frac{1}{N} \frac{\sin^2(\pi N f/r)}{\sin^2(\pi f/r)} \right] \qquad (5.37)$$

$$G_{X_\infty}(f) = r|\Phi(f)|^2 \left[\overline{|A|^2} - |\mu_A|^2 + r|\mu_A|^2 \sum_{n=-\infty}^{\infty} \delta(f - nr) \right] \qquad (5.38)$$

For the independent case, where the mean amplitude μ_A is zero, the simpler result follows:

$$G_X(ND, f) = G_{X_\infty}(f) = r\overline{|A|^2} |\Phi(f)|^2 = r\sigma_A^2 |\Phi(f)|^2 \qquad (5.39)$$

5.2.2 Examples and Spectral Issues for Communication Systems

The above theory has direct application to communication of information via signaling waveforms, as the power spectral density contains the following information. First, whether there are signal components in the transmitted signal which do not convey information. Such components are periodic, show up as impulses in the power spectral density, and indicate inefficient signaling. Second, how spectrally efficient the signaling scheme is in terms of the level of information transmitted in the frequency band containing the majority of signal energy. The usual measure here is the number of bits of information per Hz of bandwidth. A greater degree of spectral efficiency allows a greater number of signal or information channels in a specified frequency band. Third, the degree of spectral rolloff associated with the residual signal energy outside of the band used to measure spectral efficiency. The degree of spectral rolloff is a measure of the spectral spread and such spread impairs the ability of a receiver associated with an adjacent signal channel to recover a signal in that adjacent channel.

The following examples give some insight into these issues, although they primarily illustrate the evaluation of the power spectral density of a signaling random process.

5.2.2.1 Example: Power Spectral Density of a Return to Zero Signal

Consider the case of a signaling random process, defined for the interval

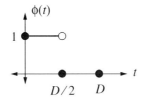

Figure 5.2 Pulse waveform.

$[0, ND]$ by the ensemble

$$E_Y = \left\{ y(\gamma_1, \ldots, \gamma_N, t) = \sum_{i=1}^{N} A_{\gamma_i} \phi(t - (i-1)D), \ \gamma_i \in \{1, 2\}, A_1 = 0, A_2 = A \right\}$$
(5.40)

and the pulse waveform ϕ has the form shown in Figure 5.2. Signaling with such a waveform is called "return to zero" (RZ) signaling. The Fourier transform of ϕ is

$$\Phi(f) = \frac{1}{2r} \mathrm{sinc}\left(\frac{f}{2r}\right) e^{-j\pi f/2r} \qquad r = 1/D$$
(5.41)

Assuming independent and equally probable amplitudes, such that $\mu_A = A/2$, $\overline{A^2} = A^2/2$, $\sigma_A^2 = A^2/4$, and $R_{A_1 A_{1+i}} = |\mu_A|^2$ for $i \geq 1$, it then follows from Eqs. (5.37) and (5.38), that the power spectral density is given by

$$G_Y(ND, f) = \frac{A^2}{16r} \mathrm{sinc}^2\left(\frac{f}{2r}\right) \left[1 + \frac{1}{N} \frac{\sin^2(\pi N f/r)}{\sin^2(\pi f/r)} \right]$$
(5.42)

$$G_{Y_\infty}(f) = \frac{A^2}{16r} \mathrm{sinc}^2\left(\frac{f}{2r}\right) + \frac{A^2}{16} \mathrm{sinc}^2\left(\frac{f}{2r}\right) \sum_{n=-\infty}^{\infty} \delta(f - nr)$$
(5.43)

The power spectral density is plotted in Figure 5.3 for the case of $N = 256$, $D = 1$, $r = 1$, and $A = 1$. Clearly evident in this figure is the continuous sinc squared form and the discrete "impulsive" components. For the case where $A = 1$, the power in the impulsive components at frequencies $0, r, 2r, 3r, \ldots$ is $1/16, 0, 1/4\pi^2, 0, 1/36\pi^2, \ldots$ In Figure 5.4, the power spectral density for the infinite interval is plotted using logarithmic scaling.

The following can be inferred from these power spectral density graphs. First, the impulses in the spectrum are wasted power as far as communication of information is concerned, and thus, RZ signaling is inefficient signaling. The impulsive components, however, may facilitate synchronization and data recovery at the receiver. Second, the signaling pulse is relatively narrow, which

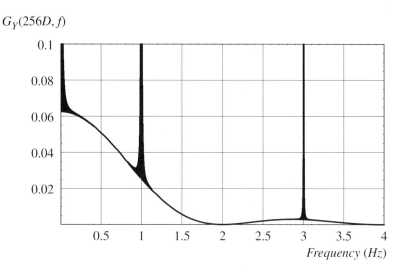

Figure 5.3 Power spectral density of a RZ signal when $D = 1$, $r = 1$, $A = 1$, and $N = 256$.

implies a relatively broad spectrum, which, in turn, implies relatively poor spectral efficiency. The main lobe of the power spectral density is from $-2r$ to $2r$ Hz, that is, a bandwidth of $2r$ Hz. This implies a spectral efficiency of 0.5 bit/Hz, which is low when compared with, for example, signaling with raised cosine pulses as shown in the next example. Third, the envelope of the power spectral density rolls off at a $1/f^2$ rate which is generally inadequate for most communication systems.

5.2.2.2 Example: Power Spectral Density of a Bipolar Signal
Consider the case of a signaling random process, characterized on the interval $[0, ND]$ by the ensemble

$$E_X = \left\{ x(\gamma_1, \ldots, \gamma_N, t) = \sum_{i=1}^{N} \phi(\gamma_i, t - (i-1)D),\ \phi \in E_\Phi, \gamma_i \in \{-1, 0, 1\} \right\} \quad (5.44)$$

where

$$E_\Phi = \{\phi(\gamma_i, t): \phi(\gamma_i, t) = \gamma_i p(\beta, t - D/2),\ \gamma_i \in \{-1, 0, 1\}\} \quad (5.45)$$

$$p(\beta, t) = A \operatorname{sinc}\left(\frac{t}{D}\right) \frac{\cos(\pi \beta t / D)}{1 - (2\beta t / D)^2} \quad (5.46)$$

Here, $p(\beta, t)$ is the inverse Fourier transform of a raised cosine spectrum, which

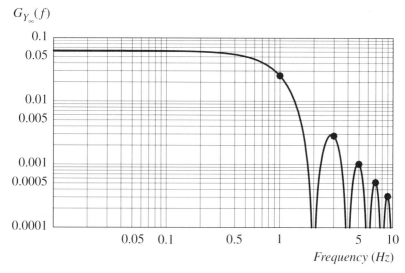

Figure 5.4 Power spectral density of a RZ signal evaluated on the infinite interval when $D = 1$, $r = 1$, and $A = 1$. The dots represent the power in impulsive components. The power in the impulse at 0 Hz is 0.0625.

is defined according to (Carlson, 1986 p. 406; Proakis, 1995 p. 546)

$$P(\beta, f) = \begin{cases} \dfrac{A}{r} & |f| < \dfrac{r}{2}(1 - \beta) \\ \dfrac{A}{r} \cos^2\left[\dfrac{\pi}{2} \dfrac{|f| - r(1-\beta)/2}{\beta r}\right] & \dfrac{r}{2}(1-\beta) \leqslant |f| < \dfrac{r}{2}(1+\beta) \\ 0 & |f| \geqslant \dfrac{r}{2}(1+\beta) \end{cases} \quad (5.47)$$

where $r = 1/D$. P is shown in Figure 5.5 for the cases of $\beta = 0.5$ and $\beta = 1.0$. The parameter β is the rolloff factor and is such that $0 < \beta \leqslant 1$. The graph of $p(\beta, t)$ is shown in Figure 5.6 for the cases of $\beta = 0.5$ and $\beta = 1.0$.

A bipolar signal is generated when binary data is encoded with no waveform corresponding to a logic 0, and with logic 1 being encoded alternatively with p and $-p$ as illustrated in Figure 5.7. This encoding leads to correlation between adjacent signaling waveforms, but ensures the signaling set has zero mean for the case of equally probable data.

Consistent with this encoding, it follows from Eq. (5.26) that the power spectral density of the signaling random process is

$$G_X(ND, f) = r\overline{|\Phi(f)|^2} + 2r\left[1 - \dfrac{1}{N}\right] \text{Re}[e^{j2\pi Df} R_{\Phi_1 \Phi_2}(f)] \quad (5.48)$$

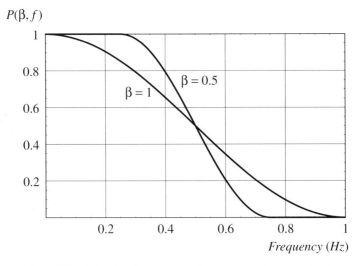

Figure 5.5 Raised cosine spectrum for the case where $r = A = 1$.

where

$$\overline{|\Phi(f)|^2} = \sum_{\gamma=-1}^{1} p_\gamma |\Phi(\gamma, f)|^2$$

$$R_{\Phi_1 \Phi_2}(f) = \sum_{\gamma_1=-1}^{1} \sum_{\gamma_2=-1}^{1} p_{\gamma_1 \gamma_2} \Phi(\gamma_1, f) \Phi^*(\gamma_2, f)$$

(5.49)

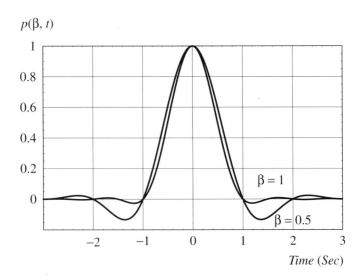

Figure 5.6 Inverse Fourier transform of raised cosine spectrum for the case where $D = A = 1$.

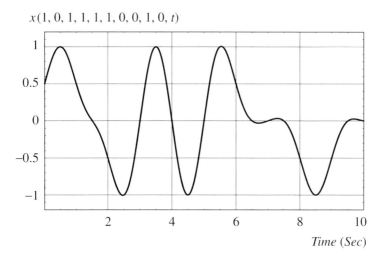

Figure 5.7 Bipolar signaling waveform where pulses associated with a raised cosine spectrum have been used. $\beta = 1$, $D = 1$, $A = 1$, and data is $\{1, 0, 1, 1, 1, 1, 0, 0, 1, 0\}$.

In these equations, $p_{-1} = p_1 = 0.25$ and $p_0 = 0.5$. By considering possible data and the corresponding signaling waveforms in two consecutive signaling intervals, it follows that the probabilities of two consecutive signaling waveforms, that is $P[\phi(\gamma_1, t), \phi(\gamma_2, t)] = p_{\gamma_1 \gamma_2}$, are as tabulated in Table 5.1. Using the results in this table, it follows that

$$\overline{|\Phi(f)|^2} = 0.5|P(\beta, f)|^2 \quad \text{and} \quad R_{\Phi_1 \Phi_2}(f) = -0.25|P(\beta, f)|^2$$

and hence,

$$G_X(ND, f) = 0.5r|P(\beta, f)|^2 \left\{ 1 - \left[1 - \frac{1}{N}\right]\cos(2\pi f/r) \right\} \tag{5.50}$$

$$G_{X_\infty}(f) = r\sin^2(\pi f/r)|P(\beta, f)|^2 \tag{5.51}$$

where the relationship $\sin^2(A) = 0.5 - 0.5\cos(2A)$ has been used. The power

TABLE 5.1 Possible Outcomes for Two Consecutive Signaling Intervals

Data	Signaling Waveforms	Probabilities
00	0, 0	$p_{00} = 0.25$
01	0, $\phi(-1, t)$ or 0, $\phi(1, t)$	$p_{0,-1} = p_{01} = 0.125$
10	$\phi(-1, t)$, 0 or $\phi(1, t)$, 0	$p_{-1,0} = p_{10} = 0.125$
11	$\phi(-1, t), \phi(1, t)$ or $\phi(1, t), \phi(-1, t)$	$p_{-1,1} = p_{1,-1} = 0.125$

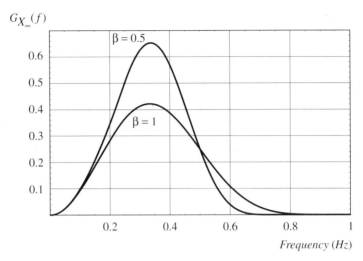

Figure 5.8 Power spectral density of a signaling random process using pulses associated with a raised cosine spectrum, bipolar coding and with $D = r = A = 1$.

spectral density is plotted in Figure 5.8 for the case of $N \to \infty$ when $\beta = 0.5$ and $\beta = 1.0$.

In comparison with RZ signaling, the following can be noted. First, the bipolar coding ensures that the signaling random process has a zero mean, and therefore, there are no impulses and no redundant signal components in the power spectral density. Second, the signaling is more spectrally efficient. For example, with $\beta = 1$, the spectrum is bandlimited to r Hz which implies 1 bit/Hz (twice as efficient as RZ signaling). Third, on the infinite interval with no truncation of the signaling pulses defined in Eq. (5.46), the spectral rolloff is infinite (there is no spectral spread). In practice, the signaling pulses are truncated and this results in spectral spread which can be readily determined. Finally, the encoding ensures that the power spectral density is zero at zero frequency, ensuring that a bipolar signal can be passed by a linear system whose transfer function has zero response at dc.

5.3 DIGITAL TO ANALOGUE CONVERTER QUANTIZATION

Increasingly, information signals are generated via a digital processor that generates very accurate sample values, and these are put to a M bit digital to analogue converter (DAC), at a constant rate of $r = 1/D$ samples/sec. To ascertain the power spectral density of the generated signal, consider a M bit DAC with 2^M equally spaced levels between and including $\pm A$. The difference between DAC levels is denoted Δ, where $\Delta = 2A/(2^M - 1)$. Associated with the ith sample value x_i, is a quantization error ε_i, as illustrated in Figure 5.9, such

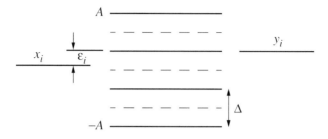

Figure 5.9 Illustration of quantization error with a 2-bit DAC (4 levels).

that in the ith sample interval $[(i-1)D, iD]$, the constant level $y_i = x_i + \varepsilon_i$ is generated. The model is one of an additive error to an ideal signal. In general, the actual levels in a DAC will vary from device to device because of manufacturing tolerances and will vary with device age, etc. Accordingly, it is appropriate to consider an infinite ensemble of DACs, where each is driven by the same sample values, such that in the ith sample interval ε_i is independent of x_i when considered across the ensemble. From the nature of quantization, it follows that ε_i takes on values with a uniform distribution, from the interval $[-\Delta/2, \Delta/2]$. A further assumption is that the DAC resolution and rate of signal change are such that the quantization errors from one sample interval to the next are uncorrelated. With such assumptions, the ensemble of DAC output signals for the interval $[0, ND]$, which define a random process Y, is

$$E_Y = \begin{cases} y(\varepsilon_1, \ldots, \varepsilon_N, t) = \sum_{i=1}^{N} (x_i + \varepsilon_i)\phi(t - (i-1)D) \\ \qquad = x_S(t) + \sum_{i=1}^{N} \varepsilon_i \phi(t - (i-1)D) \end{cases} \quad \varepsilon_i \in \left[-\frac{\Delta}{2}, \frac{\Delta}{2}\right] \quad (5.52)$$

where ϕ is a pulse function defined according to

$$\phi(t) = \begin{cases} 1 & 0 \leq t < D \\ 0 & \text{elsewhere} \end{cases} \qquad |\Phi(f)| = \frac{1}{r}\left|\operatorname{sinc}\left(\frac{f}{r}\right)\right| \quad (5.53)$$

and $x_S(t)$ is a step approximation to the desired signal x, that is,

$$x_S(t) = \sum_{i=1}^{N} x_i \phi(t - (i-1)D) \quad (5.54)$$

The probabilities associated with the quantization error are such that

$$P[\varepsilon_i \in [\varepsilon_o, \varepsilon_o + d\varepsilon]] = f_\varepsilon(\varepsilon_o)\, d\varepsilon$$

with $f_\varepsilon(\varepsilon) = 1/\Delta$ for $-\Delta/2 \leqslant \varepsilon < \Delta/2$ and $f_\varepsilon(\varepsilon) = 0$ elsewhere. Clearly, the mean of ε_i is zero and the variance of ε_i is given by $\sigma_{\varepsilon_i}^2 = \Delta^2/12$ (Papoulis, 2002 p. 165).

The random process Y can be considered to be the summation of a degenerate random process, defined by the signal x_S and a random process E associated with the quantization error, and defined by the ensemble

$$E_E = \left\{ e(\varepsilon_1, \ldots, \varepsilon_N, t) = \sum_{i=1}^{N} \varepsilon_i \phi(t - (i-1)D) \right\} \quad (5.55)$$

As E has zero mean, it follows from Theorem 4.6 that the power spectral densities of x_S and E add, that is,

$$G_Y(T, f) = G_{X_S}(T, f) + G_E(T, f) \quad (5.56)$$

From Eq. (5.39) it follows that the power spectral density of E is

$$G_E(ND, f) = \frac{r\Delta^2 |\Phi(f)|^2}{12} = \frac{\Delta^2 \operatorname{sinc}^2(f/r)}{12r} = \frac{(2A)^2 \operatorname{sinc}^2(f/r)}{12(2^M - 1)^2 r} \quad (5.57)$$

A normalized power spectral density, with normalization in respect of the DAC range of $2A$ and the output rate r, can be defined as

$$G_n(ND, f) = \frac{r G_E(ND, rf)}{(2A)^2} = \frac{\operatorname{sinc}^2(f)}{12(2^M - 1)^2} \quad (5.58)$$

This power spectral density is plotted in Figure 5.10 for DACs with 8, 10, 12, 14, and 16 bits.

5.3.1 Notes

First, for a fixed DAC range, the power spectral density due to quantization is inversely proportional to the number of levels and inversely proportional to the output rate. Second, in the frequency range of $-r/2 \leqslant f < r/2$, where the spectrum of a generated signal is located, it is common to approximate the power spectral density by the constant level of

$$G_E(ND, 0) = \frac{(2A)^2}{12(2^M - 1)^2 r} \quad (5.59)$$

As a measure of the signal to noise ratio performance that is achievable with a M bit DAC, consider the case where a sinusoid with amplitude of A volts is being generated, and the DAC output is filtered such that the effective quantization noise power is consistent with the level given by Eq. (5.59) in an

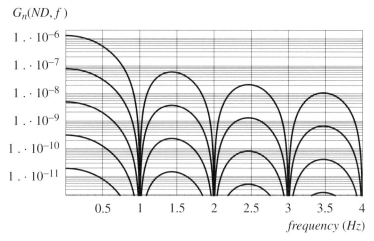

Figure 5.10 *Normalized power spectral density of a DAC due to quantization. Upper to lower curves, respectively, are for 8-, 10-, 12-, 14-, and 16-bit DACs.*

ideal bandwidth of $r/2$ Hz. The signal to noise ratio achievable is (Franco, 2002 p. 565)

$$\text{SNR}(M) = (A/\sqrt{2})^2 \left[\frac{1}{2(r/2)} \frac{12(2^M - 1)^2 r}{(2A)^2} \right] = 1.5(2^M - 1)^2 \approx 1.5(2^{2M}) \quad (5.60)$$

$$\text{SNR}_{\text{dB}}(M) = 10 \log[\text{SNR}(M)] \approx 1.76 + 6.02M \quad \text{dB} \quad (5.61)$$

For 8-, 10-, 12-, 14-, and 16-bit DACs, the respective achievable signal to noise ratios are 50 dB, 62 dB, 74 dB, 86 dB, and 98 dB.

One example of signal generation of a bandpass communications signal, with a digital signal processor and DAC, is discussed in Rensen (1999).

5.4 JITTER

The additive noise on a signal will introduce variations in the time instants a signal crosses a set threshold level. Consequently, a signal generated on the basis of the time an input signal crosses a set threshold will exhibit variations, denoted jitter, from the ideal zero noise case. Jitter arises in many practical applications, including synchronization of signals, hard limiting of signals, and digital circuitry. The archetypical jitter case is that of a periodic pulse train, which is corrupted by noise prior to the input of a comparator, as shown in Figure 5.11. The noise will alter both the start and finish times of a given output pulse shape, as shown in Figure 5.12.

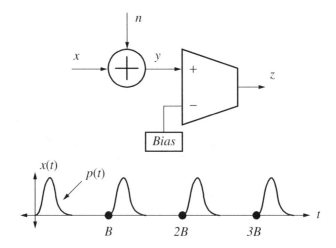

Figure 5.11 Schematic diagram of periodic signal, that is corrupted by noise and input into a comparator.

The signals x and z defined in these figures can be modeled on the interval $[0, NB]$, according to

$$x(t) = \sum_{i=1}^{N} p(t - (i-1)B) \tag{5.62}$$

$$z(\gamma_1, \ldots, \gamma_N, t) = \sum_{i=1}^{N} \phi(\gamma_i, t - (i-1)B) \tag{5.63}$$

where p is the pulse waveform defining the input pulse train, and z is one waveform from a random process Z, defined by the ensemble E_Z,

$$E_Z = \left\{ z(\gamma_1, \ldots, \gamma_N, t) = \sum_{i=1}^{N} \phi(\gamma_i, t - (i-1)B),\ \gamma_i = (a_i, d_i, w_i) \in S_\Gamma,\ \phi \in E_\Phi \right\} \tag{5.64}$$

Here, $S_\Gamma = S_A \times S_D \times S_W$, where S_A, S_D, and S_W are sample spaces of random variables A, D, and W with respective outcomes a, d, and w, and respective probability density functions f_A, f_D, and f_W. The set of signaling waveforms, E_Φ is defined according to (assuming $A_o \neq 0$),

$$E_\Phi = \left\{ \phi(\gamma, t) = A_o(1+a)r\left(\frac{t - \mu_D - d}{\mu_W + w}\right),\ \gamma = (a, d, w),\ a \in S_A, d \in S_D, w \in S_W \right\} \tag{5.65}$$

Here, A_o is the amplitude of the comparator output pulse for the ideal zero noise case, r is a normalized pulse waveform defined in Figure 5.13, a accounts for variations in the comparator output amplitude, d accounts for the advance/

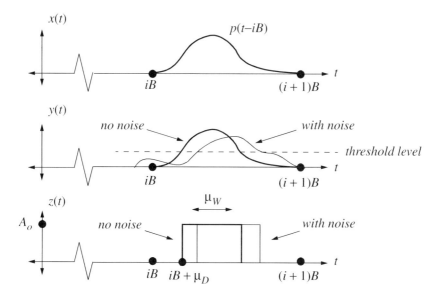

Figure 5.12 Illustration of how noise alters the ith comparator output pulse shape.

delay of the commencement of the output pulse, w accounts for the variation in the width of the output pulse, and μ_D and μ_W, respectively are the mean delay and mean width of the output pulse. By definition, the random variables A, D, and W have zero mean.

To facilitate analysis, the assumption is made that the noise is uncorrelated over a time interval consistent with the duration of the comparator output pulse. The implication of this assumption is that the delay d_i, of the ith comparator output pulse is independent of the width, as specified by w_i. Further, the delay and width of the ith comparator output pulse are assumed to be independent of the delay and width of any other output pulse, and the pulse amplitude is assumed to be independent of the pulse delay and width. With such assumptions, it follows that

$$P[\phi(\gamma, t)|_{a \in I_A, d \in I_D, w \in I_W}] = \int_{I_A} f_A(a)\, da \int_{I_D} f_D(\delta)\, d\delta \int_{I_W} f_W(w)\, dw \quad (5.66)$$

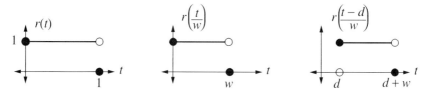

Figure 5.13 Definition of the rectangle function r.

consistent with the probability density function of the random variable Γ, whose outcomes are denoted γ, being such that

$$f_\Gamma(\gamma) = f_A(a) f_D(d) f_W(w) \qquad \gamma = (a, d, w) \tag{5.67}$$

Clearly, Z is a signaling random process. As detailed in Appendix 2, previously derived results for the power spectral density of such a random process can be used to derive the power spectral density of Z. The result is given in the following theorem.

THEOREM 5.2. POWER SPECTRAL DENSITY — JITTERED SIGNAL   *The power spectral density of the random process Z characterizing jitter and modeled by the ensemble and associated signaling set, as per Eqs. (5.64) and (5.65), is*

$$G_Z(NB, f) = rA_o^2(1 + \bar{A}^2) \int_{-\infty}^{\infty} (\mu_W + w)^2 |R[(\mu_W + w)f]|^2 f_W(w) \, dw$$

$$+ rA_o^2 |F_D(f)|^2 \left| \int_{-\infty}^{\infty} (\mu_W + w) R[(\mu_W + w)f] f_W(w) \, dw \right|^2$$

$$\times \left[\frac{1}{N} \frac{\sin^2(\pi Nf/r)}{\sin^2(\pi f/r)} - 1 \right] \tag{5.68}$$

$$G_{Z_\infty}(f) = rA_o^2(1 + \bar{A}^2) \int_{-\infty}^{\infty} (\mu_W + w)^2 |R[(\mu_w + w)f]|^2 f_W(w) \, dw$$

$$+ rA_o^2 |F_D(f)|^2 \left| \int_{-\infty}^{\infty} (\mu_W + w) R[(\mu_W + w)f] f_W(w) \, dw \right|^2$$

$$\times \left[r \sum_{i=-\infty}^{\infty} \delta(f - ir) - 1 \right] \tag{5.69}$$

where $r = 1/B$, F_D is the Fourier transform of f_D, and R is the Fourier transform of r, that is,

$$F_D(f) = \int_{-\infty}^{\infty} f_D(\delta) e^{-j2\pi f\delta} \, d\delta \qquad R(f) = \int_{-\infty}^{\infty} r(t) e^{-j2\pi ft} \, dt = \text{sinc}(f) e^{-j\pi f}$$

$$\tag{5.70}$$

and

$$\bar{A}^2 = \int_{-\infty}^{\infty} a^2 f_A(a) \, da$$

Proof. The proof of this theorem is given in Appendix 2.

5.4.1 Case 1—Constant Amplitude

The common case is where the amplitude is constant, that is, $\bar{A}^2 = 0$. For this case, the expressions for the power spectral density given in Eqs. (5.68) and (5.69), readily simplify.

5.4.2 Case 2—Zero Mean Amplitude

For the special case where the mean of the amplitude is zero, that is, $A_o = 0$, the signaling set is

$$E_\Phi = \left\{ \phi(\gamma, t) = ar\left(\frac{t - \mu_D - d}{\mu_W + w}\right), \gamma = (a, d, w), a \in S_A, d \in S_D, w \in S_W \right\} \quad (5.71)$$

For this case, the power spectral density takes on the simpler form,

$$G_Z(NB, f) = G_{Z_x}(f) = r\bar{A}^2 \int_{-\infty}^{\infty} (\mu_W + w)^2 |R[(\mu_W + w)f]|^2 f_W(w)\, dw \quad (5.72)$$

5.4.3 Example—Jitter of a Pulse Train with Gaussian Variations

Consider the case where a comparator is driven by a periodic pulse train with period $B = 1/r$ and a pulse width μ_W. Further, assume the pulse train is corrupted by noise such that the delay and width density functions, f_D and f_W, are Gaussian with variances σ_D^2 and σ_W^2, that is, $f_\Gamma(\gamma) = e^{-\gamma^2/2\sigma_\Gamma^2}/\sqrt{2\pi}\,\sigma_\Gamma$ for $\Gamma \in \{D, W\}$. The comparator output pulses are assumed to be of constant height A_o, and have a mean width μ_W. As shown in Appendix 3, the power spectral density of the comparator output random process is

$$G_Z(NB, f) = \frac{rA_o^2}{2\pi^2 f^2}[1 - \cos(2\pi\mu_W f)e^{-2\pi^2 f^2 \sigma_W^2}]$$

$$+ \frac{rA_o^2 e^{-4\pi^2 \sigma_D^2 f^2}}{4\pi^2 f^2}[1 + e^{-4\pi^2 f^2 \sigma_W^2} - 2\cos(2\pi\mu_W f)e^{-2\pi^2 f^2 \sigma_W^2}]$$

$$\times \left[\frac{1}{N}\frac{\sin^2(\pi Nf/r)}{\sin^2(\pi f/r)} - 1\right] \quad (5.73)$$

$$G_{Z_x}(f) = \frac{rA_o^2}{2\pi^2 f^2}[1 - \cos(2\pi\mu_W f)e^{-2\pi^2 f^2 \sigma_W^2}]$$

$$+ \frac{rA_o^2 e^{-4\pi^2 \sigma_D^2 f^2}}{4\pi^2 f^2}[1 + e^{-4\pi^2 f^2 \sigma_W^2} - 2\cos(2\pi\mu_W f)e^{-2\pi^2 f^2 \sigma_W^2}]$$

$$\times \left[r \sum_{i=-\infty}^{\infty} \delta(f - ir) - 1\right] \quad (5.74)$$

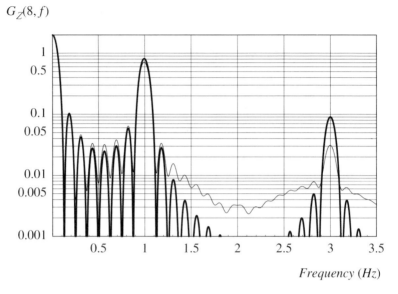

Figure 5.14 Effect of jitter on the power spectral density of a pulse train, evaluated on [0, 8], for the case where $B = r = A_o = 1$. The thicker line is for the zero jitter case.

The power spectral density, $G_Z(8, f)$, for the case where the mean comparator output pulse width is $\mu_W = B/2$, is plotted in Figure 5.14 for the ideal case, and for the case of $\sigma_W^2 = \sigma_D^2 = (0.05B)^2$. Clearly, the effect of jitter is to lead to spectral spread and to reduce the peak height of the harmonic components.

Note that the zero jitter case yields a periodic pulse train, as shown in Figure 4.7, where $W = T_p/2$ and $T_p = 1$. Accordingly, the power spectral density shown for the zero jitter case is the same as that shown in Figure 4.9 for a periodic pulse train.

5.5 SHOT NOISE

Shot noise occurs in many physical processes, including current flow in active electronic devices. Accordingly, a derivation of the power spectral density of such a process is important. To this end, consider an interval $[0, T]$ that is quantized into M intervals of duration Δt sec. Such quantization defines the set of times: $\{0, \Delta t, 2\Delta t, \ldots, (M-1)\Delta t\}$. Clearly, $M\Delta t = T$.

Next, consider the following experiment:

1. A time on the interval $[0, T]$ is chosen at random, with a uniform probability density function, and quantized to yield a number from the set $\{0, \Delta t, 2\Delta t, \ldots, (M-1)\Delta t\}$. It then follows that $P[i\Delta t] = 1/M$.
2. Step 1 is repeated N times.

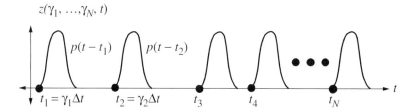

Figure 5.15 Illustration of a signal from a shot noise process.

A shot noise process is one where a "pulse" waveform is associated with each of the N times defined by the above experiment, and an example of a shot noise waveform is shown in Figure 5.15.

Consistent with the above description, a shot noise process Z is defined by the ensemble

$$E_Z = \left\{ z(\gamma_1, \ldots, \gamma_N, t) = \sum_{i=1}^{N} p(t - \gamma_i \Delta t), \gamma_i \in \{0, 1, \ldots, M-1\}, P[\gamma_i] = 1/M \right\}$$
(5.75)

where $p \in L$. The power spectral density of a shot noise process is detailed in the following theorem.

THEOREM 5.3. POWER SPECTRAL DENSITY OF A SHOT NOISE PROCESS *With the assumption that the interval $[0, T]$ is sufficiently long, such that the effect of including the contribution of pulse waveforms outside of this interval is negligible, and with the limit of $\Delta t \to 0$, the power spectral density of a shot noise process is*

$$G_Z(T, f) = \lambda |P(f)|^2 + \lambda^2 |P(f)|^2 \left(1 - \frac{1}{\lambda T}\right) T \operatorname{sinc}^2(fT) \quad (5.76)$$

$$G_{Z_z}(f) = \lambda |P(f)|^2 + \lambda^2 |P(0)|^2 \delta(f) \quad (5.77)$$

where $\lambda = N/T$ is the average number of waveforms/sec.

Proof. The proof of this theorem is given in Appendix 4.

5.5.1 Shot Noise due to Electrons Crossing a Barrier

Consider a classical description where electrons are moving through an entity due to some mechanism, and at random times, are crossing a boundary $x = x_o$ as illustrated in Figure 5.16.

With a classical description, the electrons behave as particles with finite dimensions, and an electron will take dt sec to cross a boundary. The charge

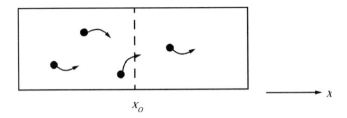

Figure 5.16 Electron movement.

q_i, passing a boundary due to the ith electron is as illustrated in Figure 5.17, where $q = 1.6 \cdot 10^{-19} C$. Also shown in this figure is the current flow i_i through the boundary due to the ith electron. The relationship $i(t) = dq/dt$ implies that

$$\int_0^\infty i_i(t)\, dt = -q$$

It is convenient to define a normalized pulse function h, according to

$$h(t) = \frac{-i_i(t + t_i)}{q} \qquad H(0) = \int_0^\infty h(t)\, dt = 1 \qquad (5.78)$$

where H is the Fourier transform of h. Assuming all electrons behave in a similar manner, the current generated by electrons passing the boundary can be written as

$$i(t) = \sum_{i=1}^{\infty} -qh(t - t_i) \qquad (5.79)$$

If, on average, there are λ electrons/sec passing the boundary, it then follows from Theorem 5.3 that the power spectral density of the random process associated with such a current flow is

$$G_\infty(f) = q^2\lambda |H(f)|^2 + q^2\lambda^2 |H(0)|^2 \delta(f) \qquad (5.80)$$

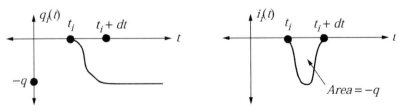

Figure 5.17 Illustration of a charge q crossing a boundary in dt sec and the resulting contribution to current flow through the boundary.

As $\bar{I} = q\lambda$ is the magnitude of the mean current flowing through the boundary, and $H(0) = 1$, it follows that the power spectral density can be written as

$$G_\infty(f) = q\bar{I}|H(f)|^2 + \bar{I}^2\delta(f) \qquad (5.81)$$

If the transition time for electrons is short relative to the measurement response time, then for frequencies less than the bandwidth of the measuring system $|H(f)|^2 \approx H^2(0) = 1$ and the most commonly used result

$$G_\infty(f) \approx q\bar{I} + \bar{I}^2\delta(f) \qquad (5.82)$$

is obtained. Ignoring the impulse at zero frequency, Schottky's formula $G_\infty(f) = q\bar{I}$ results (Davenport, 1958 p. 123).

Appropriate references for shot noise are Davenport (1958 ch. 7) and Rice (1944).

5.5.2 Shot Noise with Dead Time

Consider a shot noise process, where the occurrence of a pulse precludes the occurrence of a second pulse for a time t_Z as illustrated in Figure 5.18. This exclusion time t_Z, represents a "dead time" or "dead zone." Underlying such a random process is a point random process where, for the interval $[0, T]$, the outcomes are sets of times $\{t_1, \ldots, t_{N_m}\}$. Such a set of times is generated by the following experiment:

A number γ_1 is chosen at random from the sample space S_Γ, defined by a random variable Γ, with a density function f_Γ, and is added to the dead time t_Z to create the first time t_1. The density function is such that $f_\Gamma(\gamma) = 0$ for $\gamma < 0$. The second time, t_2 is given by t_1 plus the dead time t_Z, plus another number γ_2, chosen at random from S_Γ. This is repeated and the ith time is given by

$$t_i = \sum_{k=1}^{i} t_Z + \gamma_k = it_Z + \sum_{k=1}^{i} \gamma_k \qquad \gamma_k \in S_\Gamma \qquad (5.83)$$

This process is stopped when $t_{N_m+1} \geq T$ and $t_{N_m} < T$.

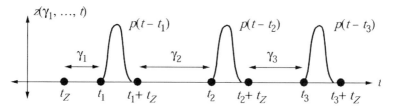

Figure 5.18 Illustration of a signal from a shot noise process with a dead time t_Z.

This experiment generates N_m numbers $\gamma_1, \ldots, \gamma_{N_m}$ that have a probability of occurrence consistent with

$$P[(\gamma_1, \ldots, \gamma_{N_m})|_{\gamma_i \in I_i}] = \prod_{i=1}^{N_m} \int_{I_i} f_\Gamma(\gamma)\, d\gamma \qquad (5.84)$$

Associated in a one-to-one manner with these numbers, is a set of times $t_1, \ldots, t_{N_m}$ as defined by Eq. (5.83). Hence,

$$P[(t_1, \ldots, t_{N_m})|_{\gamma_i \in I_i}] = \prod_{i=1}^{N_m} \int_{I_i} f_\Gamma(\gamma)\, d\gamma \qquad t_i = it_Z + \sum_{k=1}^{i} \gamma_k \qquad (5.85)$$

When a pulse function is associated with these times, a shot noise process Z with a dead time t_Z, is defined. Before formally defining such a random process, it is useful to consider the variation in N_m for a fixed interval $[0, T]$, a fixed dead time t_Z, and with the mean of the random variable Γ denoted μ_Γ.

5.5.2.1 Variation in Number of Pulses

Consider the random variable Γ_N, defined as the sum of N identically distributed and independent random variables $\Gamma_1, \ldots, \Gamma_N$, plus the time of N dead zones, that is,

$$\Gamma_N = Nt_Z + \sum_{i=1}^{N} \Gamma_i \qquad (5.86)$$

The mean and variance of Γ_N, are $N(t_Z + \mu_\Gamma)$ and $N\sigma_\Gamma^2$, respectively, where μ_Γ and σ_Γ^2 are the mean and variance of Γ_i. It follows from the central limit theorem (Grimmett, 1992 p. 175; Larson, 1986 p. 322), with probability 0.95, that an outcome of Γ_N is within $1.96\sqrt{N}\sigma_\Gamma$ of the mean $N(t_Z + \mu_\Gamma)$ as N becomes increasingly large. Hence, the relative variation in an outcome from Γ_N, as given by

$$\frac{1.96\sqrt{N}\sigma_\Gamma}{N(t_Z + \mu_\Gamma)} = \frac{1.96\sigma_\Gamma}{\sqrt{N}(t_Z + \mu_\Gamma)} \qquad (5.87)$$

clearly approaches zero as N increases. Accordingly, a reasonable approximation is to use a fixed number of outcomes $N = T/(t_Z + \mu_\Gamma)$ in the interval $[0, T]$, rather than a variable number N_m.

5.5.2.2 Power Spectral Density

A shot noise process Z, with a dead time t_Z, can be defined for the interval $[0, T]$ by the ensemble

$$E_Z = \left\{ z(\gamma_1, \ldots, \gamma_N, t) = \sum_{i=1}^{N} p(t - t_i),\ t_i = it_Z + \sum_{k=1}^{i} \gamma_k,\ \gamma_k \in S_\Gamma \right\} \qquad (5.88)$$

where $p \in L$, $N = T/(t_Z + \mu_\Gamma)$, and $P[z(\gamma_1, \ldots, \gamma_N, t)] = P[\gamma_1, \ldots, \gamma_N]$. The power spectral density of this shot noise process is detailed in the following theorem.

THEOREM 5.4. POWER SPECTRAL DENSITY OF SHOT NOISE PROCESS WITH DEAD TIME *The power spectral density on the intervals $[0, T]$ and $[0, \infty]$, of a shot noise process with a dead time t_Z is*

$$G_Z(T, f) = \lambda |P(f)|^2 \left\{ 1 + 2\text{Re}\left[\sum_{k=1}^{N-1} \left(1 - \frac{k}{\lambda T}\right) e^{j2\pi f k t_Z} [F_\Gamma^k(-f)] \right] \right\} \quad (5.89)$$

$$G_{Z_\infty}(f) = \lambda |P(f)|^2 \left\{ 1 + 2\text{Re}\left[\sum_{k=1}^{\infty} e^{j2\pi f k t_Z} [F_\Gamma^k(-f)] \right] \right\} \quad (5.90)$$

where F_Γ is the Fourier transform of the density function f_Γ, $N = T/(t_Z + \mu_\Gamma)$, μ_Γ is the mean of the random variable Γ, and $\lambda = N/T = 1/(t_Z + \mu_\Gamma)$ is the average number of pulses/sec.

Proof. The proof of this theorem is given in Appendix 5.

5.5.3 Example

Consider the case where the probability density function f_Γ is uniform on the interval $[0, \tau]$, that is, $f_\Gamma(\gamma) = 1/\tau$ for $0 \leq \gamma < \tau$ and is zero elsewhere. Then, from Theorem 2.33

$$F_\Gamma(f) = \text{sinc}(\tau f) e^{-j\pi f \tau} \qquad \mu_\Gamma = \tau/2 \quad (5.91)$$

and

$$G_Z(T, f) = \lambda |P(f)|^2 \left\{ 1 + 2 \sum_{k=1}^{N-1} \left(1 - \frac{k}{\lambda T}\right) \cos\left[2\pi f k \left(t_Z + \frac{\tau}{2}\right)\right] \text{sinc}^k(\tau f) \right\}$$

(5.92)

where $N = T/(t_Z + \tau/2)$. This result is plotted in Figure 5.19 along with the power spectral density of a regular shot noise process, for the case where the pulse function is rectangular, that is, $p(t) = 1$ for $0 \leq t < t_p$ and is zero elsewhere, whereupon $|P(f)|^2 = t_p^2 \text{sinc}^2(t_p f)$. The values used are $t_p = 1$, $t_Z = 5$, $\tau = 10$, $\lambda = 0.1$, and $T = 1000$.

Not unexpectedly, the power spectral densities are similar. This is because the mean time between pulses $t_Z + \tau/2 = 10$, is double the dead zone time $t_Z = 5$, and the dead zone has only a moderate influence. Clearly, the power spectral densities of shot noise processes with and without a dead zone, will become increasingly different when the mean time between pulses, for the shot noise process, becomes less than the dead zone time.

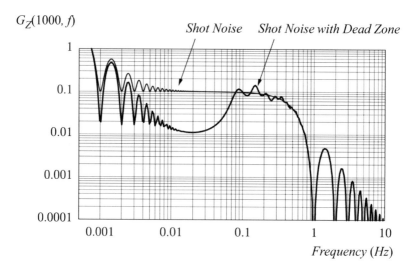

Figure 5.19 Power spectral density of a shot noise process with and without a dead time.

For the case where the dead zone time is much longer than the mean μ_Γ of Γ, the random process approaches that of a periodic signal with period close to t_Z, that is, a jittered periodic signal. For example, with $T = 8$, $t_Z = 1$, $\mu_\Gamma = 0.01$, $t_p = 0.5$, $N = 8$, and $\lambda = 1$, the power spectral density given by Eq. (5.92) can be shown to approach that of Figure 4.9, which is for a periodic pulse train.

5.6 GENERALIZED SIGNALING PROCESSES

Combining the characteristics of a signaling random process and a shot noise process, the generalized signaling random process W, can be defined on $[0, T]$ by the ensemble

$$E_W = \left\{ w(\gamma_1, \ldots, \gamma_N, t) = \sum_{i=1}^{N} \phi(\lambda_i, t - q_i),\ \gamma_i = (\lambda_i, q_i),\ \phi \in E_\Phi,\ \lambda_i \in S_\Lambda,\ q_i \in S_Q \right\}$$

(5.93)

where S_Λ and S_Q are the respective sample spaces for the independent random variables Λ and Q, with outcomes λ and q and with probability density functions f_Λ and f_Q. For the case where λ_i is independent of λ_j for $i \neq j$ and q_i is independent of q_j for $i \neq j$, the probability associated with outcomes of W are such that

$$P[w(\gamma_1, \ldots, \gamma_N, t)|_{\lambda_i \in I_i, q_i \in J_i}] = \sum_{i=1}^{N} \int_{I_i} f_\Lambda(\lambda)\, d\lambda \int_{J_i} f_Q(q)\, dq \quad (5.94)$$

The signal set E_Φ is defined as

$$E_\Phi = \{\phi(\lambda, t): \lambda \in S_\Lambda\} \tag{5.95}$$

where

$$P[\phi(\lambda, t)|_{\lambda \in [\lambda_o, \lambda_o + d\lambda]}] = f_\Lambda(\lambda_o)\, d\lambda \tag{5.96}$$

The following theorem details the power spectral density of this generalized signaling random process.

THEOREM 5.5. POWER SPECTRAL DENSITY OF A GENERALIZED SIGNALING PROCESS *Assuming the effect of including components of the signaling waveforms outside the interval $[0, T]$ is negligible, and the ith signaling waveform is independent of the jth waveform when $i \neq j$, the power spectral density of the random process W, defined by the ensemble as per Eq. (5.93), is*

$$G_W(T, f) = r\overline{|\Phi(f)|^2} + r^2\left(1 - \frac{1}{rT}\right)T|F_Q(f)|^2|\mu_\Phi(f)|^2 \tag{5.97}$$

where $r = N/T$ is the average waveform rate, F_Q is the Fourier transform of the density function f_Q, and

$$\mu_\Phi(f) = \int_{-\infty}^{\infty} \Phi(\lambda, f) f_\Lambda(\lambda)\, d\lambda \qquad \overline{|\Phi(f)|^2} = \int_{-\infty}^{\infty} |\Phi(\lambda, f)|^2 f_\Lambda(\lambda)\, d\lambda \tag{5.98}$$

Here, Φ is the Fourier transform of a waveform from the signaling set, and evaluated over the interval $(-\infty, \infty)$, that is,

$$\Phi(\lambda, f) = \int_{-\infty}^{\infty} \phi(\lambda, t) e^{-j2\pi f t}\, dt \tag{5.99}$$

Proof. The proof of this theorem is given in Appendix 6.

5.6.1 Uniform Distribution of Times

For the usual case of a uniform distribution of times on the interval $[0, T]$, consistent with $f_Q(q) = 1/T$ for $0 \leq q < T$ and zero elsewhere, it follows that $F_Q(f) = \mathrm{sinc}(fT)e^{-j\pi fT}$, and hence,

$$G_W(T, f) = r\overline{|\Phi(f)|^2} + r^2\left(1 - \frac{1}{rT}\right)T\,\mathrm{sinc}^2(fT)|\mu_\Phi(f)|^2 \tag{5.100}$$

$$G_{W_z}(f) = r\overline{|\Phi(f)|^2} + r^2|\mu_\Phi(0)|^2\delta(f) \tag{5.101}$$

Not surprisingly, by comparison with the results for the shot noise case, that is, Eqs. (5.76) and (5.77), this result is a straightforward generalization of that case.

APPENDIX 1: PROOF OF THEOREM 5.1

Theorem 4.7 states that the power spectral density of the sum of N random processes $X_1, \ldots, X_N$, is given by

$$G_Z(T, f) = \sum_{i=1}^{N} G_i(T, f) + \sum_{i=1}^{N} \sum_{\substack{k=1 \\ k \neq i}}^{N} G_{ik}(T, f) \qquad (5.102)$$

With $T = ND$, this result can be used directly with the ith random process X_i being defined by the ensemble

$$E_{X_i} = \{x_i(\gamma_i, t) = \phi(\gamma_i, t - (i-1)D), \; \gamma_i \in S_\Gamma, \phi \in E_\Phi\} \qquad (5.103)$$

where E_Φ is defined by Eq. (5.3). Using the definitions for the power spectral density for the countable and uncountable cases, it follows, assuming contributions of the signaling waveforms outside of the interval $[0, T]$ can be included, that

$$G_i(T, f) = \frac{1}{T} \sum_{\gamma=1}^{\infty} p_\gamma |\Phi(\gamma, f)|^2 \qquad G_i(T, f) = \frac{1}{T} \int_{-\infty}^{\infty} |\Phi(\gamma, f)|^2 f_\Gamma(\gamma) \, d\gamma \qquad (5.104)$$

Hence, using the definitions for $\overline{|\Phi(f)|^2}$, as per Eq. (5.17), it follows that $G_i(T, f) = \overline{|\Phi(f)|^2}/T$ for both cases. Further, when $i \neq k$, the respective results follow for the countable and uncountable cases,

$$G_{ik}(T, f) = \frac{1}{T} \left[\sum_{\gamma_i=1}^{\infty} \sum_{\gamma_k=1}^{\infty} p_{\gamma_i \gamma_k} \Phi(\gamma_i, f) \Phi^*(\gamma_k, f) \right] e^{-j2\pi(i-1)Df} e^{j2\pi(k-1)Df}$$

$$= \frac{1}{T} e^{-j2\pi(i-k)Df} R_{\Phi_i \Phi_k}(f) \qquad (5.105)$$

$$G_{ik}(T, f) = \frac{1}{T} \left[\int_{-\infty}^{\infty} \int_{-\infty}^{\infty} \Phi(\gamma_i, f) \Phi^*(\gamma_k, f) f_{\Gamma_i \Gamma_k}(\gamma_i, \gamma_k) \, d\gamma_i \, d\gamma_k \right] e^{-j2\pi(i-1)Df} e^{j2\pi(k-1)Df}$$

$$= \frac{1}{T} e^{-j2\pi(i-k)Df} R_{\Phi_i \Phi_k}(f) \qquad (5.106)$$

Hence, using the definition for $R_{\Phi_i \Phi_k}(f)$, as per Eq. 5.18, it follows for both the countable and uncountable cases, that $G_{ik}(T, f) = e^{-j2\pi(i-k)Df} R_{\Phi_i \Phi_k}(f)/T$.

Thus, with $r = N/T = 1/D$, it follows that

$$G_X(ND, f) = r\overline{|\Phi(f)|^2} + \frac{1}{T} \sum_{i=1}^{N} \sum_{\substack{k=1 \\ k \neq i}}^{N} e^{-j2\pi(i-k)Df} R_{\Phi_i \Phi_k}(f) \qquad (5.107)$$

Further simplification relies on simplying the second term in this equation, denoted T_2. This is done for the countable case — the uncountable case follows in an analogous manner. By definition

$$R_{\Phi_i \Phi_k}(f) = \sum_{\gamma_i = 1}^{\infty} \sum_{\gamma_k = 1}^{\infty} p_{\gamma_i \gamma_k} \Phi_{ik} \qquad \Phi_{ik} = \Phi(\gamma_i, f)\Phi^*(\gamma_k, f) \qquad \Phi_{ki} = \Phi_{ik}^* \qquad (5.108)$$

and

$$T_2 = \frac{e^{j2\pi Df}}{T} \sum_{\gamma_1=1}^{\infty} \sum_{\gamma_2=1}^{\infty} p_{\gamma_1 \gamma_2} \Phi_{12} + \frac{e^{-j2\pi Df}}{T} \sum_{\gamma_2=1}^{\infty} \sum_{\gamma_1=1}^{\infty} p_{\gamma_2 \gamma_1} \Phi_{21}$$

$$+ \frac{e^{j2\pi 2Df}}{T} \sum_{\gamma_1=1}^{\infty} \sum_{\gamma_3=1}^{\infty} p_{\gamma_1 \gamma_3} \Phi_{13}$$

$$+ \frac{e^{-j2\pi 2Df}}{T} \sum_{\gamma_3=1}^{\infty} \sum_{\gamma_1=1}^{\infty} p_{\gamma_3 \gamma_1} \Phi_{31} + \cdots \qquad (5.109)$$

To further simplify this equation, first note that there are $N - 1$ summations where $k = i \pm 1$, $N - 2$ summations where $k = i \pm 2, \ldots$, and 1 summation where $k = i \pm (N - 1)$. Second, note the assumption $p_{\gamma_i \gamma_k} = p_{\gamma_{q+i} \gamma_{q+k}}$, that is, correlations only depend on the difference between the location of signaling intervals and not on their absolute location. Third, note it is the case that $p_{\gamma_i \gamma_k} = p_{\gamma_k \gamma_i}$ which follows because $p_{\gamma_i \gamma_k}$ is the probability of $\phi(\gamma_i, t)$ in the ith interval, and $\phi(\gamma_k, t)$ in the kth interval, while $p_{\gamma_k \gamma_i}$ is the probability of $\phi(\gamma_k, t)$ in the kth interval, and $\phi(\gamma_i, t)$ in the ith interval. It then follows that

$$T_2 = 2r\left[1 - \frac{1}{N}\right] \text{Re}\left[e^{j2\pi Df} \sum_{\gamma_1=1}^{\infty} \sum_{\gamma_2=1}^{\infty} p_{\gamma_1 \gamma_2} \Phi(\gamma_1, f)\Phi^*(\gamma_2, f)\right]$$

$$+ 2r\left[1 - \frac{2}{N}\right] \text{Re}\left[e^{j2\pi 2Df} \sum_{\gamma_1=1}^{\infty} \sum_{\gamma_3=1}^{\infty} p_{\gamma_1 \gamma_3} \Phi(\gamma_1, f)\Phi^*(\gamma_3, f)\right] + \cdots$$

$$+ 2r\left[1 - \frac{N-1}{N}\right] \text{Re}\left[e^{j2\pi(N-1)Df} \sum_{\gamma_1=1}^{\infty} \sum_{\gamma_N=1}^{\infty} p_{\gamma_1 \gamma_N} \Phi(\gamma_1, f)\Phi^*(\gamma_N, f)\right]$$

$$(5.110)$$

which can be rewritten as

$$T_2 = 2r \sum_{i=1}^{N-1}\left[1-\frac{i}{N}\right]\text{Re}\left[e^{j2\pi iDf}\sum_{\gamma_1=1}^{\infty}\sum_{\gamma_{1+i}=1}^{\infty}p_{\gamma_1\gamma_{1+i}}\Phi(\gamma_1,f)\Phi^*(\gamma_{1+i},f)\right]$$

(5.111)

Adding and subtracting the term $p_{\gamma_1}p_{\gamma_{1+i}}\Phi(\gamma_1,f)\Phi^*(\gamma_{1+i},f)$ in the inner summation yields

$$T_2 = r|\mu_\Phi(f)|^2\left\{2\sum_{i=1}^{N-1}\left[1-\frac{i}{N}\right]\cos(2\pi iDf)\right\} + 2r\sum_{i=1}^{N-1}\left[1-\frac{i}{N}\right]$$
$$\times \text{Re}\left\{e^{j2\pi iDf}\sum_{\gamma_1=1}^{\infty}\sum_{\gamma_{1+i}=1}^{\infty}\times[p_{\gamma_1\gamma_{1+i}}-p_{\gamma_1}p_{\gamma_{1+i}}]\Phi(\gamma_1,f)\Phi^*(\gamma_{1+i},f)\right\}$$

(5.112)

Using the definition for $R_{\Phi_i\Phi_k}(f)$, a result implicit in Theorem 2.32, namely

$$2\sum_{i=1}^{N-1}\left[1-\frac{i}{N}\right]\cos[2\pi iDf] = \frac{1}{N}\frac{\sin^2(\pi NfD)}{\sin^2(\pi fD)} - 1 \qquad (5.113)$$

and assuming $p_{\gamma_1\gamma_{1+i}} = p_{\gamma_1}p_{\gamma_{1+i}}$ for $i > m$, it follows that

$$T_2 = r|\mu_\Phi(f)|^2\left[\frac{1}{N}\frac{\sin^2(\pi NfD)}{\sin^2(\pi fD)} - 1\right]$$
$$+ 2r\sum_{i=1}^{m}\left[1-\frac{i}{N}\right]\text{Re}[e^{j2\pi iDf}(R_{\Phi_i\Phi_{1+i}}(f) - |\mu_\Phi(f)|^2)]$$

(5.114)

Substituting this result into Eq. (5.107) yields

$$G_X(ND,f) = r\overline{|\Phi(f)|^2} - r|\mu_\Phi(f)|^2 + r|\mu_\Phi(f)|^2\left[\frac{1}{N}\frac{\sin^2(\pi NfD)}{\sin^2(\pi fD)}\right]$$
$$+ 2r\sum_{i=1}^{m}\left[1-\frac{i}{N}\right]\text{Re}[e^{j2\pi iDf}(R_{\Phi_i\Phi_{1+i}}(f) - |\mu_\Phi(f)|^2)]$$

(5.115)

Again, using a result from Theorem 3.32, the result for the infinite interval follows:

$$G_{X_\infty}(f) = r\overline{|\Phi(f)|^2} - r|\mu_\Phi(f)|^2 + r^2|\mu_\Phi(f)|^2 \sum_{n=-\infty}^{\infty} \delta(f - nr)$$

$$+ 2r \sum_{i=1}^{m} \text{Re}[e^{j2\pi i D f}(R_{\Phi_i \Phi_{1+i}}(f) - |\mu_\Phi(f)|^2)] \quad (5.116)$$

APPENDIX 2: PROOF OF THEOREM 5.2

As the noise is assumed to be uncorrelated over a time interval consistent with the comparator output pulse, it follows that $R_{\Phi_i \Phi_{1+i}}(f)$, as defined by Eq. (5.18), equals $|\mu_\Phi(f)|^2$. Hence, from Theorem 5.1, the power spectral density of Z is given by

$$G_Z(NB, f) = r\overline{|\Phi(f)|^2} + r|\mu_\Phi(f)|^2 \left[\frac{1}{N}\frac{\sin^2(\pi N f/r)}{\sin^2(\pi f/r)} - 1\right] \quad (5.117)$$

where $r = 1/B$, and it remains to determine $\mu_\Phi(f)$ and $\overline{|\Phi(f)|^2}$. As

$$\phi(\gamma, t) = A_o(1 + a)r\left(\frac{t - \mu_D - d}{\mu_W + w}\right)$$

with $\gamma = (a, d, w)$, and

$$r(t) \leftrightarrow R(f) \Rightarrow r\left(\frac{t}{\alpha}\right) \leftrightarrow \alpha R(\alpha f) \Rightarrow r\left(\frac{t - \beta}{\alpha}\right) \leftrightarrow \alpha R(\alpha f)e^{-j2\pi f\beta} \quad \alpha > 0$$

(5.118)

it follows that

$$\Phi(\gamma, f) = A_o(1 + a)(\mu_W + w)R[(\mu_W + w)f]e^{-j2\pi f(\mu_D + d)} \quad (5.119)$$

Then, as $f_\Gamma(\gamma) = f_A(a) f_D(d) f_W(w)$, and $\mu_\Phi(f) = \int \Phi(\gamma, f) f_\Gamma(\gamma) \, d\gamma$, it follows that

$$\mu_\Phi(f) = \int_{-\infty}^{\infty} \int_{-\infty}^{\infty} \int_{-\infty}^{\infty} A_o(1 + a)(\mu_W + w)$$

$$\times R[(\mu_W + w)f]e^{-j2\pi f(\mu_D + \delta)} f_A(a) f_D(\delta) f_W(w) \, da \, d\delta \, dw$$

$$= A_o e^{-j2\pi f \mu_D} \int_{-\infty}^{\infty} e^{-j2\pi f \delta} f_D(\delta) \, d\delta \int_{-\infty}^{\infty} (\mu_W + w) R[(\mu_W + w)f] f_W(w) \, dw$$

(5.120)

where the independence and the zero mean property of the random variables has been used. With the Fourier transform definition

$$F_D(f) = \int_{-\infty}^{\infty} e^{-j2\pi f \delta} f_D(\delta) \, d\delta \tag{5.121}$$

it follows that

$$|\mu_\Phi(f)| = A_o |F_D(f)| \left| \int_{-\infty}^{\infty} (\mu_W + w) R[(\mu_W + w)f] f_W(w) \, dw \right| \tag{5.122}$$

Similarly, from the definition of $\overline{|\Phi(f)|^2} = \int |\Phi(\gamma, f)|^2 f_\Gamma(\gamma) \, d\gamma$, it follows that

$$\overline{|\Phi(f)|^2} = \int_{-\infty}^{\infty} \int_{-\infty}^{\infty} \int_{-\infty}^{\infty} A_o^2 (1 + a)^2 (\mu_W + w)^2$$

$$\times |R[(\mu_W + w)f]|^2 f_A(a) f_D(\delta) f_W(w) \, da \, d\delta \, dw$$

$$= A_o^2 (1 + \bar{A}^2) \int_{-\infty}^{\infty} (\mu_W + w)^2 |R[(\mu_W + w)f]|^2 f_W(w) \, dw \tag{5.123}$$

Hence,

$$G_Z(NB, f) = rA_o^2(1 + \bar{A}^2) \int_{-\infty}^{\infty} (\mu_W + w)^2 |R[(\mu_W + w)f]|^2 f_W(w) \, dw$$

$$+ rA_o^2 |F_D(f)|^2 \left| \int_{-\infty}^{\infty} (\mu_W + w) R[(\mu_W + w)f] f_W(w) \, dw \right|^2$$

$$\times \left[\frac{1}{N} \frac{\sin^2(\pi Nf/r)}{\sin^2(\pi f/r)} - 1 \right] \tag{5.124}$$

and

$$G_{Z\infty}(f) = rA_o^2(1 + \bar{A}^2) \int_{-\infty}^{\infty} (\mu_W + w)^2 |R[(\mu_W + w)f]|^2 f_W(w) \, dw$$

$$+ rA_o^2 |F_D(f)|^2 \left| \int_{-\infty}^{\infty} (\mu_W + w) R[(\mu_W + w)f] f_W(w) \, dw \right|^2$$

$$\times \left[r \sum_{i=-\infty}^{\infty} \delta(f - ir) - 1 \right] \tag{5.125}$$

APPENDIX 3: PROOF OF EQUATION 5.73

With $\bar{A}^2 = 0$, Gaussian probability density functions for W and D and $R(f) = \text{sinc}(f)e^{-j\pi f}$, it follows from Eq. (5.68) that

$$G_Z(NB, f) = rA_o^2 \int_{-\infty}^{\infty} (\mu_W + w)^2 \, \text{sinc}^2[(\mu_W + w)f] \frac{e^{-w^2/2\sigma_W^2}}{\sqrt{2\pi}\,\sigma_W} dw$$

$$+ rA_o^2 e^{-4\pi^2 \sigma_D^2 f^2} \left[\frac{1}{N} \frac{\sin^2(\pi Nf/r)}{\sin^2(\pi f/r)} - 1 \right]$$

$$\times \left| \int_{-\infty}^{\infty} (\mu_W + w) \, \text{sinc}[(\mu_W + w)f] e^{-j\pi(\mu_W + w)f} \frac{e^{-w^2/2\sigma_W^2}}{\sqrt{2\pi}\,\sigma_W} dw \right|^2$$

(5.126)

where the Fourier transform result (McGillem, 1991 p. 168)

$$e^{-b^2 t^2} \leftrightarrow \frac{\sqrt{\pi}}{b} e^{-\pi^2 f^2/b^2} \quad \Rightarrow \quad \frac{e^{-d^2/2\sigma_D^2}}{\sqrt{2\pi}\,\sigma_D} \leftrightarrow e^{-2\pi^2 \sigma_D^2 f^2} \quad (5.127)$$

has been used to evaluate $F_D(f)$. Using the definition for the sinc function, it follows that

$$G_Z(NB, f) = \frac{rA_o^2}{\pi^2 f^2} \int_{-\infty}^{\infty} \sin^2[\pi(\mu_W + w)f] \frac{e^{-w^2/2\sigma_W^2}}{\sqrt{2\pi}\,\sigma_W} dw$$

$$+ \frac{rA_o^2 e^{-4\pi^2 \sigma_D^2 f^2}}{\pi^2 f^2} \left[\frac{1}{N} \frac{\sin^2(\pi Nf/r)}{\sin^2(\pi f/r)} - 1 \right]$$

$$\times \left| \int_{-\infty}^{\infty} \sin[\pi(\mu_W + w)f] e^{-j\pi(\mu_W + w)f} \frac{e^{-w^2/2\sigma_W^2}}{\sqrt{2\pi}\,\sigma_W} dw \right|^2 \quad (5.128)$$

Using the identity

$$2\sin^2(A) = 1 - \cos(2A),$$

the standard expansion for $\cos(A + B)$, the fact that the integral of a density function is unity, the integral of an even and odd function is zero, and the integral result (Spiegel, 1968 p. 98)

$$\int_0^{\infty} \cos(bx) e^{-ax^2} dx = \frac{1}{2}\sqrt{\frac{\pi}{a}}\, e^{-b^2/4a}$$

it follows that

$$\int_{-\infty}^{\infty} \cos(2\pi w f) \frac{e^{-w^2/2\sigma_w^2}}{\sqrt{2\pi}\,\sigma_w}\, dw = e^{-2\pi^2 \sigma_w^2 f^2} \tag{5.129}$$

and the first term in Eq. (5.128) simplifies to

$$\frac{rA_o^2}{2\pi^2 f^2}[1 - \cos(2\pi\mu_w f)e^{-2\pi^2 f^2 \sigma_w^2}] \tag{5.130}$$

Consider the second integral in Eq. (5.128) with the sin function written in its equivalent exponential form:

$$I_2 = \frac{1}{2j}\int_{-\infty}^{\infty}[e^{j\pi(\mu_w + w)f} - e^{-j\pi(\mu_w + w)f}]e^{-j\pi(\mu_w + w)f}\frac{e^{-w^2/2\sigma_w^2}}{\sqrt{2\pi}\,\sigma_w}\, dw$$

$$= \frac{1}{2j}\left[1 - e^{-j2\pi\mu_w f}\int_{-\infty}^{\infty} e^{-j2\pi w f}\frac{e^{-w^2/2\sigma_w^2}}{\sqrt{2\pi}\,\sigma_w}\, dw\right] \tag{5.131}$$

Writing the complex exponential term in its equivalent trigonometric form, and noting that one of the resulting integrals is of an even and odd function which integrates to zero, it follows from Eq. (5.129) that

$$I_2 = \frac{1}{2j}[1 - e^{-j2\pi\mu_w f}\, e^{-2\pi^2 f^2 \sigma_w^2}] \tag{5.132}$$

Substitution of these integral results yields the required form

$$G_Z(NB, f) = \frac{rA_o^2}{2\pi^2 f^2}[1 - \cos(2\pi\mu_w f)e^{-2\pi^2 f^2 \sigma_w^2}]$$

$$+ \frac{rA_o^2 e^{-4\pi^2 \sigma_D^2 f^2}}{4\pi^2 f^2}[1 + e^{-4\pi^2 f^2 \sigma_w^2} - 2\cos(2\pi\mu_w f)e^{-2\pi^2 f^2 \sigma_w^2}]$$

$$\times \left[\frac{1}{N}\frac{\sin^2(\pi N f/r)}{\sin^2(\pi f/r)} - 1\right] \tag{5.133}$$

Using a limit result from Theorem 2.32 for the last term in this equation yields the result for the infinite interval.

APPENDIX 4: PROOF OF THEOREM 5.3

Theorem 4.7 states that the power spectral density of the sum of N random processes $X_1, \ldots, X_N$, is given by

$$G_Z(T, f) = \sum_{i=1}^{N} G_i(T, f) + \sum_{i=1}^{N}\sum_{\substack{k=1 \\ k \neq i}}^{N} G_{ik}(T, f) \tag{5.134}$$

APPENDIX 4: PROOF OF THEOREM 5.3

This result can be used directly with the ith random process X_i being defined by the ensemble

$$E_{X_i} = \{x_i(\gamma_i, t) = p(t - \gamma_i \Delta t), \gamma_i \in \{0, \ldots, M-1\}, P[\gamma_i] = 1/M\} \quad (5.135)$$

whereupon, it follows that

$$G_i(T, f) = \frac{1}{T}\sum_{\gamma_i=0}^{M-1} \frac{1}{M} |P(f)e^{-j2\pi\gamma_i \Delta t f}|^2 = \frac{|P(f)|^2}{T} \quad (5.136)$$

To establish an expression for the cross power spectral density between X_i and X_k for $i \neq k$, note that these random processes are independent and identical. From Eq. (4.52) it then follows that

$$G_{ik}(T, f) = \frac{\bar{X}_i(T,f)\bar{X}_k^*(T,f)}{T} = \frac{|\bar{X}_i(T,f)|^2}{T} = \frac{1}{T}\left|\sum_{\gamma_i=0}^{M-1}\frac{1}{M}P(f)e^{-j2\pi\gamma_i\Delta tf}\right|^2$$

$$= \frac{|P(f)|^2}{T}\frac{\sin^2(\pi M \Delta t f)}{M^2 \sin^2(\pi \Delta t f)} \quad (5.137)$$

where the last result is from Theorem 2.32. Using these results, it then follows that the power spectral density is given by

$$G_Z(T, f) = \frac{N|P(f)|^2}{T} + \frac{(N^2 - N)|P(f)|^2}{T}\frac{\sin^2(\pi M \Delta t f)}{M^2 \sin^2(\pi \Delta t f)} \quad (5.138)$$

For the finite interval $[0, T]$ with N and T fixed, it follows for any fixed frequency range $f \in [-f_x, f_x]$, that there will exist a $\Delta t \to 0$ and a $M \to \infty$ with $\Delta t M = T$, such that $\sin(\pi \Delta t f) \approx \pi \Delta t f$, and hence,

$$G_Z(T, f) \approx \frac{N|P(f)|^2}{T} + \frac{N^2|P(f)|^2}{T^2}\left(1 - \frac{1}{N}\right)\frac{T \sin^2(\pi M \Delta t f)}{(\pi M \Delta t f)^2}$$

$$= \lambda|P(f)|^2 + \lambda^2|P(f)|^2\left(1 - \frac{1}{\lambda T}\right)T\,\text{sinc}^2(fT) \quad (5.139)$$

where $\lambda = N/T$ is the average number of waveforms/sec. This is the required result for the finite interval. For the infinite interval, a result from Theorem 2.32 yields the required result, namely,

$$G_{Z_\infty}(f) = \lim_{T \to \infty} G(T, f) = \lambda|P(f)|^2 + \lambda^2|P(0)|^2\delta(f) \quad (5.140)$$

APPENDIX 5: PROOF OF THEOREM 5.4

As γ_i is independent of γ_k for $i \neq k$ the power spectral density of the random process Z, with an ensemble given by Eq. (5.88), is

$$G_Z(T, f) = \frac{1}{T} \int_{-\infty}^{\infty} \cdots \int_{-\infty}^{\infty} f_{\Gamma}(\gamma_1) \cdots f_{\Gamma}(\gamma_N) \left| \sum_{i=1}^{N} P(f) e^{-j2\pi f t_i} \right|^2 d\gamma_1 \cdots d\gamma_N \quad (5.141)$$

Substitution of the result $t_i = it_Z + \sum_{k=1}^{i} \gamma_k$ yields

$$G_Z(T, f) = \frac{|P(f)|^2}{T} \int_{-\infty}^{\infty} \cdots \int_{-\infty}^{\infty} f_{\Gamma}(\gamma_1) \cdots f_{\Gamma}(\gamma_N)$$
$$\times \left| \sum_{i=1}^{N} \exp\left[-j2\pi f \left(it_Z + \sum_{k=1}^{i} \gamma_k \right) \right] \right|^2 d\gamma_1 \cdots d\gamma_N \quad (5.142)$$

Further simplification relies on the following result:

$$\left| \sum_{i=1}^{N} e^{jh(i)} \right|^2 = N + \sum_{p=1}^{N} \sum_{q=1, q \neq p}^{N} e^{jh(p)} e^{-jh(q)}$$
$$= N + 2\text{Re}\left[\sum_{p=1}^{N} \sum_{q=1, q>p}^{N} e^{j[h(p) - h(q)]} \right] \quad (5.143)$$

With

$$h(p) = -2\pi f p t_Z - 2\pi f \sum_{k=1}^{p} \gamma_k \quad (5.144)$$

and $q > p$, it follows that

$$h(p) - h(q) = 2\pi f(q - p) t_Z + 2\pi f \sum_{k=p+1}^{q} \gamma_k \quad (5.145)$$

Hence,

$$G_Z(T, f) = \frac{|P(f)|^2}{T} \left[N + 2\text{Re} \int_{-\infty}^{\infty} \cdots \int_{-\infty}^{\infty} f_{\Gamma}(\gamma_1) \cdots f_{\Gamma}(\gamma_N) \right.$$
$$\left. \times \sum_{p=1}^{N} \sum_{\substack{q=1 \\ q>p}}^{N} \exp[j2\pi f(q-p) t_Z] \exp\left(j2\pi f \sum_{k=p+1}^{q} \gamma_k \right) d\gamma_1 \cdots d\gamma_N \right]$$
$$(5.146)$$

Interchanging the order of summation and integration in the second term in this equation yields, for the argument of the Re operator,

$$\sum_{\substack{p=1 \\ q>p}}^{N} \sum_{q=1}^{N} e^{j2\pi(q-p)ftz} \int_{-\infty}^{\infty} \cdots \int_{-\infty}^{\infty} \exp\left[j2\pi f \sum_{k=p+1}^{q} \gamma_k\right] f_\Gamma(\gamma_1) \cdots f_\Gamma(\gamma_N) d\gamma_1 \cdots d\gamma_N$$

$$= \sum_{\substack{p=1 \\ q>p}}^{N} \sum_{q=1}^{N} e^{j2\pi(q-p)ftz} F_\Gamma^{q-p}(-f) \quad (5.147)$$

where F_Γ is the Fourier transform of the density function f_Γ. In the double summation in this equation there are $N-1$ terms, where $q = p+1$; $N-2$ terms, where $q = p+2,\ldots$; and 1 term where $q = p+(N-1)$. Thus, this double summation can be written as

$$\sum_{\substack{p=1 \\ q>p}}^{N} \sum_{q=1}^{N} e^{j2\pi(q-p)ftz} F_\Gamma^{q-p}(-f) = \sum_{k=1}^{N-1} [N-k] e^{j2\pi fktz} F_\Gamma^k(-f) \quad (5.148)$$

With this result, it follows that

$$G_Z(T, f) = \frac{N|P(f)|^2}{T}\left\{1 + 2\text{Re}\left[\sum_{k=1}^{N-1}\left(1 - \frac{k}{N}\right) e^{j2\pi fktz} F_\Gamma^k(-f)\right]\right\} \quad (5.149)$$

With $\lambda = N/T$ the required results follow, that is,

$$G_Z(T, f) = \lambda |P(f)|^2 \left\{1 + 2\text{Re}\left[\sum_{k=1}^{N-1}\left(1 - \frac{k}{\lambda T}\right) e^{j2\pi kftz} F_\Gamma^k(-f)\right]\right\} \quad (5.150)$$

$$G_{Z_\infty}(f) = \lambda |P(f)|^2 \left\{1 + 2\text{Re}\left[\sum_{k=1}^{\infty} e^{j2\pi kftz} F_\Gamma^k(-f)\right]\right\} \quad (5.151)$$

APPENDIX 6: PROOF OF THEOREM 5.5

Theorem 4.7 states that the power spectral density of the sum of N random processes $W_1,\ldots,W_N$, is given by

$$G(T, f) = \sum_{i=1}^{N} G_i(T, f) + \sum_{i=1}^{N} \sum_{\substack{j=1 \\ j\neq i}}^{N} G_{ij}(T, f) \quad (5.152)$$

This result can be used directly with the ith random process W_i being defined by the ensemble

$$E_{W_i} = \{w_i(\gamma_i, t) = \phi(\lambda_i, t - q_i), \gamma_i = (\lambda_i, q_i), \lambda_i \in S_\Lambda, q_i \in S_Q, \phi \in E_\Phi\} \quad (5.153)$$

Using the definition for the power spectral density for the uncountable case, it follows, assuming contributions of the signaling waveforms outside of the interval $[0, T]$ can be included, that

$$G_i(T, f) = \frac{1}{T} \int_{S_\Lambda} \int_{S_Q} |\Phi(\lambda, f) e^{-j2\pi f q}|^2 f_\Lambda(\lambda) f_Q(q)\, dq\, d\lambda = \frac{\overline{|\Phi(f)|^2}}{T} \quad (5.154)$$

where, by definition

$$\overline{|\Phi(f)|^2} = \int_{S_\Lambda} |\Phi(\lambda, f)|^2 f_\Lambda(\lambda)\, d\lambda \quad (5.155)$$

Further, when $i \neq k$,

$$G_{ik}(T, f) = \frac{1}{T} \int_{S_\Lambda} \int_{S_\Lambda} \Phi(\lambda_i, f) \Phi^*(\lambda_k, f) f_\Lambda(\lambda_i) f_\Lambda(\lambda_k)\, d\lambda_i\, d\lambda_k$$

$$\times \int_{S_Q} \int_{S_Q} e^{-j2\pi f q_i} e^{j2\pi f q_k} f_Q(q_i) f_Q(q_k)\, dq_i\, dq_k \quad (5.156)$$

$$= \frac{|F_Q(f)|^2}{T} \left| \int_{S_\Lambda} \Phi(\lambda, f) f_\Lambda(\lambda)\, d\lambda \right|^2 = \frac{|F_Q(f)|^2}{T} |\mu_\Phi(f)|^2$$

where

$$\mu_\Phi(f) = \int_{S_\Lambda} \Phi(\lambda, f) f_\Lambda(\lambda)\, d\lambda \quad (5.157)$$

and F_Q is the Fourier transform of the density function f_Q. With $r = N/T$, it follows that

$$G_W(ND, f) = r \overline{|\Phi(f)|^2} + r(N-1)|F_Q(f)|^2 |\mu_\Phi(f)|^2 \quad (5.158)$$

which is the required result.

6

Power Spectral Density of Standard Random Processes — Part 2

6.1 INTRODUCTION

This chapter continues the discussion of standard random processes commenced in Chapter 5. Specifically, the power spectral density associated with sampling, quadrature amplitude modulation, and a random walk, are discussed. It is shown that a $1/f$ power spectral density is consistent with a summation of bounded random walks.

6.2 SAMPLED SIGNALS

Sampling of signals is widespread with the increasing trend towards processing signals digitally. One goal is to establish, from samples of the signal, the Fourier transform of the signal. Consider a signal x, that is piecewise smooth on $[0, ND]$, as illustrated in Figure 6.1. One approach for establishing the Fourier transform of such a signal is to use a Riemann sum (Spivak, 1994 p. 279) to approximate the integral defining the Fourier transform, that is,

$$\int_0^{ND} x(t)e^{-j2\pi ft}\, dt \approx D\left[\frac{x(0^+)}{2} + \sum_{p=1}^{N-1} x(pD)e^{-j2\pi pDf} + \frac{x(ND^-)e^{-j2\pi NDf}}{2}\right] \quad (6.1)$$

If x is piecewise smooth on $[0, T]$, then from Theorem 2.7 it has bounded variation on this interval. It then follows from Theorem 2.19, that this

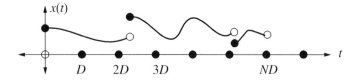

Figure 6.1 Piecewise smooth function on [0, ND].

approximation can be made arbitrarily accurate by increasing the number of samples taken. The following theorem establishes an exact relationship between this Riemann sum and the Fourier transform of x. This relationship facilitates evaluation of the power spectral density of a sampled signal.

THEOREM 6.1. SAMPLING RELATIONSHIP *Consider $N+1$ samples, taken at $0, D, \ldots, ND$ sec with a sampling frequency $f_S = 1/D$ Hz, of a piecewise smooth signal x (see Figure 6.1). If X is the Fourier transform of x, and*

$$\lim_{M \to \infty} \sum_{k=-M}^{M} X(ND, f - kf_S)$$

converges for all $f \in \mathbf{R}$, then

$$f_S \sum_{k=-\infty}^{\infty} X(ND, f - kf_S) = \frac{x(0^+)}{2} + \sum_{p=1}^{N-1} \frac{x(pD^-) + x(pD^+)}{2} e^{-j2\pi pDf} + \frac{x(ND^-) e^{-j2\pi NDf}}{2} \quad (6.2)$$

A sufficient condition for $\sum_{k=-M}^{M} X(ND, f - kf_S)$ to converge as $M \to \infty$, is the existence of $k_o, \alpha > 0$, such that $|X(ND, f)| < k_o/|f|^{1+\alpha}$ for $f \in \mathbf{R}$.

Proof. The proof of this result is given in Appendix 1.

6.2.0.1 Example Consider the function

$$x(t) = \begin{cases} 1 & 0 \leq t < ND \\ 0 & \text{elsewhere} \end{cases}$$

whose Fourier transform (see Theorem 2.33) is

$$X(ND, f) = (N/f_S) \operatorname{sinc}(Nf/f_S) e^{-j\pi fN/f_S}$$

and which does not satisfy the requirement that there exists $k_o, \alpha > 0$, such that

$|X(ND, f)| < k_o/|f|^{1+\alpha}$. However, for $f = if_S$, with $i \in \mathbf{Z}$, the summation

$$\lim_{M \to \infty} \sum_{k=-M}^{M} X(ND, f - kf_S)$$

converges and is equal to the ith term N/f_S. Equation (6.2) is then easily proved as both sides are equal to N.

When $f \neq if_S$, with $i \in \mathbf{Z}$, it follows, that after standard manipulation that

$$f_S \sum_{k=-M}^{M} X(ND, f - kf_S) = N \operatorname{sinc}\left(\frac{Nf}{f_S}\right) e^{-j\pi Nf/f_S} \left[1 + 2 \sum_{k=1}^{M} \frac{1}{\left(1 - \dfrac{k^2}{f^2/f_S^2}\right)}\right]$$

(6.3)

which clearly converges as $M \to \infty$, provided $f/f_S \notin \mathbf{Z}$. For example, if $f = f_S/4$ and $N = 2$, it follows from the result (Gradshteyn, 1980 p. 8)

$$\sum_{k=1}^{\infty} \frac{1}{(1 - 4k)(1 + 4k)} = -\frac{1}{2} + \frac{\pi}{8}$$

that

$$f_S \sum_{k=-\infty}^{\infty} X\left(2D, \frac{f_S}{4} - kf_S\right) = -j \qquad (6.4)$$

This result agrees with the Riemann sum for the case where $N = 2$ as

$$\frac{x(0^+)}{2} + \sum_{p=1}^{1} x(pD)e^{-j2\pi pDf} + \frac{x(ND^-)e^{-j2\pi NDf}}{2}\bigg|_{f = f_S/4} = 0.5 - j - 0.5 = -j$$

(6.5)

6.2.1 Power Spectral Density of Sampled Signal

Consider a signal x, as illustrated in Figure 6.1, which is piecewise smooth on $[0, ND]$ and is sampled at a rate $f_S = 1/D$ by a sampling signal S_Δ, defined according to

$$S_\Delta(t) = \sum_{k=-\infty}^{\infty} \delta_\Delta(t - kD) \qquad (6.6)$$

where δ_Δ is defined by the graph of S_Δ shown in Figure 6.2. On the interval $[0, ND)$ the signal y_Δ, as a consequence of sampling the signal x, is defined

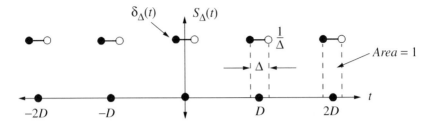

Figure 6.2 Sampling signal.

according to

$$y_\Delta(t) = \begin{cases} x(t)S_\Delta(t) = x(t)\delta_\Delta(t) + \sum_{k=1}^{N-1} x(t)\delta_\Delta(t - kD) + x(t)\delta_\Delta(t - ND) \\ 0 \quad t \notin [0, ND) \end{cases} \quad (6.7)$$

The Fourier transform and power spectral density of y_Δ as Δ approaches zero, are specified in the following theorem.

THEOREM 6.2. FOURIER TRANSFORM AND POWER SPECTRAL DENSITY AFTER SAMPLING *If x is piecewise smooth on $[0, ND]$, is sampled at a rate $f_S = 1/D$, and is such that $\lim_{M \to \infty} \sum_{k=-M}^{M} X(ND, f - kf_S)$ converges for all $f \in \mathbf{R}$, then with Y_Δ as the Fourier transform of y_Δ, it follows that*

$$Y(ND, f) = \lim_{\Delta \to 0} Y_\Delta(ND, f) = f_S \sum_{k=-\infty}^{\infty} X(ND, f - kf_S) \quad (6.8)$$

$$G_Y(ND, f) = \lim_{\Delta \to 0} G_{Y_\Delta}(ND, f) = f_S^2 \sum_{k=-\infty}^{\infty} G_X(ND, f - kf_S)$$

$$+ \frac{f_S^2}{ND} \sum_{k=-\infty}^{\infty} \sum_{\substack{n=-\infty \\ n \neq k}}^{\infty} X(ND, f - kf_S) X^*(ND, f - nf_S) \quad (6.9)$$

Proof. The proof of this theorem is given in Appendix 2.

6.2.1.1 Notes If it is the case that

$$|X(ND, f)| \gg \left| \sum_{k \neq 0} X(ND, f - kf_S) \right| \quad \text{for } f \in (-f_S/2, f_S/2)$$

then

$$G_Y(ND, f) \approx f_S^2 G_X(ND, f) \quad f \in (-f_S/2, f_S/2) \quad (6.10)$$

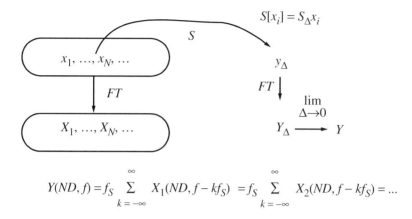

Figure 6.3 Illustration of sampling relationships.

and sampling has produced a scaled version of the true power spectral density in the frequency interval $[-f_S/2, f_S/2]$.

Figure 6.3 illustrates the relationship between the set of signals $\{x_1, \ldots, x_N, \ldots\}$, that are identical on arbitrarily small neighborhoods of the points $0^+, D, \ldots, ND^-$, and the Fourier transform of the sampled signal Y_Δ.

Clearly, sampling results in the Fourier transform and the power spectral density being repeated at integer multiples of the sampling frequency. To illustrate this, the power spectral density of a sampled 4 Hz sinusoid $A\sin(2\pi f_c t)$ is shown in Figure 6.4, where the sampling rate is 20 Hz and the

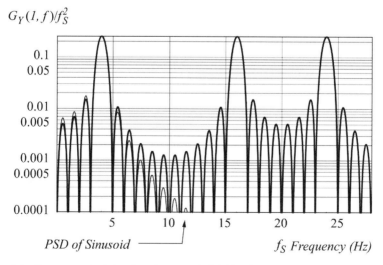

Figure 6.4 Power spectral density of a sampled 4 Hz sinusoid with unity amplitude. The sampling rate is 20 Hz and samples are from a 1 sec interval.

measurement interval is 1 sec. The power spectral density of such a sinusoid has been detailed in Section 3.2.3.3.

References for sampling theory include, Papoulis (1977 p. 160f), Champeney (1987 p. 162f), and Higgins (1996).

6.2.2 Power Spectral Density of Sampled Random Process

Consider a random process X that is characterized by an ensemble E_X of piecewise smooth signals on $[0, ND]$,

$$E_X = \{x: S_X \times R \to C\} \tag{6.11}$$

where $S_X \subseteq Z^+$ for the countable case and $S_X \subseteq R$ for the uncountable case. Consider a specific signal $x(\gamma, t)$ from E_X. Associated with this signal is an infinite set of sampled signals, defined according to

$$\{y_\Delta(\gamma, t) = S_\Delta(t)x(\gamma, t): t \in [0, ND], \Delta \in \{\Delta_i\}\} \tag{6.12}$$

where $\{\Delta_i\}$ is a sequence that converges to zero. The power spectral density associated with the limit of this sequence is given in Eq. (6.9), that is,

$$G_Y(\gamma, ND, f) = \lim_{\Delta \to 0} G_{Y_\Delta}(\gamma, ND, f) = f_S^2 \sum_{k=-\infty}^{\infty} G_X(\gamma, ND, f - kf_S)$$
$$+ \frac{f_S^2}{ND} \sum_{k=-\infty}^{\infty} \sum_{\substack{n=-\infty \\ n \neq k}}^{\infty} X(\gamma, ND, f - kf_S)X^*(\gamma, ND, f - nf_S) \tag{6.13}$$

The power spectral density of the random process formed through sampling each signal in E_X is the weighted summation of the resulting individual power spectral densities, that is, for the countable case,

$$G_Y(ND, f) = \sum_{\gamma=1}^{\infty} p_\gamma G_Y(\gamma, ND, f) = f_S^2 \sum_{\gamma=1}^{\infty} p_\gamma \sum_{k=-\infty}^{\infty} G_X(\gamma, ND, f - kf_S)$$
$$+ \frac{f_S^2}{ND} \sum_{\gamma=1}^{\infty} p_\gamma \sum_{k=-\infty}^{\infty} \sum_{\substack{n=-\infty \\ n \neq k}}^{\infty} X(\gamma, ND, f - kf_S)X^*(\gamma, ND, f - nf_S)$$

$$\tag{6.14}$$

where $P[x(\gamma, t)] = P[\gamma] = p_\gamma$. An analogous result holds for the uncountable case.

6.3 QUADRATURE AMPLITUDE MODULATION

One of the most popular and important communication modulation formats is quadrature amplitude modulation (QAM). A QAM signal x, is defined according to

$$x(t) = i(t) \cos(2\pi f_c t) - q(t) \sin(2\pi f_c t) \qquad (6.15)$$
$$= u(t) - v(t)$$

where i and q, respectively, are denoted the "inphase" and "quadrature" signals, f_c is the carrier frequency, $u(t) = i(t) \cos(2\pi f_c t)$, and $v(t) = q(t) \sin(2\pi f_c t)$.

In the general case, the signals i and q are specific signals from ensembles of two different random processes I and Q. Consider the case where the random process I is defined by the ensemble E_I, according to

$$E_I = \{i_k : \boldsymbol{R} \to \boldsymbol{C}, k \in \boldsymbol{Z}^+, P[i_k] = p_k\} \qquad (6.16)$$

A corresponding random process U, is defined by the ensemble E_U:

$$E_U = \{u_k : \boldsymbol{R} \to \boldsymbol{C}, u_k(t) = i_k(t) \cos(2\pi f_c t), k \in \boldsymbol{Z}^+, P[u_k] = p_k\} \qquad (6.17)$$

Similarly, the random processes Q and V can be defined by the ensembles E_Q and E_V:

$$E_Q = \{q_l : \boldsymbol{R} \to \boldsymbol{C}, l \in \boldsymbol{Z}^+, P[q_l] = p_l\} \qquad (6.18)$$
$$E_V = \{v_l : \boldsymbol{R} \to \boldsymbol{C}, v_l(t) = q_l(t) \sin(2\pi f_c t), l \in \boldsymbol{Z}^+, P[v_l] = p_l\} \qquad (6.19)$$

The random process $X = U - V$ can then be defined, in a manner consistent with Eq. (6.15), by the ensemble E_X:

$$E_X = \left\{ x_{kl} : \boldsymbol{R} \to \boldsymbol{C} \begin{array}{l} x_{kl}(t) = i_k(t) \cos(2\pi f_c t) - q_l(t) \sin(2\pi f_c t), \\ k, l \in \boldsymbol{Z}^+, P[x_{kl}] = P[i_k, q_l] = p_{kl} \end{array} \right\} \qquad (6.20)$$

For practical communication systems, the energy associated with all signals is finite. Thus, according to Theorem 3.6, the power spectral density of the modulating random processes I and Q, denoted G_I and G_Q, are finite for all frequencies when evaluated over the finite interval $[0, T]$. The assumption of finite energy is implicit in the following theorem and subsequent results.

THEOREM 6.3. POWER SPECTRAL DENSITY OF U, V, AND X *The power spectral density of U, V, and X on the interval $[0, T]$, are*

$$G_U(T, f) = \frac{G_I(T, f - f_c) + G_I(T, f + f_c)}{4}$$
$$+ \frac{1}{2T} \operatorname{Re}\left[\sum_{k=1}^{\infty} p_k [I_k(T, f - f_c) I_k^*(T, f + f_c)] \right] \quad (6.21)$$

$$G_V(T, f) = \frac{G_Q(T, f - f_c) + G_Q(T, f + f_c)}{4}$$
$$- \frac{1}{2T} \operatorname{Re}\left[\sum_{l=1}^{\infty} p_l [Q_l(T, f - f_c) Q_l^*(T, f + f_c)] \right] \quad (6.22)$$

$$G_X(T, f) = \frac{G_I(T, f - f_c) + G_I(T, f + f_c)}{4}$$
$$+ \frac{1}{2T} \operatorname{Re}\left[\sum_{k=1}^{\infty} p_k [I_k(T, f - f_c) I_k^*(T, f + f_c)] \right]$$
$$+ \frac{G_Q(T, f - f_c) + G_Q(T, f + f_c)}{4}$$
$$- \frac{1}{2T} \operatorname{Re}\left[\sum_{l=1}^{\infty} p_l [Q_l(T, f - f_c) Q_l^*(T, f + f_c)] \right]$$
$$+ \frac{\operatorname{Im}[G_{IQ}(T, f - f_c)]}{2}$$
$$- \frac{1}{2T} \operatorname{Im}\left[\sum_{k=1}^{\infty} \sum_{l=1}^{\infty} p_{kl} [I_k(T, f - f_c) Q_l^*(T, f + f_c)] \right]$$
$$+ \frac{-\operatorname{Im}[G_{IQ}(T, f + f_c)]}{2}$$
$$+ \frac{1}{2T} \operatorname{Im}\left[\sum_{k=1}^{\infty} \sum_{l=1}^{\infty} p_{kl} [I_k(T, f + f_c) Q_l^*(T, f - f_c)] \right] \quad (6.23)$$

where I_k and Q_l, are respectively, the Fourier transforms of i_k and q_l.

Proof. The proof of this theorem is given in Appendix 3.

6.3.1 Case 1: Bandlimited Signals

A common practical case in communication systems is where the power spectral densities of the inphase and quadrature components are only of significant level in the frequency range $-W < f < W$, where $W \ll f_c$, as

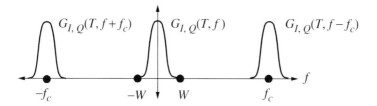

Figure 6.5 Forms for $G_I(T, f)$ and $G_Q(T, f)$ consistent with the bandlimited case.

illustrated in Figure 6.5. A general condition for the simplification that follows, is for the Fourier transforms of the inphase and quadrature signals to have negligible magnitude for frequencies greater than f_c, or less than $-f_c$.

For the case where I, Q, and the carrier frequency f_c are such that

$$\frac{2}{T}\left|\mathrm{Re}\sum_{k=1}^{\infty}p_k[I_k(T, f - f_c)I_k^*(T, f + f_c)]\right| \ll G_I(T, f - f_c) + G_I(T, f + f_c) \quad (6.24)$$

$$\frac{2}{T}\left|\mathrm{Re}\sum_{l=1}^{\infty}\frac{p_l[Q_l(T, f - f_c)Q_l^*(T, f + f_c)]}{T}\right| \ll G_Q(T, f - f_c) + G_Q(T, f + f_c)$$

$$(6.25)$$

$$\frac{1}{T}\left|\mathrm{Im}\sum_{k=1}^{\infty}\sum_{l=1}^{\infty}p_{kl}[I_k(T, f - f_c)Q_l^*(T, f + f_c)]\right|$$

$$+ \frac{1}{T}\left|\mathrm{Im}\sum_{k=1}^{\infty}\sum_{l=1}^{\infty}p_{kl}[I_k(T, f + f_c)Q_l^*(T, f - f_c)]\right| \ll G_X(T, f) \quad (6.26)$$

then the following approximation is valid:

$$G_X(T, f) \approx \frac{G_I(T, f - f_c) + G_I(T, f + f_c)}{4} + \frac{G_Q(T, f - f_c) + G_Q(T, f + f_c)}{4}$$

$$+ \frac{\mathrm{Im}[G_{IQ}(T, f - f_c)]}{2} - \frac{\mathrm{Im}[G_{IQ}(T, f + f_c)]}{2} \quad (6.27)$$

This approximate expression can be written very simply, if the definition of an equivalent low pass process, as discussed next, is used.

DEFINITION: EQUIVALENT LOW PASS RANDOM PROCESS An equivalent low pass signal w, defined according to (Proakis, 1995 p. 155),

$$w(t) = i(t) + jq(t) \quad (6.28)$$

where i and q are real signals, can be associated with a quadrature carrier signal

$$x(t) = i(t)\cos(2\pi f_c t) - q(t)\sin(2\pi f_c t) \tag{6.29}$$

as

$$x(t) = Re[w(t)e^{j2\pi f_c t}] \tag{6.30}$$

With the quadrature carrier random process X, defined by the ensemble E_X, as per Eq. (6.20), the equivalent low pass random process W can be defined by the ensemble E_W, according to

$$E_W = \{w_{kl}: \mathbf{R} \to \mathbf{C},\ w_{kl}(t) = i_k(t) + jq_l(t),\ k, l \in \mathbf{Z}^+,\ P[w_{kl}] = P[i_k, q_l] = p_{kl}\} \tag{6.31}$$

The power spectral density of W is specified in the following theorem.

THEOREM 6.4. POWER SPECTRAL DENSITY OF EQUIVALENT LOW PASS RANDOM PROCESS *If the power spectral densities of I and Q, denoted G_I and G_Q, can be validly defined, then the power spectral density of W, on the interval $[0, T]$, is*

$$\begin{aligned} G_W(T, f) &= G_I(T, f) + G_Q(T, f) + 2Im[G_{IQ}(T, f)] \\ G_W(T, -f) &= G_I(T, f) + G_Q(T, f) - 2Im[G_{IQ}(T, f)] \end{aligned} \tag{6.32}$$

Proof. The proof of the first result follows directly from Theorem 4.5, and by noting that $Re[-jG_{IQ}(T, f)] = Im[G_{IQ}(T, f)]$. The proof of the second result follows from the first result using the fact that for real signals, $X(T, -f) = X^*(T, f)$, which implies $G_X(T, -f) = G_X(T, f)$ and $G_{IQ}(T, -f) = G_{IQ}^*(T, f)$.

6.3.1.1 Notes With such a definition, it follows for the case of real bandlimited random processes, that the power spectral density of the QAM random process, as given in Eq. (6.27), can be written as

$$G_X(T, f) \approx \frac{G_W(T, f - f_c) + G_W(T, -f - f_c)}{4} \tag{6.33}$$

This simple form is one reason for the popularity of equivalent low pass random processes.

6.3.2 Case 2: Independent Inphase and Quadrature Processes

For the case where the random processes I and Q are independent, that is, $p_{kl} = p_k p_l$, the result from Section 4.5.2 for independent random processes, namely, $G_{IQ}(T, f) = \bar{I}(T, f)\bar{Q}^*(T, f)/T$, where $\bar{I}$ and $\bar{Q}$ are the respective averaged Fourier transforms of the signals defined by the random processes I and Q, yields

$$G_X(T, f) = \frac{G_I(T, f - f_c) + G_I(T, f + f_c)}{4}$$

$$+ \frac{1}{2T} \text{Re}\left[\sum_{k=1}^{\infty} p_k [I_k(T, f - f_c) I_k^*(T, f + f_c)]\right]$$

$$+ \frac{G_Q(T, f - f_c) + G_Q(T, f + f_c)}{4}$$

$$- \frac{1}{2T} \text{Re}\left[\sum_{l=1}^{\infty} p_l [Q_l(T, f - f_c) Q_l^*(T, f + f_c)]\right]$$

$$+ \frac{\text{Im}[\bar{I}(T, f - f_c)\bar{Q}^*(T, f - f_c)]}{2T} - \frac{\text{Im}[\bar{I}(T, f - f_c)\bar{Q}^*(T, f + f_c)]}{2T}$$

$$- \frac{\text{Im}[\bar{I}(T, f + f_c)\bar{Q}^*(T, f + f_c)]}{2T} + \frac{\text{Im}[\bar{I}(T, f + f_c)\bar{Q}^*(T, f - f_c)]}{2T}$$

(6.34)

For the independent and bandlimited case, the following approximation is valid:

$$G_X(T, f) \approx \frac{G_I(T, f - f_c) + G_I(T, f + f_c)}{4} + \frac{G_Q(T, f - f_c) + G_Q(T, f + f_c)}{4}$$

$$+ \frac{\text{Im}[\bar{I}(T, f - f_c)\bar{Q}^*(T, f - f_c)]}{2T} - \frac{\text{Im}[\bar{I}(T, f + f_c)\bar{Q}^*(T, f + f_c)]}{2T}$$

(6.35)

Further, if I and Q are identical random processes, then $\bar{Q}^*(T, f)$ equals the conjugate of $\bar{I}(T, f)$, and hence, $\bar{I}(T, f - f_c)\bar{Q}^*(T, f - f_c) = |\bar{I}(T, f - f_c)|^2$. Thus, for the identical, independent, and bandlimited case, it follows that

$$G_X(T, f) \approx \frac{G_I(T, f - f_c) + G_I(T, f + f_c)}{4} + \frac{G_Q(T, f - f_c) + G_Q(T, f + f_c)}{4}$$

$$- \frac{G_I(T, f - f_c) + G_I(T, f + f_c)}{2} = \frac{G_Q(T, f - f_c) + G_Q(T, f + f_c)}{2}$$

(6.36)

For the case where the random processes I and Q have constant means μ_i

and μ_q on the interval $[0, T]$, it follows from Section 4.5.2 that

$$G_{IQ}(T, f) = \bar{I}(T, f)\bar{Q}^*(T, f)/T = \mu_i \mu_q^* T \operatorname{sinc}^2(fT) \tag{6.37}$$

Hence, if the signals are real with a constant mean, then the imaginary part of $\bar{I}(T, f)\bar{Q}^*(T, f)$ is zero, and the following result holds when the signals are not necessarily bandlimited.

$$G_X(T, f) = \frac{G_I(T, f - f_c) + G_I(T, f + f_c)}{4}$$

$$+ \frac{1}{2T} \operatorname{Re}\left[\sum_{k=1}^{\infty} p_k [I_k(T, f - f_c) I_k^*(T, f + f_c)] \right]$$

$$+ \frac{G_Q(T, f - f_c) + G_Q(T, f + f_c)}{4}$$

$$- \frac{1}{2T} \operatorname{Re}\left[\sum_{l=1}^{\infty} p_l [Q_l(T, f - f_c) Q_l^*(T, f + f_c)] \right]$$

$$- \frac{\operatorname{Im}[\bar{I}(T, f - f_c)\bar{Q}^*(T, f + f_c)]}{2T} + \frac{\operatorname{Im}[\bar{I}(T, f + f_c)\bar{Q}^*(T, f - f_c)]}{2T}$$

$$\tag{6.38}$$

As shown in Appendix 4, for the independent case with real constant means, the last two terms in Eq. (6.38) can be neglected as $T \to \infty$ to yield

$$G_{X_\infty}(f) = \frac{G_{I_\infty}(f - f_c) + G_{I_\infty}(f + f_c)}{4} + \frac{G_{Q_\infty}(f - f_c) + G_{Q_\infty}(f + f_c)}{4}$$

$$+ \lim_{T \to \infty} \frac{1}{2T} \operatorname{Re}\left[\sum_{k=1}^{\infty} p_k [I_k(T, f - f_c) I_k^*(T, f + f_c)] \right] \tag{6.39}$$

$$- \lim_{T \to \infty} \frac{1}{2T} \operatorname{Re}\left[\sum_{l=1}^{\infty} p_l [Q_l(T, f - f_c) Q_k^*(T, f + f_c)] \right]$$

With the further assumption of bandlimited signals, it follows that

$$G_X(T, f) \approx \frac{1}{4}[G_I(T, f - f_c) + G_I(T, f + f_c) + G_Q(T, f - f_c) + G_Q(T, f + f_c)]$$

$$= \frac{1}{4}[G_W(T, f - f_c) + G_W(T, f + f_c)] \tag{6.40}$$

$$G_{X_\infty}(f) \approx \frac{1}{4}[G_{I_\infty}(f - f_c) + G_{I_\infty}(f + f_c) + G_{Q_\infty}(f - f_c) + G_{Q_\infty}(f + f_c)]$$

$$= \frac{1}{4}[G_{W_\infty}(f - f_c) + G_{W_\infty}(f + f_c)] \tag{6.41}$$

6.3.3 Case 3: Independent and Zero Mean Case

When the inphase and quadrature random processes are independent, and one or both of them have a zero mean, it follows that

$$G_X(T, f) = \frac{G_I(T, f - f_c) + G_I(T, f + f_c)}{4}$$

$$+ \frac{1}{2T} \text{Re}\left[\sum_{k=1}^{\infty} p_k[I_k(T, f - f_c)I_k^*(T, f + f_c)]\right]$$

$$+ \frac{G_Q(T, f - f_c) + G_Q(T, f + f_c)}{4} \quad (6.42)$$

$$- \frac{1}{2T} \text{Re}\left[\sum_{l=1}^{\infty} p_l[Q_l(T, f - f_c)Q_l^*(T, f + f_c)]\right]$$

For the case of bandlimited signals,

$$G_X(T, f) = \frac{1}{4}[G_I(T, f - f_c) + G_I(T, f + f_c) + G_Q(T, f - f_c) + G_Q(T, f + f_c)]$$

$$= \frac{1}{4}[G_W(T, f - f_c) + G_W(T, f + f_c)] \quad (6.43)$$

$$G_{X_\infty}(f) = \frac{1}{4}[G_{I_\infty}(f - f_c) + G_{I_\infty}(f + f_c) + G_{Q_\infty}(f - f_c) + G_{Q_\infty}(f + f_c)]$$

$$= \frac{1}{4}[G_{W_\infty}(f - f_c) + G_{W_\infty}(f + f_c)] \quad (6.44)$$

6.3.4 Example

For communication systems, I and Q are usually signaling random processes with power spectral densities given by Theorem 5.1. For example, consider the quadrature amplitude modulation random process X, where I and Q are independent and have identical RZ signaling random processes with power spectral densities as per Eqs. (5.42) and (5.43), and as shown in Figure 5.3. With f_c appropriately chosen, the bandlimited approximation, as per Eq. (6.36), is valid, that is,

$$G_X(T, f) \approx \frac{G_I(T, f - f_c) + G_I(T, f + f_c)}{2} = \frac{G_Q(T, f - f_c) + G_Q(T, f + f_c)}{2}$$

$$(6.45)$$

This power spectral density is shown in Figure 6.6 for the case where $f_c = 10$.

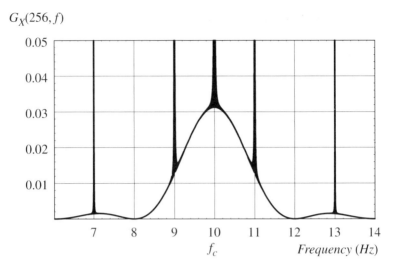

Figure 6.6 Power spectral density of a QAM signal where both the inphase and quadrature random processes are RZ random processes with power spectral densities as shown in Figure 5.3.

6.4 RANDOM WALKS

The quintessential nonstationary random process is a random walk, and such a random process has been extensively studied (for example, see Feller, 1957 ch. 3). The limit of a random walk in terms of an increasingly small step size and step interval, yields the Wiener process or Brownian motion (Grimmett, 1992 p. 342; Gillespie, 1996).

A random walk is clearly nonstationary, however, this does not present a problem for the power spectral density evaluated on an interval $[0, T]$ because it has its basis in the average power on this interval. The average power, and hence, the power spectral density, will change with the interval length and appropriate care must be taken when interpreting the power spectral density.

The model used for a random walk leads to a model for a bounded random walk which has a signaling random process form. Such a process has constant average power after an initial transient period. Bounded random walks provide a basis for synthesizing a $1/f$ power spectral density form. A synthesis is given for this form in the next section, and such a synthesis is consistent with a simple model for $1/f$ noise.

6.4.1 Modeling of a Random Walk

DEFINITION: SIGNAL DEFINITION FOR A RANDOM WALK A random walk is a signal that exhibits a step jump every D sec, with a step size randomly chosen with equal probability from the set $\{\pm A\}$. The signal is constant between jumps and initially is zero for the first interval of D sec.

A random walk random process, or random walk for short, consists of the ensemble of individual random walks defined by a set step interval, step jump, and step probabilities.

Consistent with the above definition, an individual random walk can be modeled on the interval $[0, T]$ as a summation of step waveforms consistent with those shown in Figure 6.7. With such a model, a random walk random process X, can be modeled on the interval $[0, T]$ by the ensemble E_X,

$$E_X = \left\{ x(\gamma_1, \ldots, \gamma_N, t) = A \sum_{i=1}^{N} \gamma_i u(t - iD), \gamma_i \in \{-1, 1\} \right\} \quad (6.46)$$

where $P[\gamma_i = \pm 1] = 0.5$, u is the unit step function, and $T = (N + 1)D$.

For the more general case, the step size takes on values from a zero mean continuous random variable Γ, with a density function f_Γ, and sample space S_Γ, that is,

$$E_X = \left\{ x(\gamma_1, \ldots, \gamma_N, t) = \sum_{i=1}^{N} \gamma_i u(t - iD), \gamma_i \in S_\Gamma \right\} \quad (6.47)$$

where $P[\gamma \in (\gamma_o, \gamma_o + d\gamma)] = f_\Gamma(\gamma_o) d\gamma$. For independent step sizes, it is the case that

$$P[x(\gamma_1, \ldots, \gamma_N, t)|_{\gamma_i \in I_i}] = \prod_{i=1}^{N} P[\gamma_i \in I_i] = \prod_{i=1}^{N} \int_{I_i} f_\Gamma(\gamma) d\gamma \quad (6.48)$$

Two random walks from the ensemble E_X, for the case of equally probable step sizes, $S_\Gamma = \{\pm 1\}$ and $D = 1$, are shown in Figure 6.8 for $t \in [0, 100]$.

6.4.2 Power Spectral Density of a Random Walk

By defining a random process X_i, on the interval $[0, T]$, by the ensemble

$$E_{X_i} = \{x(\gamma_i, t) = \gamma_i u(t - iD), \gamma_i \in S_\Gamma\} \quad (6.49)$$

it follows that the random process X, for $[0, T]$, is the sum of the N random

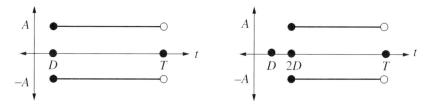

Figure 6.7 Possible waveforms associated with the first (left graph) and second (right graph) steps of a random walk.

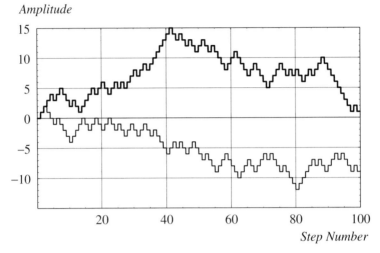

Figure 6.8 Two random walks on the interval [0, 100], for the case where $S_\Gamma = \{\pm 1\}$ and $D = 1$.

processes $X_1, \ldots, X_N$. The power spectral density of X_i is given by

$$G_{X_i}(T, f) = \frac{\sigma_\Gamma^2 |U_i(T, f)|^2}{T} \qquad U_i(T, f) = \frac{e^{-j2\pi fT} - e^{-j2\pi fiD}}{-j2\pi f} \qquad (6.50)$$

where $U_i(T, f)$ is the Fourier transform of $u(t - iD)$ evaluated over the interval $[0, T]$, σ_Γ^2 is the variance of Γ, and Γ has a zero mean. As the random processes $X_1, \ldots, X_N$ are independent and have zero mean, it follows from Theorem 4.6 that the power spectral density of X is the sum of the individual power spectral densities according to

$$G_X(T, f) = \sum_{i=1}^{N} G_{X_i}(T, f) \qquad (6.51)$$

Evaluation of the appropriate Fourier transforms yields

$$\begin{aligned} G_X(T, f) &= \frac{\sigma_\Gamma^2}{\pi^2 f^2 T} \sum_{i=1}^{N} \sin^2(i\pi fD) \qquad T = (N+1)D \\ &= \frac{\sigma_\Gamma^2 r}{2\pi^2 f^2} \left[1 - \frac{1}{2N+2}\right]\left[1 - \frac{\sin((2N+1)\pi fD)}{(2N+1)\sin(\pi fD)}\right] \end{aligned} \qquad (6.52)$$

where $r = 1/D$, and the last result follows from writing $\sin^2$ in terms of complex exponentials and using standard results for geometric series (Gradshteyn, 1980 p. 30).

By integrating the power spectral density, it follows that the average power in the random process is

$$\bar{P} = \int_{-\infty}^{\infty} G_X(T, f)\, df = \frac{\sigma_\Gamma^2 N}{2} = \frac{\sigma_\Gamma^2}{2}\left[\frac{T}{D} - 1\right] \quad (6.53)$$

Clearly, for $T \gg D$ the average power increases linearly with T, that is, the rms value of a random walk increases according to $\sqrt{T}$.

The power spectral density is shown in Figure 6.9 for the case of $D = r = 1$, $\bar{P} = 1$, and $N = 10$ which is consistent with $\sigma_\Gamma^2 = 0.2$. For the case where $f \ll 1/T$ and $N \gg 1$, the power spectral density approaches the constant value

$$G_X(T, f) \approx \frac{\sigma_\Gamma^2 N^2 D}{3} \qquad f \ll 1/T \quad (6.54)$$

The term

$$1 - \frac{\sin((2N + 1)\pi f D)}{(2N + 1)\sin(\pi f D)}$$

in Eq. (6.52), has the form shown in Figure 6.10, and hence, for $f \gg 1/T$ and $N \gg 1$ the power spectral density can be approximated as follows:

$$G_X(T, f) \approx \begin{cases} \dfrac{\sigma_\Gamma^2 r}{2\pi^2 f^2} & f \gg 1/T, f \neq kr, k \in \mathbf{Z} \\ 0 & f = kr, k \in \mathbf{Z} \end{cases} \quad (6.55)$$

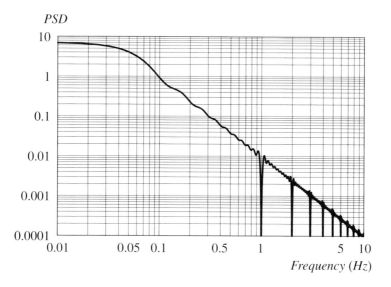

Figure 6.9 Power spectral density of a 10-step random walk with unity power and $D = r = 1$.

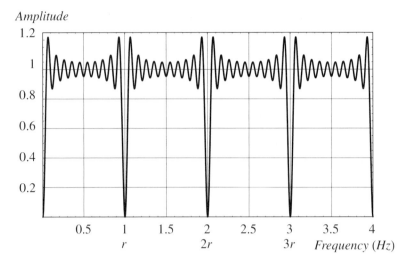

Figure 6.10 Plot of $1 - \sin((2N + 1)\pi fD)/(2N + 1)\sin(\pi fD)$ as a function of frequency f for the case of $r = D = 1$ and $N = 10$. The ripple around the level of 1 decreases as N increases.

6.4.3 Bounded Random Walk

One model for a random walk X bounded on the interval $[0, T]$, where $T = (N + 1)D$, is defined by the ensemble

$$E_X = \left\{ x(\gamma_1, \ldots, \gamma_N, t) = \sum_{i=1}^{N} \gamma_i \phi(t - iD), \gamma_i \in S_\Gamma \right\} \quad (6.56)$$

where ϕ has the form shown in Figure 6.11. For the case where $T_b = bD$, $b \in Z^+$, the random walk is correlated for T_b sec and, if $\gamma_i \in \{\pm 1\}$, it is bounded above and below by the levels $\pm b$. For the interval $[0, T_b]$, the random walk is that of a standard random walk. An example of a bounded random walk is shown in Figure 6.12 for the case where $S_\Gamma = \{\pm 1\}$, $P[\gamma_i = \pm 1] = 0.5$, $D = 1$, and $b = 10$.

The bounded random walk X, as defined by Eq. (6.56), is a signaling random process with zero mean. According to Theorem 5.1, its power spectral

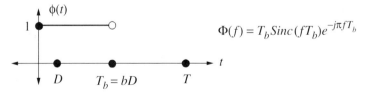

Figure 6.11 Pulse function for bounded random walk.

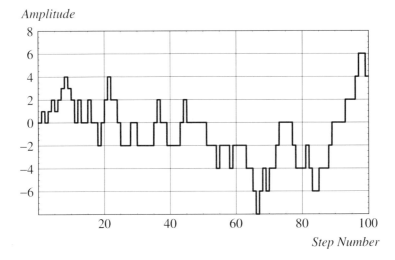

Figure 6.12 A bounded random walk with bounds of ± 10 ($b = 10$), $D = 1$, and $S_r = \{\pm 1\}$.

density is given by

$$G_X(T, f) = \frac{r\sigma_r^2}{1 + 1/N}|\Phi(f)|^2 = \frac{bT_b\sigma_r^2}{1 + 1/N}\text{sinc}^2(fT_b) \qquad T \gg T_b \quad (6.57)$$

where $r = 1/D$, Φ is the Fourier transform of ϕ, and the factor $1 + 1/N$ arises from the fact that $N/T = r/(1 + 1/N)$ for the case where $T = (N + 1)D$. The assumption made in this expression, is that the energy associated with the windowing effect of the interval $[0, T]$ on pulse functions is negligible. This is the case for $T \gg T_b$ or equivalently, $b \ll N$.

The average power in such a random process, obtained by integrating the power spectral density, and noting that the integral of $\sin^2(px)/x^2$ over the interval $(-\infty, \infty)$ equals πp (Spiegel, 1968 p. 96), is

$$\bar{P} = \int_{-\infty}^{\infty} G_X(T, f)\, df = \frac{r\sigma_r^2 T_b}{1 + 1/N} = \frac{b\sigma_r^2}{1 + 1/N} \qquad (6.58)$$

With this result, the power spectral density can be written as

$$G_X(T, f) = \bar{P}T_b\,\text{sinc}^2(fT_b) \qquad T \gg T_b \qquad (6.59)$$

The power spectral density is shown in Figure 6.13 for the case of $D = r = 1$, $\bar{P} = 1$, $b = 10$, and $N = 100$.

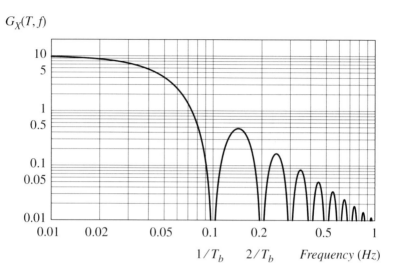

Figure 6.13 Power spectral density of a 100-step bounded random walk with unity power, correlation time of 10 steps, and $D = r = 1$.

6.5 1/f NOISE

Random processes that exhibit a power spectral density of the form $1/f^\alpha$ over a finite frequency range, where α is close to unity, are ubiquitous and, for example, such a form has been associated with economic data, traffic flow, annual rainfall, and noise in resistors, metals, and semiconductor devices [see, for example, Keshner (1982), Buckingham (1983), and Stephany (1998)]. Such noise, denoted $1/f$ or flicker noise, has been the subject of thorough investigation and modeling.

Research to explain $1/f$ noise has been along two lines. First, to ascertain physical attributes and origin of $1/f$ noise in a given entity (Buckingham, 1983; Hooge, 1981; Bell, 1980; Stephany, 1998). Second, research has been conducted to propose models, that is, random processes, that exhibit $1/f$ noise [see, for example, Hooge (1997), Kaulakys (1998), and Howard (2000)]. Many modeling approaches have been used including use of random walks [see, for example, Jantsch (1987) and Tunaley (1976)]. In the following section, a specific model for $1/f$ noise, based on bounded random walks, is demonstrated.

6.5.1 Synthesis of a 1/f Power Spectral Density Using Bounded Random Walks

A $1/f$ power spectral density form can be synthesized over a finite frequency range from a summation of distinct power spectral densities. For independent random processes with zero means, the goal of synthesis is to find N practical random processes, with respective power spectral densities $G_1, \ldots, G_N$, such that the $1/f$ form is approximated over a set frequency range $[f_L, f_U]$, that is,

$$\sum_{i=1}^{N} G_i(T, f) \approx \frac{k}{f} \quad f_L < f < f_U \quad \text{or} \quad \int_{f_L}^{f_U} \left| \sum_{i=1}^{N} G_i(T, f) - \frac{k}{f} \right| df \ll \int_{f_L}^{f_U} \frac{k}{f} df \quad (6.60)$$

To demonstrate such a synthesis, consider the power spectral density of the summation of $N + 1$ bounded random walks,

$$G(T, f) = \sum_{i=0}^{N} G_i(T, f) \quad (6.61)$$

where $G_i(T, f)$ is the power spectral density of a bounded random walk, as given by Eq. (6.59), with $T_b = bD_i$ and $D_i = 2^i D$. For equal powers in all random processes, with unity total power, $D = 1$ and $b = 10$, the power spectral density of G is shown in Figure 6.14, along with the ideal $1/f$ form. The step durations in the individual random processes are $1, 2, 4, 8, \ldots, 1024$ sec. The summation of bounded random walks clearly approximates the $1/f$ form over a restricted frequency range. A smoother approximation to the $1/f$ form can be achieved through a distribution of step durations or step rates, such that the average power in random processes with step durations between D and $2D$, is the same

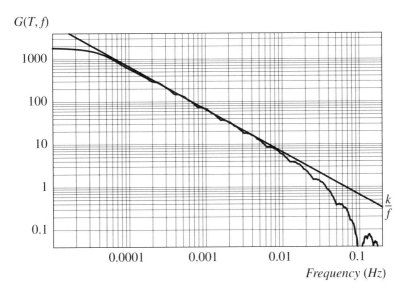

Figure 6.14 Power spectral density from the summation of 11 independent bounded random walks with step duration of $1, 2, 4, 8, \ldots, 1024$ sec.

as that for random processes with step durations between 2D and 4D, and so on. The lower limit to the $1/f$ form decreases as the step duration and interval duration increases.

This synthesis shows that a first order model consistent with $1/f$ noise, is a sum of equal power bounded random walks, with step durations that form a geometric series with a ratio of 2. Such a description provides a simple model and explanation for $1/f$ noise.

APPENDIX 1: PROOF OF THEOREM 6.1

If x is piecewise smooth on $[0, ND]$, then $X(ND, f)$ is finite for all f. Further, if there exists $k_o, \alpha > 0$, such that $|X(ND, f)| < k_o/|f|^{1+\alpha}$ for $f \in \mathbf{R}$, then, from the integral test (Knopp, 1956 pp. 64–65), it follows that

$$\lim_{M \to \infty} \sum_{k=-M}^{M} X(ND, f - kf_S) \qquad (6.62)$$

converges for all f and is bounded above. This condition is a sufficient condition. In general, if this summation converges, and this is assumed below, it is valid to define Y according to

$$Y(ND, f) = f_S \sum_{k=-\infty}^{\infty} X(ND, f - kf_S) \qquad f \in \mathbf{R}, f_S \in \mathbf{R}^+ \qquad (6.63)$$

Clearly, Y is periodic with respect to f with period f_S. Hence, on any interval of the form $[f_x, f_x + f_S]$, where $f_x \in \mathbf{R}$, Y can be written as an exponential Fourier series, according to

$$Y(ND, f) = \sum_{n=-\infty}^{\infty} c_n e^{j2\pi nDf} \qquad (6.64)$$

where $D = 1/f_S$ and the nth coefficient c_n, is given by

$$c_n = \frac{1}{f_S} \int_{f_x}^{f_x + f_S} \left[f_S \sum_{k=-\infty}^{\infty} X(ND, f - kf_S) \right] e^{-j2\pi nDf} df = \int_{f_x}^{f_x + f_S} \lim_{M \to \infty} g(M, f) df \qquad (6.65)$$

Here,

$$g(M, f) = \left[\sum_{k=-M}^{M} X(ND, f - kf_S) \right] e^{-j2\pi nDf} \qquad (6.66)$$

As it has been assumed that $\lim_{M \to \infty} g(M, f)$ is finite for $f \in [f_x, f_x + f_S]$, the interchange of limit and integral operations in this equation is valid. Thus,

$$c_n = \sum_{k=-\infty}^{\infty} \int_{f_x}^{f_x+f_S} X(ND, f - kf_S) e^{-j2\pi nDf} df \tag{6.67}$$

A change of variable, $\lambda = f - kf_S$, yields

$$c_n = \sum_{k=-\infty}^{\infty} e^{-j2\pi nk} \int_{f_x - kf_S}^{f_x - (k-1)f_S} X(ND, \lambda) e^{-j2\pi nD\lambda} d\lambda \tag{6.68}$$

Since the exponential term outside of the integral is unity for all values of the index k, and x is piecewise smooth, it follows from Theorem 2.30 that

$$c_n = \int_{-\infty}^{\infty} X(ND, \lambda) e^{j2\pi(-nD)\lambda} d\lambda = \begin{cases} 0 & -nD \notin [0, ND] \\ x(0^+)/2 & n = 0 \\ \dfrac{x(-nD^-) + x(-nD^+)}{2} & -nD \in (0, ND) \\ x(ND^-)/2 & -nD = ND \end{cases}$$

(6.69)

Thus,

$$Y(ND, f) = \frac{x(0^+)}{2} + \sum_{p=1}^{N-1} \frac{x(pD^-) + x(pD^+)}{2} e^{-j2\pi pDf} + \frac{x(ND^-) e^{-j2\pi NDf}}{2}$$

(6.70)

which is the required result.

APPENDIX 2: PROOF OF THEOREM 6.2

The Fourier transform of y_Δ, as defined by Eq. (6.7), is

$$Y_\Delta(ND, f) = \frac{1}{\Delta} \int_0^{\Delta/2} x(t) e^{-j2\pi ft} dt + \sum_{k=1}^{N-1} \frac{1}{\Delta} \int_{kD - \Delta/2}^{kD + \Delta/2} x(t) e^{-j2\pi ft} dt + \frac{1}{\Delta} \int_{ND - \Delta/2}^{ND} x(t) e^{-j2\pi ft} dt \tag{6.71}$$

For any $\varepsilon > 0$ and for any frequency range $[-f_z, f_z]$, there will exist a $\Delta > 0$, such that over an interval of measure Δ, centered at kD, both $x(t)$ and

$e^{-j2\pi ft}$ are such that

$$\begin{aligned}
-\varepsilon < |x(kD^-) - x(t)| < \varepsilon & \qquad kD - \Delta/2 < t < kD \\
-\varepsilon < |x(t) - x(kD^+)| < \varepsilon & \qquad kD < t < kD + \Delta/2 \\
-\varepsilon < |e^{-j2\pi ft} - e^{-j2\pi fkD}| < \varepsilon & \qquad kD - \Delta/2 < t < kD + \Delta/2
\end{aligned} \qquad (6.72)$$

These bounds are guaranteed by piecewise smoothness. Hence, for a frequency range $[-f_z, f_z]$, there will exist a Δ, such that $Y_\Delta(ND, f)$ can be approximated by

$$\frac{x(0^+)}{2} + \sum_{k=1}^{N-1} \frac{x(kD^-) + x(kD^+)}{2} e^{-j2\pi fkD} + \frac{x(ND^-)e^{-j2\pi fND}}{2} \qquad (6.73)$$

with an arbitrarily small error. Using the results from Theorem 6.1, it follows that

$$Y(ND, f) = \lim_{\Delta \to 0} Y_\Delta(ND, f) = f_S \sum_{k=-\infty}^{\infty} X(ND, f - kf_S) \qquad (6.74)$$

where $f_S = 1/D$. Further, since $\lim_{\Delta \to 0} Y_\Delta(ND, f)$ is bounded for $f \in R$, it follows that $|Y(ND, f)|^2 = \lim_{\Delta \to 0} |Y_\Delta(ND, f)|^2$ is finite for all frequencies, and hence,

$$G_Y(ND, f) = \lim_{\Delta \to 0} \frac{|Y_\Delta(ND, f)|^2}{ND} = \frac{f_S^2}{ND} \left| \sum_{k=-\infty}^{\infty} X(ND, f - kf_S) \right|^2 \qquad (6.75)$$

Expanding the summation within the modulus sign, yields the required result, namely,

$$\begin{aligned}
G_Y(ND, f) = & f_S^2 \sum_{k=-\infty}^{\infty} G_X(ND, f - kf_S) \\
& + \frac{f_S^2}{ND} \sum_{k=-\infty}^{\infty} \sum_{\substack{n=-\infty \\ n \ne k}}^{\infty} X(ND, f - kf_S) X^*(ND, f - nf_S)
\end{aligned} \qquad (6.76)$$

APPENDIX 3: PROOF OF THEOREM 6.3

A.3.1 Power Spectral Density of U and V

By definition, the power spectral densities of I and U on the interval $[0, T]$, are respectively,

$$G_I(T, f) = \sum_{k=1}^{\infty} p_k \frac{|I_k(T, f)|^2}{T} \qquad G_U(T, f) = \sum_{k=1}^{\infty} p_k \frac{|U_k(T, f)|^2}{T} \qquad (6.77)$$

where

$$I_k(T, f) = \int_0^T i_k(t) e^{-j2\pi ft} \, dt$$

$$U_k(T, f) = \int_0^T i_k(t) \cos(2\pi f_c t) e^{-j2\pi ft} \, dt \quad (6.78)$$

$$= \frac{I_k(T, f - f_c) + I_k(T, f + f_c)}{2}$$

Here, the relationship $\cos(\theta) = 0.5[e^{j\theta} + e^{-j\theta}]$ has been used to obtain the result for U_k. It then follows that

$$G_U(T, f) = \sum_{k=1}^{\infty} p_k \frac{|I_k(T, f - f_c) + I_k(T, f + f_c)|^2}{4T} = \frac{G_I(T, f - f_c) + G_I(T, f + f_c)}{4}$$

$$+ \frac{1}{2T} \mathrm{Re}\left[\sum_{k=1}^{\infty} p_k [I_k(T, f - f_c) I_k^*(T, f + f_c)] \right] \quad (6.79)$$

Similarly, since $v_l(t) = q_l(t) \sin(2\pi f_c t)$, the result $\sin(\theta) = 0.5j[-e^{j\theta} + e^{-j\theta}]$ implies that

$$V_l(T, f) = \frac{-jQ_l(T, f - f_c) + jQ_l(T, f + f_c)}{2} \quad (6.80)$$

where Q_l and V_l, are respectively, the Fourier transforms of q_l and v_l evaluated on the interval $[0, T]$. It then follows that

$$G_V(T, f) = \frac{G_Q(T, f - f_c) + G_Q(T, f + f_c)}{4}$$

$$- \frac{1}{2T} \mathrm{Re}\left[\sum_{l=1}^{\infty} p_l [Q_l(T, f - f_c) Q_l^*(T, f + f_c)] \right] \quad (6.81)$$

A.3.2 Power Spectral Density of Quadrature Amplitude Modulation Signal

Since $X = U - V$, it follows from Theorem 4.5 that

$$G_X(T, f) = G_U(T, f) + G_V(T, f) - 2\mathrm{Re}[G_{UV}(T, f)] \quad (6.82)$$

where G_U and G_V are given in Eqs. (6.79) and (6.81) and

$$G_{UV}(T, f) = \frac{1}{T} \sum_{k=1}^{\infty} \sum_{l=1}^{\infty} p_{kl} U_k(T, f) V_l^*(T, f)$$

$$= \frac{1}{4T} \sum_{k=1}^{\infty} \sum_{l=1}^{\infty} p_{kl} \begin{bmatrix} [I_k(T, f - f_c) + I_k(T, f + f_c)] \\ \times [jQ_l^*(T, f - f_c) - jQ_l^*(T, f + f_c)] \end{bmatrix}$$ (6.83)

With the result

$$\sum_{k=1}^{\infty} \sum_{l=1}^{\infty} \frac{p_{kl} I_k(T, f - f_c) jQ_l^*(T, f - f_c)}{T} = jG_{IQ}(T, f - f_c)$$ (6.84)

it then follows that

$$G_X(T, f) = G_U(T, f) + G_V(T, f) - \frac{1}{2} \text{Re}[jG_{IQ}(T, f - f_c) - jG_{IQ}(T, f + f_c)]$$

$$+ \frac{-1}{2T} \text{Re}\left[-j \sum_{k=1}^{\infty} \sum_{l=1}^{\infty} p_{kl} I_k(T, f - f_c) Q_l^*(T, f + f_c)\right]$$ (6.85)

$$+ \frac{-1}{2T} \text{Re}\left[j \sum_{k=1}^{\infty} \sum_{l=1}^{\infty} p_{kl} I_k(T, f + f_c) Q_l^*(T, f - f_c)\right]$$

As $\text{Re}[jG_{IQ}] = -\text{Im}[G_{IQ}]$, the required result follows.

APPENDIX 4: PROOF OF EQUATION 6.39

For the case where the random processes I and Q have constant means on the interval $[0, T]$, that is, $\mu_i(T, t) = \mu_i$ and $\mu_q(T, t) = \mu_q$, it follows that

$$\bar{I}(T, f) = \mu_i \int_0^T e^{-j2\pi ft} dt = \begin{cases} \mu_i[1 - e^{-j2\pi fT}]/j2\pi f & f \neq 0 \\ \mu_i T & f = 0 \end{cases}$$ (6.86)

and similarly for $Q(T, f)$. Hence,

$$\bar{I}(T, f - f_c)\bar{Q}^*(T, f + f_c) = \begin{cases} \dfrac{\mu_i \mu_q^*[1 - e^{-j2\pi(f - f_c)T}][1 - e^{j2\pi(f + f_c)T}]}{4\pi^2(f^2 - f_c^2)} & f \neq \pm f_c \\ \dfrac{\mu_i \mu_q^*[1 - e^{j2\pi(2f_c)T}]T}{-j2\pi(2f_c)} & f = \pm f_c \end{cases}$$

(6.87)

For $f \neq \pm f_c$, and for any $\varepsilon > 0$, there will exist a T, such that

$$\left|\frac{\bar{I}(T, f - f_c)\bar{Q}^*(T, f + f_c)}{2T}\right| \leq \frac{4|\mu_i \mu_q^*|}{8\pi^2 |f^2 - f_c^2| T} < \varepsilon \qquad (6.88)$$

For $f = \pm f_c$, and for the case where μ_i and μ_q are both real, it follows that

$$\frac{\text{Im}[\bar{I}(T, f - f_c)\bar{Q}^*(T, f + f_c)]}{2T} = \frac{\mu_i \mu_q [1 - \cos(2\pi(2f_c)T)]}{8\pi f_c} \qquad (6.89)$$

which is clearly finite for all values of T. However, the integral of this component is zero, that is, it is a component associated with zero energy. Accordingly, this component can be neglected as $T \to \infty$.

7

Memoryless Transformations of Random Processes

7.1 INTRODUCTION

This chapter uses the fact that a memoryless nonlinearity does not affect the disjointness of a disjoint random process to illustrate a procedure for ascertaining the power spectral density of a signaling random process after a memoryless transformation. Several examples are given, including two illustrating the application of this approach to frequency modulation (FM) spectral analysis. Alternative approaches are given in Davenport (1958 ch. 12) and Thomas (1969 ch. 6).

7.2 POWER SPECTRAL DENSITY AFTER A MEMORYLESS TRANSFORMATION

The approach given in this chapter relies on a disjoint partition of signals on a fixed interval. The following section gives the relevant results.

7.2.1 Decomposition of Output Using Input Time Partition

Consider a signal f which, based on a set of disjoint time intervals $\{I_1, \ldots, I_N\}$, can be written as a summation of disjoint waveforms according to

$$f(t) = \sum_{i=1}^{N} f_i(t) \qquad f_i(t) = \begin{cases} f(t) & t \in I_i \\ 0 & \text{elsewhere} \end{cases} \qquad (7.1)$$

POWER SPECTRAL DENSITY AFTER A MEMORYLESS TRANSFORMATION

If such a signal is input into a memoryless nonlinearity characterized by an operator G, then the output signal $g = G(f)$ can be written as a summation of disjoint waveforms according to

$$g(t) = \sum_{i=1}^{N} g_i(t) \qquad g_i(t) = \begin{cases} g(t) & t \in I_i \\ 0 & t \notin I_i \end{cases} \qquad (7.2)$$

where, as detailed in Section 2.3.3,

$$g_i(t) = \begin{cases} G(f_i(t)) & t \in I_i \\ 0 & t \notin I_i \end{cases} \qquad (7.3)$$

7.2.1.1 Implication If all signals from a signaling random process can be written as a summation of disjoint signals, then this result can be used to define each of the corresponding output signals after a memoryless transformation and hence, define a signaling random process for the output random process. As the power spectral density of a signaling random process is well defined (see Theorem 5.1), such an approach allows the output power spectral density to be readily evaluated.

Clearly, the applicability of this approach depends on the extent to which signals from a signaling random processes can be written as a summation of disjoint waveforms, that is, to the extent a signaling random process can be written as a disjoint signaling random process, which is defined as follows.

DEFINITION: DISJOINT SIGNALING RANDOM PROCESS A disjoint signaling random process X, with a signaling period D, is a signaling random process where each waveform in the signaling set is zero outside the interval $[0, D]$. The ensemble E_X characterizing such a random process for the interval $[0, ND]$ is

$$E_X = \left\{ x(\gamma_1, \ldots, \gamma_N, t) = \sum_{i=1}^{N} \phi(\gamma_i, t - (i-1)D), \gamma_i \in S_\Gamma, \phi \in E_\Phi \right\} \qquad (7.4)$$

where S_Γ is the sample space of the index random variable Γ, and is such that $S_\Gamma \subseteq \mathbf{Z}^+$ for the countable case, and $S_\Gamma \subseteq \mathbf{R}$ for the uncountable case. The set of signaling waveforms, E_Φ, is defined according to

$$E_\Phi = \{\phi(\gamma, t): \gamma \in S_\Gamma, \phi(\gamma, t) = 0, t < 0, t \geq D\} \qquad (7.5)$$

7.2.1.2 Equivalent Disjoint Signaling Random Process Consider a signaling random process X, defined by the ensemble

$$E_X = \left\{ x(\zeta_1, \ldots, \zeta_N, t) = \sum_{i=1}^{N} \psi(\zeta_i, t - (i-1)D), \zeta_i \in S_Z, \psi \in E_\Psi \right\} \qquad (7.6)$$

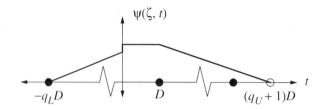

Figure 7.1 Illustration of signaling waveform.

where S_Z is the sample space of the index random variable Z and the set of signaling waveforms, E_Ψ, is defined according to

$$E_\Psi = \{\psi(\zeta, t): \zeta \in S_Z\} \tag{7.7}$$

Further, assume, as illustrated in Figure 7.1, that all signaling waveforms are nonzero only on a finite number of signaling intervals. It then follows that if a waveform in the random process starts with the signals associated with data in $[0, D]$, $[D, 2D]$, ... then a transient waveform exists in the interval $[0, q_U D]$. This transient is avoided for $t \geq 0$ if signals associated with data in the interval $[-q_U D, -(q_U - 1)D]$ and subsequent intervals are included.

The following theorem states that the random process defined in Eq. (7.6) can be written as a disjoint signaling random process with an appropriate disjoint signaling set. A likely, but not necessary consequence of this alternative characterization of a random process is the correlation between signaling waveforms in adjacent signaling intervals.

THEOREM 7.1. EQUIVALENT DISJOINT SIGNALING RANDOM PROCESS *If all signaling waveforms in the signaling set E_Ψ, associated with a signaling random process X, are zero outside $[-q_L D, (q_U + 1)D]$, where $q_L, q_U \in \{0\} \cup \mathbf{Z}^+$, then, for the steady state case, the signaling random process can be written on the interval $[0, ND]$, as a disjoint signaling random process with an ensemble*

$$E_X = \left\{ x(\gamma_1, \ldots, \gamma_N, t) = \sum_{i=1}^{N} \phi(\gamma_i, t - (i-1)D), \ \phi \in E_\Phi, \ \gamma_i \in S_\Gamma \right\} \tag{7.8}$$

The associated signaling set E_Φ is defined as

$$E_\Phi = \left\{ \phi(\gamma, t): \gamma \in S_\Gamma = S_Z \times \cdots \times S_Z, \gamma = (\zeta_{-q_U}, \ldots, \zeta_{q_L}), \zeta_{-q_U}, \ldots, \zeta_{q_L} \in S_Z \right\} \tag{7.9}$$

where

$$\phi(\gamma, t) = \begin{cases} \psi(\zeta_{-q_U}, t - (-q_U)D) + \cdots + \psi(\zeta_{q_L}, t - q_L D) & 0 \leq t < D \\ 0 & \text{elsewhere} \end{cases} \tag{7.10}$$

Proof. The proof of this result is given in Appendix 1.

7.2.1.3 Notes All waveforms in E_Φ are zero outside the interval $[0, D]$. The probability of each waveform and the correlation between waveforms, can be readily inferred from the original signaling random process. For the finite case where there are M independent signaling waveforms in E_Ψ, potentially there are $M^{q_L + q_U + 1}$ waveforms in E_Φ. In most instances the waveforms from different signaling intervals will be correlated.

7.2.2 Power Spectral Density After a Nonlinear Memoryless Transformation

Consider a disjoint signaling random process characterized over the interval $[0, ND]$ by the ensemble E_X and associated signaling set as per Eqs. (7.4) and (7.5). If waveforms from such a random process are passed through a memoryless nonlinearity, characterized by an operator G, then the corresponding output random process Y is characterized by the ensemble E_Y and associated signaling set E_Ψ, where

$$E_Y = \left\{ y(\gamma_1, \ldots, \gamma_N, t) = \sum_{i=1}^{N} \psi(\gamma_i, t - (i-1)D), \gamma_i \in S_\Gamma, \psi \in E_\Psi \right\} \quad (7.11)$$

and

$$E_\Psi = \{ \psi : \psi(\gamma, t) = G[\phi(\gamma, t)], \gamma \in S_\Gamma, \phi \in E_\Phi \} \quad (7.12)$$

Here, $P[\psi(\gamma, t)] = P[\phi(\gamma, t)] = P[\gamma]$. Clearly, the memoryless nonlinearity does not alter the signaling random process form, and the following result from Theorem 5.1 can be directly used to ascertain the power spectral density of the output random process,

$$G_Y(ND, f) = r\overline{|\Psi(f)|^2} - r|\mu_\Psi(f)|^2 + r|\mu_\Psi(f)|^2 \left[\frac{1}{N} \frac{\sin^2(\pi N f/r)}{\sin^2(\pi f/r)} \right] \quad (7.13)$$

$$+ 2r \sum_{i=1}^{m} \left[1 - \frac{i}{N} \right] \text{Re}[e^{j2\pi i D f} (R_{\Psi_1 \Psi_{1+i}}(f) - |\mu_\Psi(f)|^2)]$$

$$G_{Y_\infty}(f) = r\overline{|\Psi(f)|^2} - r|\mu_\Psi(f)|^2 + r^2|\mu_\Psi(f)|^2 \sum_{n=-\infty}^{\infty} \delta(f - nr) \quad (7.14)$$

$$+ 2r \sum_{i=1}^{m} \text{Re}[e^{j2\pi i D f} (R_{\Psi_1 \Psi_{1+i}}(f) - |\mu_\Psi(f)|^2)]$$

where $r = 1/D$ and μ_Ψ, $\overline{|\Psi(f)|^2}$, and $R_{\Psi_1 \Psi_{1+i}}$ are defined consistent with the definitions given in Theorem 5.1. For example, for the countable case $P[\gamma] = p_\gamma$ and

$$\mu_\Psi(f) = \sum_{\gamma=1}^{\infty} p_\gamma \Psi(\gamma, f) \qquad \overline{|\Psi(f)|^2} = \sum_{\gamma=1}^{\infty} p_\gamma |\Psi(\gamma, f)|^2 \qquad (7.15)$$

$$R_{\Psi_1 \Psi_{1+i}}(f) = \sum_{\gamma_1=1}^{\infty} \sum_{\gamma_{1+i}=1}^{\infty} p_{\gamma_1 \gamma_{1+i}} \Psi(\gamma_1, f) \Psi^*(\gamma_{1+i}, f) \qquad (7.16)$$

where $\Psi(\gamma, f) = \int_0^D \psi(\gamma, t) e^{-j2\pi ft} \, dt$.

7.2.3 Extension to Nonmemoryless Systems

It is clearly useful if the above approach can be extended to nonmemoryless systems. To facilitate this, it is useful to define a signaling invariant system.

7.2.3.1 Definition—Signaling Invariant System
A system is a signaling invariant system, if the output random process, in response to an input signaling random process is also a signaling random process and there is a one-to-one correspondence between waveforms in the signaling sets associated with the input and output random processes, that is, if $E_\Phi = \{\phi_i\}$ and $E_\Psi = \{\psi_i\}$ are, respectively, the input and output signaling sets, then there exists an operator G, such that $\psi_i = G[\phi_i]$.

A simple example of a signaling varying system is one where the output y, in response to an input x is defined as, $y(t) = x(t) + x(\pi t/4)$. For the case where the input is a waveform from a signaling random process the output is the summation of two signaling waveforms whose signaling intervals have an irrational ratio.

7.2.3.2 Implication
If a system is a signaling invariant system and is driven by a signaling random process, then the output is also a signaling random process whose power spectral density can be readily ascertained through use of Eqs. (7.13) and (7.14).

7.2.3.3 Signaling Invariant Systems
A simple example of a nonmemoryless, but signaling invariant system, is a system characterized by a delay, t_D. In fact, all linear time invariant systems are signaling invariant, as can be readily seen from the principle of superposition. However, the results of Chapter 8 yield a simple method for ascertaining the power spectral density of the output of a linear time invariant system, in terms of the input power spectral density, and the "transfer function" of the system.

7.3 EXAMPLES

The following sections give several examples of the above theory related to nonlinear transformations of random processes.

7.3.1 Amplitude Signaling through Memoryless Nonlinearity

Consider the case where the input random process X to a memoryless nonlinearity is a disjoint signaling random process, characterized on the interval $[0, ND]$, by the ensemble E_X:

$$E_X = \left\{ x(a_1, \ldots, a_N, t) = \sum_{i=1}^{N} \phi(a_i, t - (i-1)D), \, a_i \in S_A, \, \phi \in E_\Phi \right\} \quad (7.17)$$

where

$$E_\Phi = \left\{ \phi(a, t) = ap(t), \, a \in S_A, \, p(t) = \begin{cases} 1 & 0 \leq t < D \\ 0 & \text{elsewhere} \end{cases} \right\} \quad (7.18)$$

and $P[\phi(a, t)|_{a \in [a_o, a_o + da]}] = P[a \in [a_o, a_o + da]] = f_A(a_o)\,da$. Here, f_A is the density function of a random process A with outcomes a and sample space S_A. Assuming the signaling amplitudes are independent from one signaling interval to the next, it follows that the power spectral density of X is

$$G_X(ND, f) = r\overline{|\Phi(f)|^2} - r|\mu_\Phi(f)|^2 + r|\mu_\Phi(f)|^2 \left[\frac{1}{N}\frac{\sin^2(\pi Nf/r)}{\sin^2(\pi f/r)}\right] \quad (7.19)$$

$$G_{X_\infty}(f) = r\overline{|\Phi(f)|^2} - r|\mu_\Phi(f)|^2 + r^2|\mu_\Phi(f)|^2 \sum_{n=-\infty}^{\infty} \delta(f - nr) \quad (7.20)$$

where $r = 1/D$, and

$$\mu_\Phi(f) = P(f)\int_{-\infty}^{\infty} a f_A(a)\,da = \mu_A P(f) \qquad \mu_A = \int_{-\infty}^{\infty} a f_A(a)\,da$$

$$\overline{|\Phi(f)|^2} = |P(f)|^2 \int_{-\infty}^{\infty} a^2 f_A(a)\,da = \bar{A}^2 |P(f)|^2 \qquad \bar{A}^2 = \int_{-\infty}^{\infty} a^2 f_A(a)\,da \quad (7.21)$$

If signals from X are passed through a memoryless nonlinearity G, then, because of the disjointness of the input components of the signaling waveform, the output ensemble of the output random process Y, is

$$E_Y = \left\{ y(a_1, \ldots, a_N, t) = \sum_{i=1}^{N} \psi(a_i, t - (i-1)D), \, a_i \in S_A, \, \psi \in E_\Psi \right\} \quad (7.22)$$

where

$$E_\Psi = \{\psi(a, t) = G(a)p(t),\ a \in S_A\} \quad (7.23)$$

and

$$P[\psi(a, t)|_{a \in [a_o, a_o + da]}] = P[\phi(a, t)|_{a \in [a_o, a_o + da]}] = f_A(a_o)\, da \quad (7.24)$$

It then follows that the power spectral density of the output random process is

$$G_Y(ND, f) = r\overline{|\Psi(f)|^2} - r|\mu_\Psi(f)|^2 + r|\mu_\Psi(f)|^2 \left[\frac{1}{N}\frac{\sin^2(\pi N f/r)}{\sin^2(\pi f/r)}\right] \quad (7.25)$$

$$G_{Y_\infty}(f) = r\overline{|\Psi(f)|^2} - r|\mu_\Psi(f)|^2 + r^2|\mu_\Psi(f)|^2 \sum_{n=-\infty}^{\infty} \delta(f - nr) \quad (7.26)$$

where

$$\mu_\Psi(f) = \mu_G P(f) = P(f) \int_{-\infty}^{\infty} G(a) f_A(a)\, da$$

$$\overline{|\Psi(f)|^2} = \overline{G^2}|P(f)|^2 = |P(f)|^2 \int_{-\infty}^{\infty} G^2(a) f_A(a)\, da \quad (7.27)$$

where the following definitions have been used:

$$\mu_G = \int_{-\infty}^{\infty} G(a) f_A(a)\, da \qquad \overline{G^2} = \int_{-\infty}^{\infty} G^2(a) f_A(a)\, da \quad (7.28)$$

To illustrate these results, consider a square law device, that is, $G(a) = a^2$, and a Gaussian distribution of amplitudes according to

$$f_A(a) = e^{-a^2/2\sigma_A^2}/\sqrt{2\pi}\,\sigma_A$$

whereupon it follows that $\mu_G = \sigma_A^2$ and $\overline{G^2} = 3\sigma_A^4$ (Papoulis, 2002 p. 148). Thus, with $|P(f)| = |\text{sinc}(f/r)|/r$, it follows that

$$G_X(ND, f) = G_{X_\infty}(f) = \frac{\sigma_A^2}{r}\text{sinc}^2(f/r) \quad (7.29)$$

$$G_Y(ND, f) = \frac{2\sigma_A^4}{r}\text{sinc}^2(f/r) + \frac{\sigma_A^4}{r}\text{sinc}^2(f/r)\left[\frac{1}{N}\frac{\sin^2(\pi N f/r)}{\sin^2(\pi f/r)}\right] \quad (7.30)$$

$$G_{Y_\infty}(f) = \frac{2\sigma_A^4}{r}\text{sinc}^2(f/r) + \sigma_A^4 \delta(f) \quad (7.31)$$

Clearly, for this case, and in general, for disjoint signaling waveforms with information encoded in the signaling amplitude as per Eq. (7.18), the nonlinear transformation has scaled, but not changed the shape of the power spectral density function with frequency apart from impulsive components. For the case where the mean of the Fourier transform of the output signaling set is altered, compared with the corresponding input mean, potentially there is the introduction or removal of impulsive components in the power spectral density.

7.3.2 Nonlinear Filtering to Reduce Spectral Spread

Many nonlinearities yield spectral spread, that is, a broadening of the power spectral density. However, spectral spread is not inevitable and depends on the nature of the nonlinearity and the nature of the input signal. The following is one example of nonlinear filtering where the power spectral density spread is reduced.

Consider the case where the input signaling random process X is characterized on the interval $[0, ND]$, by the ensemble

$$E_X = \left\{ x(\gamma_1, \ldots, \gamma_N, t) = \sum_{i=1}^{N} \phi(\gamma_i, t - (i-1)D), \gamma_i \in S_\Gamma, \phi \in E_\Phi \right\} \quad (7.32)$$

where $S_\Gamma = \{-1, 1\}$, $P[\gamma_i = \pm 1] = 0.5$,

$$E_\Phi = \left\{ \phi(\gamma_i, t): \phi(\gamma_i, t) = \gamma_i A \Lambda \left(\frac{t - D/2}{D/2} \right), \gamma_i \in \{-1, 1\} \right\} \quad (7.33)$$

and the waveforms in different signaling intervals are independent. Here, Λ is the triangle function defined according to

$$\Lambda(t) = \begin{cases} 1 + t & -1 \leqslant t < 0 \\ 1 - t & 0 \leqslant t < 1 \\ 0 & \text{elsewhere} \end{cases} \quad (7.34)$$

Consider a nonlinearity, defined according to

$$G(x) = \begin{cases} -A_o & x < -A \\ A_o \sin\left(\frac{\pi}{2} \frac{x}{A}\right) & -A \leqslant x < A \\ A_o & x \geqslant A \end{cases} \quad (7.35)$$

which is shown in Figure 7.2, along with input and output waveforms.

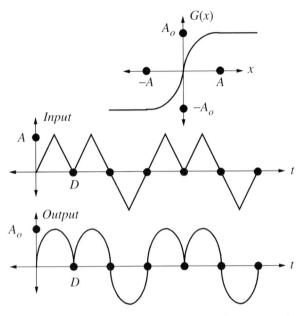

Figure 7.2 Memoryless nonlinearity and input and output waveforms.

It follows that the output signaling random process Y is characterized on the interval $[0, ND]$, by the ensemble

$$E_Y = \left\{ y(\gamma_1, \ldots, \gamma_N, t) = \sum_{i=1}^{N} \psi(\gamma_i, t - (i-1)D), \; \gamma_i \in S_\Gamma, \psi \in E_\psi \right\} \quad (7.36)$$

where

$$E_\psi = \left\{ \psi(\gamma_i, t) \colon \psi(\gamma_i, t) = \gamma_i A_o \sin\left[\frac{\pi}{2} \Lambda\left(\frac{t - D/2}{D/2}\right)\right] = \gamma_i A_o \sin\left[\frac{\pi t}{D}\right] \right\} \quad (7.37)$$
$$0 \leqslant t < D, \; \gamma_i \in S_\Gamma = \{-1, 1\}$$

Clearly, $P[\psi(\gamma_i, t)] = P[\gamma_i] = 0.5$. It follows that the power spectral density of the input and output waveforms are

$$G_X(ND, f) = G_{X_\infty}(f) = r\overline{|\Phi(f)|^2} \quad (7.38)$$

$$G_Y(ND, f) = G_{Y_\infty}(f) = r\overline{|\Psi(f)|^2} \quad (7.39)$$

where

$$\overline{|\Phi(f)|^2} = \frac{A^2}{4r^2} \operatorname{sinc}^4\left(\frac{f}{2r}\right) \qquad \overline{|\Psi(f)|^2} = \frac{4A_o^2}{\pi^2 r^2} \frac{\cos^2(\pi f/r)}{(1 - 4f^2/r^2)^2} \quad (7.40)$$

There is equal power in the input and output spectral densities when $A_o = \sqrt{2}A/\sqrt{3}$.

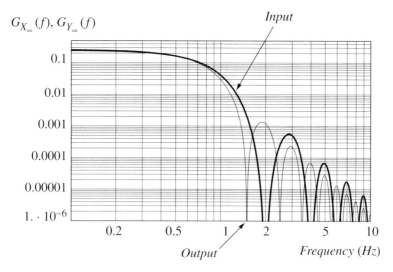

Figure 7.3 Input and output power spectral densities associated with the memoryless nonlinearity and waveforms shown in Figure 7.2.

These power spectral densities are plotted in Figure 7.3 for the case of $r = D = 1$, $A = 1$, and $A_o = \sqrt{2}/\sqrt{3}$. For this equal input and output power case, there is clear spectral narrowing consistent with the "smoothing" of the input waveform via the nonlinear transformation.

7.3.3 Power Spectral Density of Binary Frequency Shifted Keyed Modulation

As the following two examples show, signaling random process theory can readily be applied to ascertaining the power spectral density of FM random processes.

First, consider an FM signal,

$$y(t) = A \cos[x(t)] \qquad x(t) = 2\pi f_c t + \varphi(t) \qquad t \geqslant 0 \qquad (7.41)$$

where the carrier frequency f_c is an integer multiple of the signaling rate $r = 1/D$, and the binary digital modulation is such that φ has the form

$$\varphi(t) = 2\pi f_d \int_0^t \sum_{i=1}^{\infty} \gamma_i p(\lambda - (i-1)D) \, d\lambda \qquad \gamma_i \in \{-1, 1\}$$
$$= 2\pi f_d \sum_{i=1}^{\lfloor t/D \rfloor + 1} \gamma_i \int_0^t p(\lambda - (i-1)D) \, d\lambda \qquad (7.42)$$

Here, $P[\gamma_i = -1] = p_{-1}$ and $P[\gamma_i = 1] = p_1$, and the pulse function p is assumed to be such that

$$p(t) = 0 \quad t < 0, t \geqslant D \quad \int_0^D p(t)dt = D \tag{7.43}$$

which is consistent with a phase change of $\pm 2\pi(f_d/r)$ during each signaling interval of duration D sec. Clearly, $p(t)$ and $p(t - iD)$ are disjoint for $i \geqslant 1$.

With the assumptions that both f_d/r and f_c/r are integer ratios, it follows, as far as a cosine function is concerned, that the phase signal x in any interval of the form $[(i-1)D, iD]$, where $i \in Z^+$, can be written as

$$x(t) = 2\pi f_c(t - (i-1)D) + 2\pi f_d \gamma \int_0^{t-(i-1)D} p(\lambda)\, d\lambda \quad t \in [(i-1)D, iD], \gamma \in \{-1, 1\}$$

$$= \phi(\gamma, t - (i-1)D) \tag{7.44}$$

where

$$\phi(\gamma, t) = \begin{cases} 2\pi f_c t + 2\pi f_d \gamma \int_0^t p(\lambda)\, d\lambda & 0 \leqslant t < D, \gamma \in \{\pm 1\} \\ 0 & \text{elsewhere} \end{cases} \tag{7.45}$$

It then follows, for the ith signaling interval, that

$$y(t) = A\cos[x(t)] = A\cos[\phi(\gamma, t - (i-1)D)] \quad (i-1)D \leqslant t < iD \tag{7.46}$$

This formulation can be generalized to the random process case as follows: The random phase process X is defined by the ensemble E_X

$$E_X = \left\{ x(\gamma_1, \ldots, \gamma_N, t) = \sum_{i=1}^N \phi(\gamma_i, t - (i-1)D), \phi \in E_\Phi, \gamma_i \in \{\pm 1\}, t \in [0, ND] \right\} \tag{7.47}$$

where $P[\gamma_i] \in \{p_{-1}, p_1\}$ and the signaling set E_Φ is defined as

$$E_\Phi = \left\{ \phi(\gamma, t) = \begin{cases} 2\pi f_c t + 2\pi f_d \gamma \int_0^t p(\lambda)\, d\lambda & 0 \leqslant t < D, \gamma \in \{\pm 1\} \\ 0 & \text{elsewhere} \end{cases} \right\} \tag{7.48}$$

As any waveform in E_X consists of a summation of disjoint signals, a random process $Y = \cos[X]$ can be defined with an ensemble E_Y,

$$E_Y = \left\{ y(\gamma_1, \ldots, \gamma_N, t) = \sum_{i=1}^{N} \psi(\gamma_i, t - (i-1)D, \psi \in E_\Psi, \gamma_i \in \{\pm 1\}, t \in [0, ND] \right\} \quad (7.49)$$

where $y(\gamma_1, \ldots, \gamma_N, t)$ is a summation of disjoint signals from the signaling set E_Ψ,

$$E_\Psi = \left\{ \psi(\gamma, t) = \begin{cases} A \cos[\phi(\gamma, t)] & 0 \leq t < D, \gamma \in \{\pm 1\}, \phi \in E_\Phi \\ 0 & \text{elsewhere} \end{cases} \right\} \quad (7.50)$$

and $P[\psi(\gamma, t)] = P[\phi(\gamma, t)] \in \{p_{-1}, p_1\}$.

With independent data, consistent with γ_i being independent of γ_j for $i \neq j$, it follows from Eq. (7.14) that the power spectral density of Y is

$$G_{Y_x}(f) = r \sum_\gamma p_\gamma |\Psi(\gamma, f)|^2 - r \left| \sum_\gamma p_\gamma \Psi(\gamma, f) \right|^2$$

$$+ r^2 \left| \sum_\gamma p_\gamma \Psi(\gamma, f) \right|^2 \sum_{k=-\infty}^{\infty} \delta(f - kr) \qquad \gamma \in \{-1, 1\} \quad (7.51)$$

where Ψ is the Fourier transforms of ψ.

For the case where $p(t) = 1$ for $0 \leq t < D$ and zero elsewhere, that is, binary frequency shifted keyed (FSK) modulation, it follows that the signaling set is

$$E_\Psi = \begin{cases} \psi(1, t) = A \cos[2\pi(f_c + f_d)t] & 0 \leq t < D \\ \psi(-1, t) = A \cos[2\pi(f_c - f_d)t] & 0 \leq t < D \end{cases} \quad (7.52)$$

and

$$\Psi(1, f) = \frac{A}{2r} e^{j\pi(f_c + f_d - f)/r} \operatorname{sinc}\left(\frac{f_c + f_d - f}{r}\right)$$

$$+ \frac{A}{2r} e^{-j\pi(f_c + f_d + f)/r} \operatorname{sinc}\left(\frac{f_c + f_d + f}{r}\right) \quad (7.53)$$

$$\Psi(-1, f) = \frac{A}{2r} e^{j\pi(f_c - f_d - f)/r} \operatorname{sinc}\left(\frac{f_c - f_d - f}{r}\right)$$

$$+ \frac{A}{2r} e^{-j\pi(f_c - f_d + f)/r} \operatorname{sinc}\left(\frac{f_c - f_d + f}{r}\right) \quad (7.54)$$

For this case, and where $r = D = f_d = 1$, $A = \sqrt{2}$, $f_c = 10$, and $p_{-1} = p_1 = 0.5$,

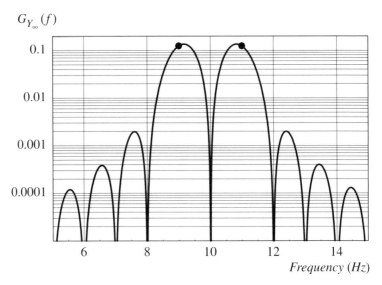

Figure 7.4 Power spectral density of a binary FSK random process with $r = D = f_d = 1$, $A = \sqrt{2}$, and $f_c = 10$. The dots represent the power in impulses.

the power spectral density, as defined by Eq. (7.51), is shown in Figure 7.4. A check on the power in the impulses can be simply undertaken by writing the FM signal $A\cos[2\pi(f_c \pm f_d)t]$ in the quadrature carrier form,

$$A\cos(2\pi f_c t)\cos(2\pi f_d t) \mp A\sin(2\pi f_c t)\sin(2\pi f_d t) \qquad (7.55)$$

The first term is periodic, and independent of the data, and yields impulses at $\pm f_c \pm f_d$ where the area under each impulse is $A^2/16$, which equals 0.125 when $A = \sqrt{2}$.

7.3.4 Frequency Modulation with Raised Cosine Pulse Shaping

Consider a FM signal with continuous phase modulation that is achieved through the use of raised cosine pulse shaping,

$$x(t) = A\sin\left[2\pi f_c t + 2\pi r \int_{-D}^{t} m(\lambda)\, d\lambda\right] \qquad t \geq 0 \qquad (7.56)$$

where $r = 1/D$ and the lower limit of $-D$ in the integral arises from the pulse waveform in the modulating signal m, which is defined according to

$$m(t) = \sum_{i=1}^{\infty} \zeta_i p(t - (i-1)D) \qquad \zeta_i \in \{\pm 0.5\}, t \geq -D \qquad (7.57)$$

EXAMPLES

Here, p is a raised cosine pulse with a duration of three signaling intervals (Proakis, 1995 p. 218), that is,

$$p(t) = \begin{cases} \frac{1}{3}\left[1 - \cos\left(\frac{2\pi(t+D)}{3D}\right)\right] & -D \leqslant t < 2D \\ 0 & \text{elsewhere} \end{cases} \quad (7.58)$$

and is shown in Figure 7.5. The integral of this raised cosine pulse shape, q, is

$$q(t) = \int_{-D}^{t} p(\lambda)\,d\lambda = \begin{cases} 0 & t < -D \\ \dfrac{t+D}{3} - \dfrac{D}{2\pi}\sin\left[\dfrac{2\pi(t+D)}{3D}\right] & -D \leqslant t < 2D \\ D & t \geqslant 2D \end{cases} \quad (7.59)$$

and the area under p is D. The value of $\zeta_i \in \{\pm 0.5\}$ in Eq. (7.57), results in each signaling waveform yielding a phase change of $\pm\pi$. The normalized integral of p, that is, $q(t/D)/D$, is shown in Figure 7.5.

As

$$\int_{-D}^{t}\left(\sum_{i=1}^{\infty}\zeta_i p(\lambda-(i-1)D)\right)d\lambda = \sum_{i=1}^{\lfloor t/D\rfloor+2}\zeta_i\int_{(i-2)D}^{t}p(\lambda-(i-1)D)\,d\lambda$$
$$= \sum_{i=1}^{\lfloor t/D\rfloor+2}\zeta_i q(t-(i-1)D) \quad t \geqslant -D \quad (7.60)$$

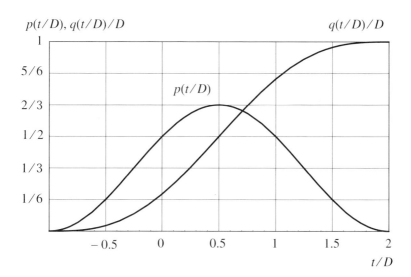

Figure 7.5 Raised cosine pulse waveform and normalized integral of such a waveform.

it follows that the FM signal defined by Eqs. (7.56) and (7.57), can be written as

$$x(t) = A \sin\left[2\pi f_c t + 2\pi r \sum_{i=1}^{\lfloor t/D \rfloor + 2} \zeta_i q(t - (i-1)D)\right] \qquad \zeta_i \in \{\pm 0.5\}, \, t \geq 0$$

(7.61)

The random process, of which this signal is one outcome, is denoted X and is defined, on the interval $[0, ND)$, by the ensemble E_X:

$$E_X = \left\{ \begin{array}{l} x(\zeta_0, \ldots, \zeta_{N+1}, t) = A \sin\left[w_c t + 2\pi r \sum_{i=0}^{\lfloor t/D \rfloor + 2} \zeta_i q(t - (i-1)D)\right] \\ w_c = 2\pi f_c, \, \zeta_i \in \{\pm 0.5\}, \, t \in [0, ND) \end{array} \right\}$$

(7.62)

where the effect of symbols in the interval $[-D, 0]$ and $[ND, (N+1)D]$, have been included to establish a steady state ensemble for $[0, ND]$. As the integral of the pulse shape is $D = 1/r$, and $\zeta_i \in \{\pm 0.5\}$, it follows that each pulse contributes a final phase shift of $\pm \pi$ radians to the argument of the sine function. Hence, each pair of symbols results in a phase shift from the set $-2\pi, 0, 2\pi$. As the sine function is periodic with period of 2π, it follows that in the ith signaling interval, $[(i-1)D, iD]$, the phase accumulation from the previous $\ldots, i-3, i-2$ symbols can be neglected. Thus, it is possible to rewrite the ensemble defining the random process X on the interval $[0, 2ND]$ in a signaling random process form, with a signaling rate of $r/2$, that is,

$$E_X = \left\{ x(\gamma_1, \ldots, \gamma_N, t) = \sum_{i=1}^{N} \phi_{\gamma_i}(t - (i-1)2D), \, \gamma_i \in \{1, \ldots, 16\}, \, \phi_{\gamma_i} \in E_\Phi, \, t \in [0, 2ND] \right\}$$

(7.63)

where the signaling set E_Φ consists of waveforms that are zero outside the interval $[0, 2D]$, and is defined according to

$$E_\Phi = \left\{ \phi_i(t) = \begin{cases} A \sin[2\pi f_c t + \varphi_i(t)] & 0 \leq t < 2D \\ 0 & \text{elsewhere} \end{cases} \quad i \in \{1, \ldots, 16\} \right\}$$ (7.64)

The waveforms in E_Φ, as well as the component phase waveforms φ_i, are detailed in Table 7.1. All waveforms in this set have equal probability, and the phase waveforms φ_i are plotted in Figure 7.6. The correlation between the signaling waveforms in adjacent signaling intervals of duration $2D$, is detailed in Table 7.2. The signaling waveforms in signaling intervals separated by at least $2D$, are independent as far as the sine operator is concerned.

EXAMPLES   221

Table 7.1   Signaling Waveforms in Signaling set

Data	Phase Waveforms $\varphi_1, \ldots, \varphi_{16}$ in [0, 2D]	Signaling Waveforms in [0, 2D]
0000	$\pi r[-q(t+D) - q(t) - q(t-D) - q(t-2D)]$	$\phi_1(t) = A\sin(2\pi f_c t + \varphi_1(t))$
0001	$\pi r[-q(t+D) - q(t) - q(t-D) + q(t-2D)]$	$\phi_2(t) = A\sin(2\pi f_c t + \varphi_2(t))$
0010	$\pi r[-q(t+D) - q(t) + q(t-D) - q(t-2D)]$	$\phi_3(t) = A\sin(2\pi f_c t + \varphi_3(t))$
0011	$\pi r[-q(t+D) - q(t) + q(t-D) + q(t-2D)]$	$\phi_4(t) = A\sin(2\pi f_c t + \varphi_4(t))$
0100	$\pi r[-q(t+D) + q(t) - q(t-D) - q(t-2D)]$	$\phi_5(t) = A\sin(2\pi f_c t + \varphi_5(t))$
0101	$\pi r[-q(t+D) + q(t) - q(t-D) + q(t-2D)]$	$\phi_6(t) = A\sin(2\pi f_c t + \varphi_6(t))$
0110	$\pi r[-q(t+D) + q(t) + q(t-D) - q(t-2D)]$	$\phi_7(t) = A\sin(2\pi f_c t + \varphi_7(t))$
0111	$\pi r[-q(t+D) + q(t) + q(t-D) + q(t-2D)]$	$\phi_8(t) = A\sin(2\pi f_c t + \varphi_8(t))$
1000	$\pi r[q(t+D) - q(t) - q(t-D) - q(t-2D)]$	$\phi_9(t) = A\sin(2\pi f_c t + \varphi_9(t))$
1001	$\pi r[q(t+D) - q(t) - q(t-D) + q(t-2D)]$	$\phi_{10}(t) = A\sin(2\pi f_c t + \varphi_{10}(t))$
1010	$\pi r[q(t+D) - q(t) + q(t-D) - q(t-2D)]$	$\phi_{11}(t) = A\sin(2\pi f_c t + \varphi_{11}(t))$
1011	$\pi r[q(t+D) - q(t) + q(t-D) + q(t-2D)]$	$\phi_{12}(t) = A\sin(2\pi f_c t + \varphi_{12}(t))$
1100	$\pi r[q(t+D) + q(t) - q(t-D) - q(t-2D)]$	$\phi_{13}(t) = A\sin(2\pi f_c t + \varphi_{13}(t))$
1101	$\pi r[q(t+D) + q(t) - q(t-D) + q(t-2D)]$	$\phi_{14}(t) = A\sin(2\pi f_c t + \varphi_{14}(t))$
1110	$\pi r[q(t+D) + q(t) + q(t-D) - q(t-2D)]$	$\phi_{15}(t) = A\sin(2\pi f_c t + \varphi_{15}(t))$
1111	$\pi r[q(t+D) + q(t) + q(t-D) + q(t-2D)]$	$\phi_{16}(t) = A\sin(2\pi f_c t + \varphi_{16}(t))$

Data of 0 and 1 correspond, respectively, to $\zeta_i = -0.5$ and $\zeta_i = 0.5$. The data in the first column are for the intervals $[-D, 0]$, $[0, D]$, $[D, 2D]$, and $[2D, 3D]$.

7.3.4.1   Determining Power Spectral Density

The power spectral density from Theorem 5.1, for a signaling random process with a rate $r_o = 1/D_o$, is

$$G_X(ND_o, f) = r_o \overline{|\Phi(f)|^2} - r_o |\mu_\Phi(f)|^2 + r_o |\mu_\Phi(f)|^2 \left[\frac{1}{N} \frac{\sin^2(\pi Nf/r_o)}{\sin^2(\pi f/r_o)} \right]$$
$$+ 2r_o \sum_{i=1}^{m} \left[1 - \frac{i}{N} \right] \text{Re}\left[e^{j2\pi i D_o f} (R_{\Phi_1 \Phi_{1+i}}(f) - |\mu_\Phi(f)|^2) \right]$$

(7.65)

where, for the case being considered, $r_o = r/2 = 1/2D$, $D_o = 2D$, $m = 1$, and

$$\mu_\Phi(f) = \sum_{i=1}^{16} p_i \Phi_i(f) \qquad \overline{|\Phi(f)|^2} = \sum_{i=1}^{16} p_i |\Phi_i(f)|^2 \qquad (7.66)$$

$$R_{\Phi_1 \Phi_2}(f) = \sum_{\gamma_1=1}^{16} \sum_{\gamma_2=1}^{16} p_{\gamma_1 \gamma_2} \Phi_{\gamma_1}(f) \Phi_{\gamma_2}^*(f) \qquad (7.67)$$

To evaluate the power spectral density, the Fourier transform of the individual waveforms in the signaling set, as defined by Eq. (7.64) and Table 7.1, are required to be evaluated. The details are given in Appendix 2. Using the results from this appendix, the power spectral density, as defined by Eqs. (7.65) to Eq. (7.67), is shown in Figure 7.7, for the case of $f_c = 10$, $r = D = 1$, $A = \sqrt{2}$, and $N \to \infty$. For the parameters used, the average power is $1V^2$ assuming a voltage signal. The power in each of the sinusoidal components with frequencies of $f_c \pm r/2$ is $0.11V^2$, and the remaining power of $0.78V^2$ is in the continuous spectrum.

222 MEMORYLESS TRANSFORMATIONS OF RANDOM PROCESSES

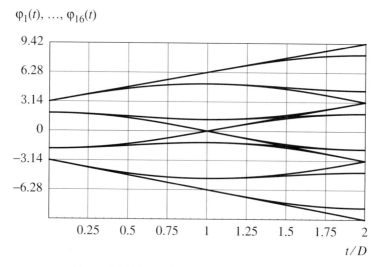

$\varphi_1(t), \ldots, \varphi_{16}(t)$

Figure 7.6 Phase signaling waveforms for [0, 2D].

The power in the impulsive components is consistent with inefficient signaling. These components can be eliminated by reducing the phase variation in each signaling waveform from π to $\pi/2$ radians. This also leads to better spectral efficiency (Proakis, 1995 p. 218). With respect to spectral efficiency, the power spectral density shown in Figure 7.7 should be compared with that

Table 7.2 Correlation between Signals in Signaling Intervals of Duration 2D

Data	Signal in ith Interval	Signal in $(i+1)$th Interval	Probability
xx0000	$\phi_1, \phi_5, \phi_9, \phi_{13}$	$\phi_1(t)$	1/64
xx0001	$\phi_1, \phi_5, \phi_9, \phi_{13}$	$\phi_2(t)$	1/64
xx0010	$\phi_1, \phi_5, \phi_9, \phi_{13}$	$\phi_3(t)$	1/64
xx0011	$\phi_1, \phi_5, \phi_9, \phi_{13}$	$\phi_4(t)$	1/64
xx0100	$\phi_2, \phi_6, \phi_{10}, \phi_{14}$	$\phi_5(t)$	1/64
xx0101	$\phi_2, \phi_6, \phi_{10}, \phi_{14}$	$\phi_6(t)$	1/64
xx0110	$\phi_2, \phi_6, \phi_{10}, \phi_{14}$	$\phi_7(t)$	1/64
xx0111	$\phi_2, \phi_6, \phi_{10}, \phi_{14}$	$\phi_8(t)$	1/64
xx1000	$\phi_3, \phi_7, \phi_{11}, \phi_{15}$	$\phi_9(t)$	1/64
xx1001	$\phi_3, \phi_7, \phi_{11}, \phi_{15}$	$\phi_{10}(t)$	1/64
xx1010	$\phi_3, \phi_7, \phi_{11}, \phi_{15}$	$\phi_{11}(t)$	1/64
xx1011	$\phi_3, \phi_7, \phi_{11}, \phi_{15}$	$\phi_{12}(t)$	1/64
xx1100	$\phi_4, \phi_8, \phi_{12}, \phi_{16}$	$\phi_{13}(t)$	1/64
xx1101	$\phi_4, \phi_8, \phi_{12}, \phi_{16}$	$\phi_{14}(t)$	1/64
xx1110	$\phi_4, \phi_8, \phi_{12}, \phi_{16}$	$\phi_{15}(t)$	1/64
xx1111	$\phi_4, \phi_8, \phi_{12}, \phi_{16}$	$\phi_{16}(t)$	1/64

The data in the first column are for the $(i-1)$th, ith and $(i+1)$th signaling intervals of duration 2D. The symbol x implies the data is arbitrary.

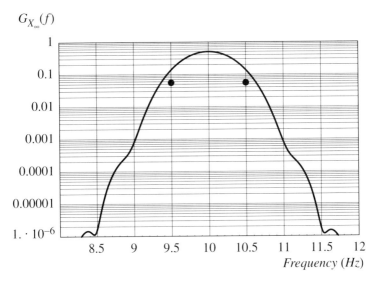

Figure 7.7 Power spectral density of a raised cosine pulse shaped FM random process with a carrier frequency of 10 Hz, $r = D = 1$, and $A = \sqrt{2}$. The dots represent the power in impulses.

shown in Figure 7.4, where pulse shaping has not been used, and the phase change for each signaling waveform is 2π radians. Finally, a further comparison of Figures 7.7 and 7.4, reveals that the pulse shaping has led to a very rapid spectral rolloff.

APPENDIX 1: PROOF OF THEOREM 7.1

Consider the steady state case and a single signaling waveform $\psi(\zeta_0, t)$ from the ensemble E_ψ, that could be associated with every signaling interval as shown in Figure 7.8. Clearly, the signal in the interval $[0, D]$ is given by

$$\psi(\zeta_0, t - (-q_U)D) + \cdots + \psi(\zeta_0, t) + \cdots + \psi(\zeta_0, t - q_L D) \qquad (7.68)$$

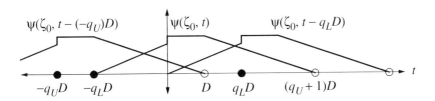

Figure 7.8 Illustration of signaling waveforms that have nonzero contributions in the interval $[0, D]$.

In general, the signal $\phi(\gamma, t)$ in the interval $[0, D]$ has the form

$$\phi(\gamma, t) = \begin{cases} \psi(\zeta_{-qU}, t-(-q_U)D) + \cdots + \psi(\zeta_0, t) + \cdots + \psi(\zeta_{qL}, t-q_L D) & 0 \leq t < D \\ 0 & \text{elsewhere} \end{cases} \quad (7.69)$$

where $\gamma = (\zeta_{-qU}, \ldots, \zeta_{qL})$ and $\zeta_{-qU}, \ldots, \zeta_{qL} \in S_Z$. By definition, $\phi(\gamma, t)$ is zero outside the interval $[0, D]$. As this interval is representative of any other interval of the form $[(i-1)D, iD]$, it follows that a signal from the random process can be written in the interval $[0, ND]$, as a sum of disjoint signals,

$$\sum_{i=1}^{N} \phi(\gamma_i, t-(i-1)D), \qquad \phi \in E_\Phi, \gamma_i \in S_\Gamma \quad (7.70)$$

where

$$E_\Phi = \{\phi(\gamma, t): \gamma \in S_\Gamma = S_Z \times \cdots \times S_Z, \gamma = (\zeta_{-qU}, \ldots, \zeta_{qL}), \zeta_{-qU}, \ldots, \zeta_{qL} \in S_Z\} \quad (7.71)$$

this is the required result.

APPENDIX 2: FOURIER RESULTS FOR RAISED COSINE FREQUENCY MODULATION

To establish the Fourier transform of each signaling waveform, explicit expressions for $q(t + D)$, $q(t)$, $q(t - D)$, and $q(t - 2D)$ are first required. Using the definition for q, as in Eq. (7.59), it follows that

$$q(t + D) = \frac{t + 2D}{3} + \frac{D}{4\pi} \sin(q_m t) + \frac{\sqrt{3}D}{4\pi} \cos(q_m t) \qquad -2D \leq t < D \quad (7.72)$$

$$q(t) = \frac{t + D}{3} + \frac{D}{4\pi} \sin(q_m t) - \frac{\sqrt{3}D}{4\pi} \cos(q_m t) \qquad -D \leq t < 2D \quad (7.73)$$

$$q(t - D) = \frac{t}{3} - \frac{2D}{4\pi} \sin(q_m t) \qquad 0 \leq t < 3D \quad (7.74)$$

$$q(t - 2D) = \frac{t - D}{3} + \frac{D}{4\pi} \sin(q_m t) + \frac{\sqrt{3}D}{4\pi} \cos(q_m t) \qquad D \leq t < 4D \quad (7.75)$$

where $q_m = 2\pi/3D$.

APPENDIX 2: FOURIER RESULTS FOR RAISED COSINE FREQUENCY MODULATION

A.2.1 Phase Waveforms for [0, D]

For $0 \leq t < D$, the phase of the waveforms, as detailed in Table 7.1, can be written as

$$\varphi_i(t) = \pi r[\gamma_{-1} q(t+D) + \gamma_0 q(t) + \gamma_1 q(t-D)] \tag{7.76}$$

where, $\gamma_{-1}, \gamma_0, \gamma_1 \in \{-1, 1\}$ depend, respectively, on data in the intervals $[-D, 0]$, $[0, D]$, and $[D, 2D]$. From Eqs. (7.72)–(7.75) it follows that

$$\varphi_i(t) = \pi r \left[\frac{D}{3}(2\gamma_{-1} + \gamma_0) + \frac{t}{3}(\gamma_{-1} + \gamma_0 + \gamma_1) \right.$$
$$\left. + \frac{D}{4\pi}(\gamma_{-1} + \gamma_0 - 2\gamma_1)\sin(q_m t) + \frac{\sqrt{3}D}{4\pi}(\gamma_{-1} - \gamma_0)\cos(q_m t) \right] \tag{7.77}$$

which can be rewritten as

$$\varphi_i(t) = q_o + q_1 t + q_2 \sin(q_m t + \theta_q) \tag{7.78}$$

where

$$q_o = \frac{\pi}{3}(2\gamma_{-1} + \gamma_0) \qquad q_1 = \frac{\pi}{3D}(\gamma_{-1} + \gamma_0 + \gamma_1) \tag{7.79}$$

$$q_2 = \frac{1}{4}\sqrt{(\gamma_{-1} + \gamma_0 - 2\gamma_1)^2 + 3(\gamma_{-1} - \gamma_0)^2} \tag{7.80}$$

$$\theta_q = \begin{cases} \tan^{-1}\left[\dfrac{\sqrt{3}(\gamma_{-1} - \gamma_0)}{\gamma_{-1} + \gamma_0 - 2\gamma_1}\right] & \gamma_{-1} + \gamma_0 - 2\gamma_1 > 0 \\ \tan^{-1}\left[\dfrac{\sqrt{3}(\gamma_{-1} - \gamma_0)}{\gamma_{-1} + \gamma_0 - 2\gamma_1}\right] + \pi & \gamma_{-1} + \gamma_0 - 2\gamma_1 < 0 \\ 0 & \gamma_{-1} - \gamma_0 = 0, \gamma_{-1} + \gamma_0 - 2\gamma_1 = 0 \end{cases} \tag{7.81}$$

A.2.2 Phase Waveforms for [D, 2D]

For $D \leq t < 2D$, the phase of the waveforms, as per Table 7.1, can be written as

$$\varphi_i(t) = \pi r[\gamma_{-1} D + \gamma_0 q(t) + \gamma_1 q(t-D) + \gamma_2 q(t-2D)] \tag{7.82}$$

where $\gamma_{-1}, \gamma_0, \gamma_1, \gamma_2 \in \{-1, 1\}$, and thus,

$$\varphi_i(t) = \pi r \left[\frac{D}{3}(3\gamma_{-1} + \gamma_0 - \gamma_2) + \frac{t}{3}(\gamma_0 + \gamma_1 + \gamma_2) \right.$$
$$\left. + \frac{D}{4\pi}(\gamma_0 - 2\gamma_1 + \gamma_2)\sin(q_m t) + \frac{\sqrt{3}D}{4\pi}(-\gamma_0 + \gamma_2)\cos(q_m t) \right] \quad (7.83)$$

This can be rewritten as

$$\varphi_i(t) = q_o + q_1 t + q_2 \sin(q_m t + \theta_q) \quad (7.84)$$

where

$$q_o = \frac{\pi}{3}(3\gamma_{-1} + \gamma_0 - \gamma_2) \qquad q_1 = \frac{\pi}{3D}(\gamma_0 + \gamma_1 + \gamma_2) \quad (7.85)$$

$$q_2 = \frac{1}{4}\sqrt{(\gamma_0 - 2\gamma_1 + \gamma_2)^2 + 3(-\gamma_0 + \gamma_2)^2} \quad (7.86)$$

$$\theta_q = \begin{cases} \tan^{-1}\left[\dfrac{\sqrt{3}(-\gamma_0 + \gamma_2)}{\gamma_0 - 2\gamma_1 + \gamma_2}\right] & \gamma_0 - 2\gamma_1 + \gamma_2 > 0 \\ \tan^{-1}\left[\dfrac{\sqrt{3}(-\gamma_0 + \gamma_2)}{\gamma_0 - 2\gamma_1 + \gamma_2}\right] + \pi & \gamma_0 - 2\gamma_1 + \gamma_2 < 0 \\ 0 & -\gamma_0 + \gamma_2 = 0, \gamma_0 - 2\gamma_1 + \gamma_2 = 0 \end{cases} \quad (7.87)$$

A.2.3 Fourier Transform of Signaling Waveforms

For the interval $[0, 2D]$, the Fourier transform of the ith waveform in the signaling set is

$$\Phi_i(f) = \int_0^{2D} A \sin(2\pi f_c t + \varphi_i(t)) e^{-j2\pi f t} \, dt$$
$$= \frac{A}{2j} \int_0^{2D} \left[e^{j[2\pi(f_c - f)t + \varphi_i(t)]} - e^{-j[2\pi(f_c + f)t + \varphi_i(t)]} \right] dt \quad (7.88)$$

Substituting for $\varphi_i(t)$, and with the definitions $u_{1n}(f) = 2\pi(f_c - f) + q_1$ and $u_{1p}(f) = 2\pi(f_c + f) + q_1$, it follows that $\Phi_i(f)$ can be written as

$$\frac{A}{2j} \int_0^D \left[e^{j[q_o + u_{1n}(f)t + q_2 \sin(q_m t + \theta_q)]} - e^{-j[q_o + u_{1p}(f)t + q_2 \sin(q_m t + \theta_q)]} \right] dt$$
$$+ \frac{A}{2j} \int_D^{2D} \left[e^{j[q_o + u_{1n}(f)t + q_2 \sin(q_m t + \theta_q)]} - e^{-j[q_o + u_{1p}(f)t + q_2 \sin(q_m t + \theta_q)]} \right] dt \quad (7.89)$$

APPENDIX 2: FOURIER RESULTS FOR RAISED COSINE FREQUENCY MODULATION

where, as is clear from the above derivation of the phase waveforms in $[0, D]$ and $[D, 2D]$, the coefficients q_o, u_{1n}, u_{1p}, q_2, and θ_q vary from $[0, D]$ to $[D, 2D]$. With the change of variable $\lambda = t + \theta_q/q_m$, and with the definitions $v_{1n}(f) = u_{1n}(f)\theta_q/q_m$, $v_{1p}(f) = u_{1p}(f)\theta_q/q_m$, it follows that $\Phi_i(f)$ can be written as

$$\Phi_i(f) = \frac{-Aje^{j[q_o - v_{1n}(f)]}}{2} \int_{\theta_q/q_m}^{D+\theta_q/q_m} e^{j[u_{1n}(f)\lambda + q_2 \sin(q_m\lambda)]} d\lambda$$

$$+ \frac{Aje^{-j[q_o - v_{1p}(f)]}}{2} \int_{\theta_q/q_m}^{D+\theta_q/q_m} e^{-j[u_{1p}(f)\lambda + q_2 \sin(q_m\lambda)]} d\lambda$$

$$+ \frac{-Aje^{j[q_o - v_{1n}(f)]}}{2} \int_{D+\theta_q/q_m}^{2D+\theta_q/q_m} e^{j[u_{1n}(f)\lambda + q_2 \sin(q_m\lambda)]} d\lambda$$

$$+ \frac{Aje^{-j[q_o - v_{1p}(f)]}}{2} \int_{D+\theta_q/q_m}^{2D+\theta_q/q_m} e^{-j[u_{1p}(f)\lambda + q_2 \sin(q_m\lambda)]} d\lambda \tag{7.90}$$

Evaluation of these integrals relies on the result,

$$I(k_o, w_C, w_m, w_A, \lambda_1, \lambda_2) = \int_{\lambda_1}^{\lambda_2} e^{jk_o[w_C\lambda + w_A \sin(w_m\lambda)]} d\lambda$$

$$= \sum_{i=0}^{\infty} (-1)^i J_i(w_A) \left[\frac{\sin((w_C - iw_m)\lambda_2) - \sin((w_C - iw_m)\lambda_1)}{w_C - iw_m} \right]$$

$$+ \sum_{i=1}^{\infty} J_i(w_A) \left[\frac{\sin((w_C + iw_m)\lambda_2) - \sin((w_C + iw_m)\lambda_1)}{w_C + iw_m} \right]$$

$$+ (-j)k_o \sum_{i=0}^{\infty} (-1)^i J_i(w_A) \left[\frac{\cos((w_C - iw_m)\lambda_2) - \cos((w_C - iw_m)\lambda_1)}{w_C - iw_m} \right]$$

$$+ (-j)k_o \sum_{i=1}^{\infty} J_i(w_A) \left[\frac{\cos((w_C + iw_m)\lambda_2) - \cos((w_C + iw_m)\lambda_1)}{w_C + iw_m} \right]$$

$$\tag{7.91}$$

where $k_o \in \{-1, 1\}$. This result arises from the standard Bessel function expansions for the terms $\cos(w_A \sin(w_m\lambda))$ and $\sin(w_A \sin(w_m\lambda))$ (Spiegel, 1968 p. 145), that is,

$$\cos(w_A \sin(w_m\lambda)) = J_0(w_A) + 2J_2(w_A) \cos(2w_m\lambda) + 2J_4(w_A) \cos(4w_m\lambda) + \cdots \tag{7.92}$$

$$\sin(w_A \sin(w_m\lambda)) = 2J_1(w_A) \sin(w_m\lambda) + 2J_3(w_A) \sin(3w_m\lambda) + \cdots \tag{7.93}$$

$\Phi_i(f)$ can be evaluated using Eq. (7.91), that is,

$$\begin{aligned}\Phi_i(f) = & \frac{-Aje^{j[q_o - v_{1n}(f)]}}{2} I(1, u_{1n}(f), q_m, q_2, \theta_q/q_m, D + \theta_q/q_m) \\ & + \frac{Aje^{-j[q_o - v_{1p}(f)]}}{2} I(-1, u_{1p}(f), q_m, q_2, \theta_q/q_m, D + \theta_q/q_m) \\ & + \frac{-Aje^{j[q_o - v_{1n}(f)]}}{2} I(1, u_{1n}(f), q_m, q_2, D + \theta_q/q_m, 2D + \theta_q/q_m) \\ & + \frac{Aje^{-j[q_o - v_{1p}(f)]}}{2} I(-1, u_{1p}(f), q_m, q_2, D + \theta_q/q_m, 2D + \theta_q/q_m)\end{aligned}$$

(7.94)

In the first two component expressions in this equation, q_o, v_{1n}, v_{1p}, q_2, and θ_q are defined for $[0, D]$, whereas in the last two component expressions, these variables are defined for $[D, 2D]$.

8
Linear System Theory

8.1 INTRODUCTION

In this chapter, the fundamental relationships between the input and output of a linear time invariant system, as illustrated in Figure 8.1, are detailed. Specifically, the relationships between the input and output time signals, Fourier transforms and power spectral densities, are established. Such relationships are fundamental to many aspects of system theory, including analysis of noise in linear systems, and low noise amplifier design.

The relationships between the parameters defined in Figure 8.1, and proved in this chapter are,

$$y(t) = \int_0^t x(\lambda)h(t-\lambda)\,d\lambda \qquad (8.1)$$

$$Y(T, f) \approx H(T, f)X(T, f) \qquad (8.2)$$

where X and Y are the respective Fourier transforms, evaluated on the interval $[0, T]$, of the signals x and y. However, as will be shown in this chapter, the relationship defined in Eq. (8.2) is an approximation. If both $x, h \in L$, then the relative error in this approximation can be made arbitrarily small by making T sufficiently large. However, stationary random signals are not Lebesgue integrable on the interval $(0, \infty)$ and hence, this convergence is not guaranteed. However, it is shown, for a broad class of signals and random processes, including periodic signals and stationary random processes, that the corresponding relationship between the input and output power spectral densities, namely,

$$G_Y(T, f) \approx |H(T, f)|^2 G_X(T, f) \qquad (8.3)$$

becomes exact as T increases without bound. Establishing the relationships, as per Eqs. (8.1)–(8.3), for a linear time invariant system requires the system impulse response to be defined, and this is the subject of the next section.

230 LINEAR SYSTEM THEORY

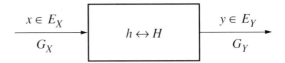

Figure 8.1 Schematic diagram of a linear system. E_X and E_Y, respectively, represent the ensemble of input and output signals. H is the Fourier transform of the impulse response function h. G_X and G_Y, respectively, are the power spectral densities of the input and output random processes.

8.2 IMPULSE RESPONSE

Fundamental to defining the impulse function of a time invariant linear system, is the function δ_Δ defined by the graph shown in Figure 8.2. The response of a linear time invariant system to the input signal δ_Δ is denoted h_Δ.

DEFINITION: IMPULSE RESPONSE By definition, the impulse response of a linear system is the output signal, in response to the input signal δ_Δ, as Δ becomes increasingly small, that is,

$$h(t) = \lim_{\Delta \to 0} h_\Delta(t) \tag{8.4}$$

8.2.1 Restrictions on Impulse Response

General requirements on the impulse function are, first, that it is integrable, that is, $h \in L[0, \infty]$, and second, that as $\Delta \to 0$, the integrated difference between h and h_Δ is negligible on sets of nonzero measure, that is, convergence in the mean over $(0, \infty)$:

$$\lim_{\Delta \to 0} \int_0^\infty |h_\Delta(t) - h(t)| dt = 0 \tag{8.5}$$

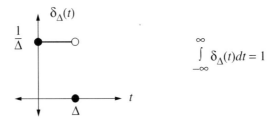

Figure 8.2 Definition of the function δ_Δ.

IMPULSE RESPONSE   231

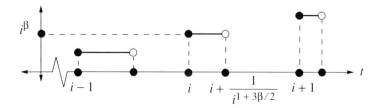

Figure 8.3 Illustration of a function that is Lebesgue integrable on [0, ∞], but is not square Lebesgue integrable on the same interval.

The following are two examples where, as $\Delta \to 0$, the integrated error between h and h_Δ is finite. First, the "identity" system where $h_\Delta(t) = \delta_\Delta(t)$ and second, the system where

$$h_\Delta(t) = \begin{cases} \Delta & t \in [\Delta, \Delta + 1/\Delta] \\ 0 & \text{elsewhere} \end{cases} \tag{8.6}$$

For both systems, and for $t \in (0, \infty)$, it follows that

$$h(t) = \lim_{\Delta \to 0} h_\Delta(t) = 0 \quad \text{but} \quad \lim_{\Delta \to 0} \int_0^\infty |h_\Delta(t) - h(t)| dt \neq 0 \tag{8.7}$$

To ensure $h \in L[0, \infty]$, and as $\Delta \to 0$ the integrated difference between h and h_Δ is negligible, the following restriction on the set of functions $\{h_\Delta\}$, denoted condition 1, is sufficient:

1. There exists a function $g \in L[0, \infty]$, such that, for all $\Delta > 0$ it is the case that $|h_\Delta(t)| \leq g(t)$ for $t \in [0, \infty]$.

The validity of this condition, in terms of guaranteeing that Eq. (8.5) holds, is given by Theorem 2.25.

Practical and stable systems are such that h_Δ is bounded and has finite energy for all values of Δ. As per Theorem 2.14 these two criteria are met by condition 1 and the following condition.

2. For $\forall \Delta > 0$, h_Δ is bounded, that is, $\forall \Delta > 0$, $|h_\Delta(t)| \leq h_{\max}$ for $t \in [0, \infty]$.

This second condition excludes a signal such as $1/\sqrt{t}$, which is integrable on $[0, \infty]$, but has infinite energy on all intervals of the form $[0, t_o]$. It also excludes signals such as the one shown in Figure 8.3, whose integral equals $\sum_{i=1}^\infty (1/i^{1+\beta/2})$, which from the comparison test (Knopp, 1956 pp. 56f), is finite for $\beta > 0$, but whose energy is given by $\sum_{i=1}^\infty (1/i^{1-\beta/2})$ and is infinite when $\beta > 0$.

8.3 INPUT–OUTPUT RELATIONSHIP

Consider the causal linear time invariant system illustrated in Figure 8.1. The well-known relationship between the input and output signals is specified in the following theorem.

THEOREM 8.1. INPUT–OUTPUT RELATIONSHIP FOR A LINEAR SYSTEM *If the input signal x to, and the system impulse response h of, a linear time invariant system are both causal, are locally integrable, and have bounded variation on all finite intervals, then the output signal, y, is given by*

$$y(t) = \int_0^t x(\lambda) h(t - \lambda) \, d\lambda \qquad t > 0 \tag{8.8}$$

Proof. The proof of this result is given in Appendix 1.

Note that this result is applicable to unstable systems where $h \notin L[0, \infty]$.

8.4 FOURIER AND LAPLACE TRANSFORM OF OUTPUT

The following theorem states the important result of the relationship between the Fourier and Laplace transforms of the input and output of a linear time invariant system.

THEOREM 8.2. TRANSFORMS OF OUTPUT SIGNAL OF A LINEAR SYSTEM *If both $x, h \in L[0, T]$, have bounded variation on $[0, T]$, and their respective Fourier transforms are denoted X and H, then the Fourier transform Y of the output signal y, evaluated on $[0, T]$, is given by*

$$Y(T, f) = \iint\limits_{\substack{0 \leq p, \lambda < T \\ p + \lambda \leq T}} x(\lambda) h(p) e^{-j2\pi f(p+\lambda)} \, d\lambda \, dp \tag{8.9}$$

$$= \tilde{Y}(T, f) - I(T, f) = X(T, f) H(T, f) - I(T, f)$$

where

$$\tilde{Y}(T, f) = X(T, f) H(T, f)$$

$$I(T, f) = \iint\limits_{\substack{0 \leq p, \lambda < T \\ p + \lambda > T}} x(\lambda) h(p) e^{-j2\pi f(p+\lambda)} \, d\lambda \, dp \tag{8.10}$$

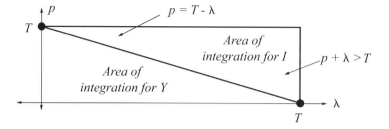

Figure 8.4 Illustration of area of integration for Y and I.

and the integration regions for both Y and I are as shown in Figure 8.4. With

$$X(T, s) = \int_0^T x(t)e^{-st}\, dt \qquad (8.11)$$

and similarly for other Laplace transformed variables, it is the case that

$$Y(T, s) = \iint_{\substack{0 \leq p, \lambda < T \\ p + \lambda \leq T}} x(\lambda)h(p)e^{-s(p+\lambda)}\, d\lambda\, dp \qquad (8.12)$$

$$= X(T, s)H(T, s) - I(T, s)$$

where

$$I(T, s) = \iint_{\substack{0 \leq p, \lambda < T \\ p + \lambda > T}} x(\lambda)h(p)e^{-s(p+\lambda)}\, d\lambda\, dp \qquad (8.13)$$

Proof. The proof of this theorem is given in Appendix 2.

For the Fourier transform case $\tilde{Y}(T, f)$, because of its simplicity, is the approximation that is normally used, and $I(T, f)$ is clearly the error between the approximate and true output Fourier transforms for a given frequency f. The next theorem gives a sufficient condition for the term I to approach zero as the interval under consideration becomes increasingly large.

THEOREM 8.3 CONVERGENCE OF APPROXIMATION *If both $x, h \in L[0, \infty]$, and have bounded variation on all closed finite intervals, then*

$$\lim_{T \to \infty} Y(T, f) = \lim_{T \to \infty} X(T, f)H(T, f) \qquad f \in \mathbf{R} \qquad (8.14)$$

$$\lim_{T \to \infty} Y(T, s) = \lim_{T \to \infty} X(T, s)H(T, s) \qquad \text{Re}[s] \geq 0 \qquad (8.15)$$

Further, if $h \in L[0, \infty]$, x is locally integrable and does not exhibit exponential increase, then $\text{Re}[s] > 0$ is a sufficient condition for

$$\lim_{T \to \infty} Y(T, s) = \lim_{T \to \infty} X(T, s)H(T, s) \tag{8.16}$$

Proof. The proof is given in Appendix 3.

8.4.1 Windowed Input and Nonwindowed Output

For completeness, the response of a linear time invariant system for the case where the input and impulse response are windowed, but the output is not windowed, as illustrated in Figure 8.5, is stated in the following theorem.

THEOREM 8.4 TRANSFORMS OF OUTPUT SIGNAL: NONWINDOWED CASE *If both x, $h \in L[0, T]$, and have bounded variation on $[0, T]$, then the Fourier and Laplace transforms $\tilde{Y}$ of the output signal y, which is not windowed, are given by*

$$\tilde{Y}(2T, f) = X(T, f)H(T, f) \tag{8.17}$$

$$\tilde{Y}(2T, s) = X(T, s)H(T, s) \tag{8.18}$$

Proof. The proof of this result is given in Appendix 4.

This result has application, when the output signal y is to be derived for the interval $[0, T]$. The procedure is as follows for the Fourier transform case. First, evaluate $X(T, f)$ and $H(T, f)$, second, evaluate $\tilde{Y}(2T, f) = X(T, f)H(T, f)$, and third, evaluate y by taking the inverse Fourier transform of $\tilde{Y}(2T, f)$. The evaluated response is valid for the interval $[0, T]$, but not $[T, 2T]$.

8.4.2 Fourier Transform of Output — Power Input Signals

Theorem 8.3 states that $\lim_{T \to \infty} \tilde{Y}(T, f) = \lim_{T \to \infty} Y(T, f)$, provided $x, h \in L$. However, for the common case of signals whose average power evaluated on $[0, T]$, does not significantly vary with T, for example, stationary or periodic

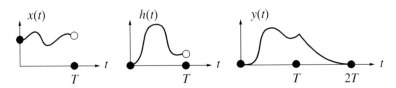

Figure 8.5 Illustration of waveforms in a linear system for the case where the impulse response and the input are windowed but the output is not.

signals, it is the case that $x \notin L$. For this situation, it can be the case that $\lim_{T \to \infty} \tilde{Y}(T, f) \neq \lim_{T \to \infty} Y(T, f)$ almost everywhere. The following example illustrates this point.

8.4.2.1 Example Consider a linear system with an impulse response and input signal, respectively, defined according to

$$h(t) = \frac{h_o e^{-t/\tau}}{\tau} \qquad t > 0, \tau > 0$$
$$x(t) = \sqrt{2} \sigma_x \sin(2\pi f_x t) \qquad t > 0 \tag{8.19}$$

For the case where $\sigma_x = 1$, $h_o = 1$, $\tau = 0.1$, $T = 1$, and $f_x = 4$, the output signal y is plotted in Figure 8.6. For these parameters, the magnitude of the true, Y, and approximate, $\tilde{Y}$, Fourier transforms, as well as the magnitude of the error, I, between these transforms, is plotted in Figure 8.7.

To establish bounds on the integral I, and hence, on how well $\tilde{Y}$ approximates Y, note that

$$|I(T, f)| = \left| \int_0^T \left[\int_{T-\lambda}^T h(p) e^{-j2\pi f p} \, dp \right] x(\lambda) e^{-j2\pi f \lambda} \, d\lambda \right|$$
$$\leq \sqrt{2} \sigma_x \int_0^T \left[\int_{T-\lambda}^T \frac{h_o e^{-p/\tau}}{\tau} \, dp \right] d\lambda = \sqrt{2} \sigma_x h_o \tau \left[1 - e^{-T/\tau} - \frac{T e^{-T/\tau}}{\tau} \right]$$
(8.20)

When T is sufficiently large, such that $T e^{-T/\tau}/\tau \ll 1$, it follows that $|I(T, f)| \leq \sqrt{2} \sigma_x h_o \tau$, which is independent of the interval length T, and only depends on

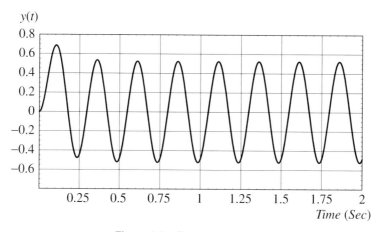

Figure 8.6 Output waveform y.

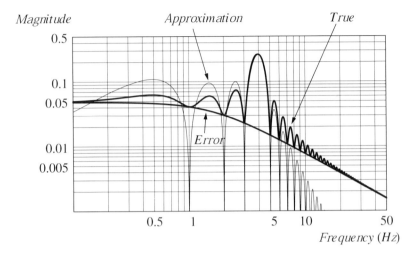

Figure 8.7 Magnitude of the true and approximate Fourier transform of the output signal as well as the magnitude of the error between these two transforms, for the case where $T = 1$.

the input signal amplitude and the system impulse response characteristics h_o and τ. For the given parameters, the bound for $|I(T, f)|$ is 0.141. From Figure 8.7, it follows that the maximum magnitude of I is 0.05, which is within this bound.

Further, the level of the error defined by $|I|$ does not increase or decrease as the interval length T increases (see Figures 8.8 and 8.9). In Figure 8.8 the

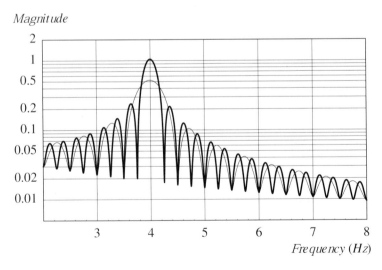

Figure 8.8 Magnitude of the true Fourier transform of the output signal for the cases where $T = 2$ (lower peak) and $T = 4$ (higher peak).

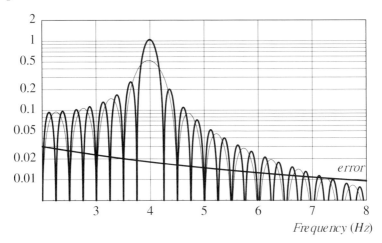

Figure 8.9 Magnitude of the approximate Fourier transform $\tilde{Y}$ of the output signal, for the cases where $T = 2$ (lower peak) and $T = 4$ (higher peak). The magnitude of the error between the true and approximate Fourier transform is identical for $T = 2$ and $T = 4$, and is the smooth curve.

magnitude of the true Fourier transform Y, is plotted for cases $T = 2$ and $T = 4$. In Figure 8.9, the magnitude of approximate Fourier transform $\tilde{Y}$, as well as the error $|I|$, are graphed for cases $T = 2$ and $T = 4$. As T increases, the lobe at the frequency of the input (4 Hz) increases in height, and decreases in width. Away from the lobe, the envelope of the magnitude of both Y and $\tilde{Y}$ remains constant as T increases and, consistent with this, I does not change with T. Clearly, for this example the approximate Fourier transform $\tilde{Y}$, does not converge to the true Fourier transform Y, defined in Eq. (8.9).

8.4.2.2 Explanation An explanation of the nonconvergence of $\tilde{Y}(T, f)$ to $Y(T, f)$ as $T \to \infty$, for signals with constant average power, can be found by noting that I can be approximated by an integral over the region defined in Figure 8.10, where t_h is a time such that $\int_{t_h}^{\infty} |h(p)|\, dp \ll \int_0^{\infty} |h(p)|\, dp$. The magnitude of this integral is relatively insensitive to an increase in the value of T. That is, as T increases the error defined by $|I|$ remains relatively static. For the case where $x \in L$, the magnitude of $\int_{T-t_h}^{T} |x(\lambda)|\, d\lambda$ decreases, in general, as T increases, and the error defined by $|I|$ converges to zero.

8.4.2.3 Power Spectral Density Clearly, on a finite interval $[0, T]$, it is the case that

$$G_Y(T, f) = \frac{|Y(T, f)|^2}{T} \neq \frac{|X(T, f)|^2 |H(T, f)|^2}{T} \quad \text{a.e.} \quad (8.21)$$

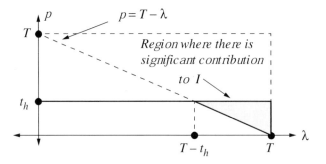

Figure 8.10 Region of integration where there is a significant contribution to the integral I. The time t_h is the time when the impulse response has negligible magnitude as defined in the text.

as $I(T, f)$ is finite. However, for the infinite interval, it follows, as $I(T, f)$ does not increase with T, that

$$\lim_{T\to\infty} G_I(T, f) = \lim_{T\to\infty} \frac{|I(T, f)|^2}{T} = 0 \qquad (8.22)$$

A consequence of this result is that

$$\lim_{T\to\infty} G_Y(T, f) = \lim_{T\to\infty} \frac{|X(T, f)|^2 |H(T, f)|^2}{T} = \lim_{T\to\infty} G_X(T, f) |H(T, f)|^2 \quad (8.23)$$

In fact, as shown in the next section, this last result holds for a broad class of signals that are not elements of $L[0, \infty]$.

8.5 INPUT–OUTPUT POWER SPECTRAL DENSITY RELATIONSHIP

Consider the case where the input random process X to a linear system, is defined on the interval $[0, T]$ by the ensemble

$$E_X = \left\{ x \colon S_\Gamma \times [0, T] \to R \quad \begin{matrix} S_\Gamma \subseteq Z^+ & \text{countable case} \\ S_\Gamma \subseteq R & \text{uncountable case} \end{matrix} \right\} \qquad (8.24)$$

where $P[x(\gamma, t)] = P[\gamma] = p_\gamma$ for the countable case, and $P[x(\gamma, t)|_{\gamma \in [\gamma_o, \gamma_o + d\gamma]}] = P[\gamma \in [\gamma_o, \gamma_o + d\gamma]] = f_\Gamma(\gamma_o) d\gamma$ for the uncountable case. Here, f_Γ is the probability density function associated with the index random variable Γ, whose sample space is S_Γ.

The output waveforms define a random process Y with an ensemble

$$E_Y = \left\{ y: S_\Gamma \times [0, T] \to \mathbf{R},\ y(\gamma, t) = \int_0^t x(\gamma, \lambda) h(t - \lambda)\, d\lambda,\ \gamma \in S_\Gamma \right\} \quad (8.25)$$

where $P[y(\gamma, t)] = P[x(\gamma, t)]$. For subsequent analysis, it is convenient to define the random process I, whose ensemble E_I is

$$E_I = \left\{ \begin{array}{l} i: S_\Gamma \times [0, T] \to \mathbf{C},\ i(\gamma, t) = \displaystyle\int_{-\infty}^{\infty} I(\gamma, T, f) e^{j2\pi f t}\, df \\[1em] \gamma \in S_\Gamma,\ I(\gamma, T, f) = \tilde{Y}(\gamma, T, f) - Y(\gamma, T, f) \end{array} \right\}$$

(8.26)

where $P[i(\gamma, t)] = P[x(\gamma, t)]$. The power spectral density of the output signal is stated in the following theorem. The subsequent theorem states the convergence of $G_Y(T, f)$ to $|H(T, f)|^2 G_X(T, f)$ as $T \to \infty$.

THEOREM 8.5. POWER SPECTRAL DENSITY OF OUTPUT RANDOM PROCESS *If $x \in E_X$ and h have bounded variation on all closed finite intervals, and x and h are locally integrable, then*

$$G_Y(T, f) = |H(T, f)|^2 G_X(T, f) - 2\mathrm{Re}[H(T, f) G_{XI}(T, f)] + G_I(T, f) \quad (8.27)$$

where

$$G_{XI}(T, f) = \begin{cases} \dfrac{1}{T} \displaystyle\sum_{\gamma=1}^{\infty} p_\gamma X(\gamma, T, f) I^*(\gamma, T, f) & \text{countable case} \\[1em] \dfrac{1}{T} \displaystyle\int_{-\infty}^{\infty} X(\gamma, T, f) I^*(\gamma, T, f) f_\Gamma(\gamma)\, d\gamma & \text{uncountable case} \end{cases} \quad (8.28)$$

Proof. Consider the countable case. By definition,

$$G_Y(T, f) = \sum_{\gamma=1}^{\infty} p_\gamma \frac{|Y(\gamma, T, f)|^2}{T} \quad (8.29)$$

Local integrability of $x \in E_X$ and h guarantee that the Fourier transforms X and H exist. The relationships $Y(\gamma, T, f) = \tilde{Y}(\gamma, T, f) - I(\gamma, T, f)$ and $\tilde{Y}(\gamma, T, f) = X(\gamma, T, f) H(T, f)$ then result in the power spectral density of Y as detailed.

THEOREM 8.6. CONVERGENCE OF OUTPUT POWER SPECTRAL DENSITY *Assume for all signals $x \in E_X$ that x is locally integrable, x has bounded variation on all closed finite intervals, and the average power of x does not increase with the* *interval length. Further, assume that h has bounded variation on all closed finite intervals and* $h \in L[0, \infty]$. *It then follows that*

$$\lim_{T \to \infty} G_Y(T, f) = \lim_{T \to \infty} |H(T, f)|^2 G_X(T, f) = \lim_{T \to \infty} G_{\tilde{Y}}(T, f) \quad (8.30)$$

or using a more convenient notation,

$$G_{Y_\infty}(f) = |H_\infty(f)|^2 G_{X_\infty}(f) \quad (8.31)$$

Further,

$$\lim_{T \to \infty} G_I(T, f) = 0 \quad (8.32)$$

If $\lim_{T \to \infty} G_{\tilde{Y}}(T, f) < \infty$ *or* $\lim_{T \to \infty} G_{\tilde{Y}}(T, f) = 0$, *then*

$$\lim_{T \to \infty} |H(T, f)| \, |G_{XI}(T, f)| = 0 \quad (8.33)$$

If $\lim_{T \to \infty} G_{\tilde{Y}}(T, f) = \infty$, *then*

$$\lim_{T \to \infty} \frac{|H(T, f)| \, |G_{XI}(T, f)|}{G_{\tilde{Y}}(T, f)} = 0 \quad (8.34)$$

Proof. The proof of this theorem is given in Appendix 5.

8.5.1 Notes

As shown in Appendix 5, there is finite energy associated with $I(T, f)$, and the power associated with $I(T, f)$ decreases to zero as $T \to \infty$. This fact results in the average power in $H(T, f) G_{XI}(T, f)$ and $G_I(T, f)$ being negligible compared with the average power in $G_{\tilde{Y}}(T, f)$ as $T \to \infty$. The required result as given by Eq. (8.30) then follows from

$$|G_Y(T, f) - G_{\tilde{Y}}(T, f)| = |-2\text{Re}[H(T, f) G_{XI}(T, f)] + G_I(T, f)| \quad (8.35)$$

To establish the rate of convergence of $G_{\tilde{Y}}(T, f)$ to $G_Y(T, f)$ when f is fixed, consider the single waveform case and a bound on the relative error between $G_{\tilde{Y}}(T, f)$ and $G_Y(T, f)$, given by

$$\varepsilon_R \leqslant \frac{2|H(T, f)| \, |G_{XI}(T, f)| + G_I(T, f)}{G_{\tilde{Y}}(T, f)} \qquad G_{\tilde{Y}}(T, f) \neq 0 \quad (8.36)$$

As $|G_{XI}(T, f)| = |X(T, f)| \, |I(T, f)|/T$, $|G_I(T, f)| = |I(T, f)|^2/T$, and $|I(T, f)|$ does not increase with T, whereas $|X(T, f)|$ generally does, it follows that a reasonable bound on the relative error is

$$\varepsilon_R \leqslant \frac{2|H(T, f)| |G_{XI}(T, f)|}{G_{\tilde{Y}}(T, f)} = \frac{2|I(T, f)|}{|H(T, f)| |X(T, f)|} \qquad G_{\tilde{Y}}(T, f) \neq 0 \quad (8.37)$$

For the case where $\lim_{T \to \infty} |X(T, f)|/\sqrt{T}$ is finite, but nonzero, the relative error is proportional to $1/\sqrt{T}$. This case is consistent with a bounded power spectral density on the infinite interval. For the case where $\lim_{T \to \infty} |X(T, f)|/T$ is finite, the relative error is proportional to $1/T$. This case is consistent with an unbounded power spectral density on the infinite interval. Such a case occurs for periodic signals at specific frequencies.

The relationship given in Eq. (8.31) underpins a significant level of analysis of noise in linear systems. One application of this result is in characterizing the noise level of an electronic circuit. Such a characterization is fundamental to low noise electronic design and is the subject of Chapter 9. The following subsection gives an important example, where the relationship given in Eq. (8.31) cannot be applied.

8.5.2 Example—Oscillator Noise

A quadrature oscillator is an entity that generates signals of the form

$$x(t) = A \cos[2\pi f_c t + \theta(t)] \qquad y(t) = A \sin[2\pi f_c t + \theta(t)] \quad (8.38)$$

where typically, $\theta' \ll 2\pi f_c$. Such signals arise from the differential equations,

$$x' = -[2\pi f_c + \theta']y \qquad x(0) = A \cos[\theta(0)] \quad (8.39)$$

$$y' = [2\pi f_c + \theta']x \qquad y(0) = A \sin[\theta(0)] \quad (8.40)$$

This result can be proved by substitution of x and y into the differential equations. For the case where the modulation θ is zero, a quadrature sinusoidal oscillator results and can be implemented as per the prototypical structure shown in Figure 8.11. In this figure, n_1 and n_2 are independent noise sources to account for the noise in the integrators and following circuitry.

With the noise sources n_1 and n_2, the differential equations characterizing the circuit of Figure 8.11 are

$$x' = -2\pi f_c(y + n_1) \qquad y' = 2\pi f_c(x + n_2) \quad (8.41)$$

Differentiation and substitution yields the following differential equation for x:

$$x'' + 4\pi^2 f_c^2 x = -2\pi f_c (n_1)' - 4\pi^2 f_c^2 n_2 \quad (8.42)$$

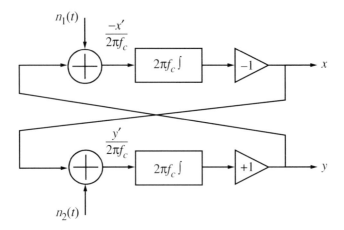

Figure 8.11 Prototypical quadrature oscillator structure.

As this is a linear differential equation, it follows that the quadrature oscillator can be modeled, as far as the output x is concerned, as shown in Figure 8.12. The impulse responses in this figure are the solutions of the differential equations,

$$x'' + 4\pi^2 f_c^2 x = -2\pi f_c \delta'(t) \qquad x'' + 4\pi^2 f_c^2 x = -4\pi^2 f_c^2 \delta(t) \qquad (8.43)$$

which equates to the solution of

$$x'' + 4\pi^2 f_c^2 x = 0 \quad t > 0 \qquad x(0) = -2\pi f_c, \; x'(0) = 0 \qquad (8.44)$$
$$x'' + 4\pi^2 f_c^2 x = 0 \quad t > 0 \qquad x(0) = 0, \; x'(0) = -4\pi^2 f_c^2 \qquad (8.45)$$

It then follows that the respective impulse responses are

$$h_1(t) = -2\pi f_c \cos(2\pi f_c t) \qquad h_2(t) = -2\pi f_c \sin(2\pi f_c t) \qquad (8.46)$$

Clearly, $h_1, h_2 \notin L[0, \infty]$, and $H_1(T, f)$ and $H_2(T, f)$ do not converge as $T \to \infty$. Consequently, Theorem 8.6 cannot be used when ascertaining the noise

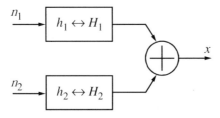

Figure 8.12 Equivalent model for the output signal x of the quadrature oscillator shown in Figure 8.11.

characteristics of an oscillator. This fact is overlooked in a significant proportion of the literature (Demir, 1998 p. 164), and alternative approaches are required to characterize the noise of an oscillator [see, for example, Demir (1998 ch. 6)].

8.6 MULTIPLE INPUT–MULTIPLE OUTPUT SYSTEMS

Two possible multiple input–multiple output (MIMO) systems are illustrated in Figures 8.13 and 8.14 where the input signals $x_1, \ldots, x_N$, respectively, are from the ensembles $E_{X_1}, \ldots, E_{X_N}$, defining the random processes $X_1, \ldots, X_N$. By definition

$$E_{X_i} = \left\{ x_i \colon S_\Gamma \times [0, T] \to R \quad \begin{matrix} S_\Gamma \subseteq Z^+ & \text{countable case} \\ S_\Gamma \subseteq R & \text{uncountable case} \end{matrix} \right\} \quad (8.47)$$

For the system shown in Figure 8.13, one signal from the ensemble for the output random process Z_u, can be written in the form

$$z_u(\zeta, t) = \sum_{i=1}^{N} w_{ui} y_i(\gamma_i, t) = \sum_{i=1}^{N} w_{ui} \int_0^t x_i(\gamma_i, \lambda) h_i(t - \lambda) \, d\lambda \qquad \gamma_i \in S_\Gamma \quad (8.48)$$

where $\zeta = (\gamma_1, \ldots, \gamma_N)$. On the interval $[0, T]$, it follows from Theorem 8.2, that

$$\begin{aligned} Z_u(\zeta, T, f) &= \sum_{i=1}^{N} w_{ui} Y_i(\gamma_i, T, f) \\ &= \sum_{i=1}^{N} w_{ui} [H_i(T, f) X_i(\gamma_i, T, f) - I_i(\gamma_i, T, f)] \end{aligned} \quad (8.49)$$

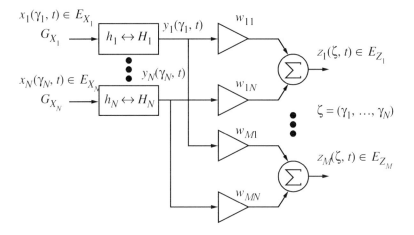

Figure 8.13 Schematic diagram of a multiple input–multiple output system.

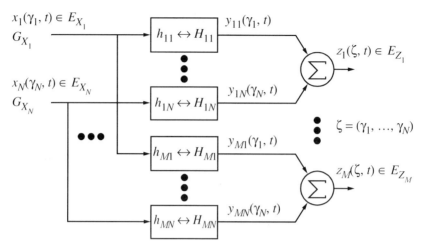

Figure 8.14 Schematic diagram of an alternative multiple input–multiple output system.

The following theorem states the relationship between the output and input power spectral densities of the system illustrated in Figure 8.13.

THEOREM 8.7. POWER SPECTRAL DENSITY RELATIONSHIPS FOR MULTIPLE INPUT–MULTIPLE OUTPUT SYSTEMS If $x_i \in E_{X_i}$ and h_i have bounded variation on all closed finite intervals, x_i is locally integrable, $h_i \in L[0, \infty]$ and the input random processes $X_1, \ldots, X_N$ are independent with zero means, then, for the infinite interval $[0, \infty]$, it is the case that

$$G_{Z_u}(f) = \sum_{i=1}^{N} |w_{ui}|^2 G_{Y_i}(f) = \sum_{i=1}^{N} |w_{ui}|^2 |H_i(f)|^2 G_{X_i}(f) \qquad (8.50)$$

where, for convenience of notation, the subscript ∞ has been dropped.

For the general case, where $X_1, \ldots, X_N$ are not necessarily independent with zero means, the following result holds for the infinite interval

$$\begin{aligned} G_{Z_u}(f) &= \sum_{i=1}^{N} |w_{ui}|^2 G_{Y_i}(f) + \sum_{i=1}^{N} \sum_{\substack{j=1 \\ j \neq i}}^{N} w_{ui} w_{uj}^* G_{Y_i Y_j}(f) \\ &= \sum_{i=1}^{N} |w_{ui}|^2 |H_i(f)|^2 G_{X_i}(f) + \sum_{i=1}^{N} \sum_{\substack{j=1 \\ j \neq i}}^{N} w_{ui} w_{uj}^* H_i(f) H_j^*(f) G_{X_i X_j}(f) \end{aligned} \qquad (8.51)$$

On the infinite interval, the cross power spectral density between the random processes X_i or Y_i and Y_j, is given by

$$G_{X_i Y_j}(f) = H_j^*(f) G_{X_i X_j}(f) \qquad (8.52)$$

$$G_{Y_i Y_j}(f) = H_i(f) H_j^*(f) G_{X_i X_j}(f) \qquad (8.53)$$

On the infinite interval, the cross power spectral density between Z_u and Z_v, is given by

$$G_{Z_u Z_v}(f) = \sum_{i=1}^{N} w_{ui} w_{vi}^* G_{Y_i}(f) + \sum_{i=1}^{N} \sum_{\substack{j=1 \\ j \neq i}}^{N} w_{ui} w_{vj}^* G_{Y_i Y_j}(f)$$

$$= \sum_{i=1}^{N} w_{ui} w_{vi}^* |H_i(f)|^2 G_{X_i}(f) + \sum_{i=1}^{N} \sum_{\substack{j=1 \\ j \neq i}}^{N} w_{ui} w_{vj}^* H_i(f) H_j^*(f) G_{X_i X_j}(f)$$

(8.54)

Proof. The proof is given in Appendix 6.

8.6.1 Alternative Multiple Input–Multiple Output System

Consider one signal from the ensemble for the output random process Z_u, defined by the structure of Figure 8.14,

$$z_u(\zeta, t) = \sum_{i=1}^{N} y_{ui}(\gamma_i, t) = \sum_{i=1}^{N} \int_0^t x_i(\gamma_i, \lambda) h_{ui}(t - \lambda) \, d\lambda \qquad \gamma_i \in S_\Gamma \quad (8.55)$$

where $\zeta = (\gamma_1, \ldots, \gamma_N)$. On an interval $[0, T]$, it follows from Theorem 8.2, that

$$Z_u(\zeta, T, f) = \sum_{i=1}^{N} Y_{ui}(\gamma_i, T, f) = \sum_{i=1}^{N} [H_{ui}(T, f) X_i(\gamma_i, T, f) - I_i(\gamma_i, T, f)] \quad (8.56)$$

The following theorem details the appropriate power spectral density relationships for this system.

THEOREM 8.8. POWER SPECTRAL DENSITY RELATIONSHIPS FOR ALTERNATIVE MULTIPLE INPUT–MULTIPLE OUTPUT SYSTEMS *If $x_i \in E_{X_i}$ and h_{ij} have bounded variation on all closed finite intervals, x_i is locally integrable and $h_{ij} \in L[0, \infty]$, then the following result holds for the infinite interval:*

$$G_{Z_u}(f) = \sum_{i=1}^{N} G_{Y_{ui}}(f) + \sum_{i=1}^{N} \sum_{\substack{j=1 \\ j \neq i}}^{N} G_{Y_{ui} Y_{uj}}(f)$$

$$= \sum_{i=1}^{N} |H_{ui}(f)|^2 G_{X_i}(f) + \sum_{i=1}^{N} \sum_{\substack{j=1 \\ j \neq i}}^{N} H_{ui}(f) H_{uj}^*(f) G_{X_i X_j}(f)$$

(8.57)

On the infinite interval, the cross power spectral density between Z_u and Z_v, is given by

$$G_{Z_u Z_v}(f) = \sum_{i=1}^{N} G_{Y_{ui} Y_{vi}}(f) + \sum_{i=1}^{N} \sum_{\substack{j=1 \\ j \neq i}}^{N} G_{Y_{ui} Y_{vj}}(f) \quad (8.58)$$

$$= \sum_{i=1}^{N} H_{ui}(f) H_{vi}^*(f) G_{X_i}(f) + \sum_{i=1}^{N} \sum_{\substack{j=1 \\ j \neq i}}^{N} H_{ui}(f) H_{vj}^*(f) G_{X_i X_j}(f)$$

Proof. The proof of this theorem is given in Appendix 7.

APPENDIX 1: PROOF OF THEOREM 8.1

If the input signal $x \in L[0, T]$ and is of bounded variation then, as per Theorem 2.19, for all $\varepsilon > 0$ there exists a $\Delta > 0$, such that x can be approximated on the interval $[0, T]$ by a step function x_Δ, defined according to

$$x_\Delta(t) = \Delta x \left[\left\lfloor \frac{t}{\Delta} \right\rfloor \Delta \right] \delta_\Delta \left[t - \left\lfloor \frac{t}{\Delta} \right\rfloor \Delta \right] = \sum_{i=0}^{\lfloor t/\Delta \rfloor} \Delta x(i\Delta) \delta_\Delta(t - i\Delta) \quad t \in [0, T] \quad (8.59)$$

such that $\int_0^T |x(t) - x_\Delta(t)| \, dt < \varepsilon$. Since the response of the system to a signal δ_Δ is h_Δ, it follows, from the causality, time invariance and linearity of the system, that the output y_Δ, at a time $t \in [0, T]$, in response to the signal x_Δ, is

$$y_\Delta(t) = \Delta x(0) h_\Delta(t) + \cdots + \Delta x \left[\left\lfloor \frac{t}{\Delta} \right\rfloor \Delta \right] h_\Delta \left[t - \left\lfloor \frac{t}{\Delta} \right\rfloor \Delta \right] \quad (8.60)$$

$$= \sum_{i=0}^{\lfloor t/\Delta \rfloor} \Delta x(i\Delta) h_\Delta(t - i\Delta)$$

It remains to show that $\lim_{\Delta \to 0} y_\Delta(t) = y(t)$ and $\lim_{\Delta \to 0} \int_0^T |y(t) - y_\Delta(t)| \, dt = 0$. To prove these results, it is convenient to define a function z according to

$$z(t, \lambda) = x(\lambda) h(t - \lambda) \quad \Rightarrow \quad y(t) = \int_0^t z(t, \lambda) \, d\lambda \quad (8.61)$$

If both x and h are causal and of bounded variation, then for any fixed value of t and consistent with Figure 8.15, z can be approximated for $\lambda \in [0, t]$ by a step function z_Δ,

$$z_\Delta(t, \lambda) = \sum_{i=0}^{\lfloor t/\Delta \rfloor} \Delta x(i\Delta) h(t - i\Delta) \delta_\Delta(\lambda - i\Delta) \quad \lambda \in [0, t] \quad (8.62)$$

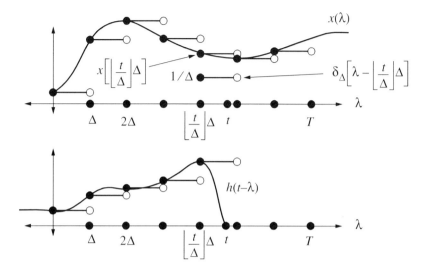

Figure 8.15 Illustration of the functions comprising z, and step approximations to them.

Theorem 2.19 then implies that for $\forall \varepsilon > 0$ there exists a $\Delta > 0$, such that

$$\int_0^t |z(t, \lambda) - z_\Delta(t, \lambda)|\, d\lambda < \varepsilon \quad \Rightarrow \quad \lim_{\Delta \to 0} \int_0^t z_\Delta(t, \lambda)\, d\lambda = \int_0^t z(t, \lambda)\, d\lambda \quad (8.63)$$

Further, it follows that

$$\begin{aligned}
\int_0^t z_\Delta(t, \lambda)\, d\lambda &= \sum_{i=0}^{\lfloor t/\Delta \rfloor} \Delta x(i\Delta) h(t - i\Delta) \int_0^t \delta(\lambda - i\Delta)\, d\lambda \\
&= \sum_{i=0}^{\lfloor t/\Delta \rfloor - 1} \Delta x(i\Delta) h(t - i\Delta) \\
&\quad + \Delta x\left[\left\lfloor \frac{t}{\Delta} \right\rfloor \Delta\right] h\left[t - \left\lfloor \frac{t}{\Delta} \right\rfloor \Delta\right] \int_{\lfloor t/\Delta \rfloor \Delta}^t \delta\left[\lambda - \left\lfloor \frac{t}{\Delta} \right\rfloor \Delta\right] d\lambda \\
&= y_\Delta(t) + \varepsilon_\Delta
\end{aligned} \quad (8.64)$$

where

$$\varepsilon_\Delta = -\Delta x\left[\left\lfloor \frac{t}{\Delta} \right\rfloor \Delta\right] h\left[t - \left\lfloor \frac{t}{\Delta} \right\rfloor \Delta\right] \int_t^{\lfloor t/\Delta \rfloor \Delta + \Delta} \delta\left[\lambda - \left\lfloor \frac{t}{\Delta} \right\rfloor \Delta\right] d\lambda \quad (8.65)$$

Clearly, $\varepsilon_\Delta \to 0$ as $\Delta \to 0$. Thus, from Eqs. (8.61), (8.63), and (8.64), it follows that $\lim_{\Delta \to 0} y_\Delta(t) = y(t)$. Finally, from Eq. (8.63), it follows that

$$\int_0^T |y(t) - y_\Delta(t)|\, dt = \int_0^T \left| \int_0^t z(t, \lambda)\, d\lambda - \int_0^t z_\Delta(t, \lambda)\, d\lambda + \varepsilon_\Delta \right| dt$$

$$\leqslant \int_0^T \left[\int_0^t |z(t, \lambda) - z_\Delta(t, \lambda)|\, d\lambda \right] dt + |\varepsilon_\Delta| T \leqslant \varepsilon T + |\varepsilon_\Delta| T$$

(8.66)

which is the final required result.

APPENDIX 2: PROOF OF THEOREM 8.2

With the stated assumptions, it follows that

$$Y(T, f) = \int_0^T y(t) e^{-j2\pi f t}\, dt = \int_0^T \left[\int_0^t x(\lambda) h(t - \lambda)\, d\lambda \right] e^{-j2\pi f t}\, dt$$

$$= \int_0^T x(\lambda) e^{-j2\pi f \lambda} \left[\int_\lambda^T h(t - \lambda) e^{-j2\pi f(t - \lambda)}\, dt \right] d\lambda$$

(8.67)

The region of integration is illustrated in Figure 8.16. A change of variable $p = t - \lambda$ in the inner integral of the last equation yields

$$Y(T, f) = \int_0^T x(\lambda) e^{-j2\pi f \lambda} \left[\int_0^{T-\lambda} h(p) e^{-j2\pi f p}\, dp \right] d\lambda$$

$$= \int_0^T x(\lambda) e^{-j2\pi f \lambda} \left[\int_0^T h(p) e^{-j2\pi f p}\, dp - \int_{T-\lambda}^T h(p) e^{-j2\pi f p}\, dp \right] d\lambda$$

(8.68)

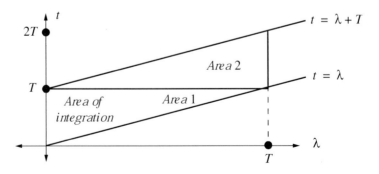

Figure 8.16 Region of integration for evaluation of Y.

which can be written in the form

$$Y(T, f) = X(T, f)H(T, f) - I(T, f) \qquad (8.69)$$

where

$$I(T, f) = \int_0^T x(\lambda)e^{-j2\pi f \lambda} \left[\int_{T-\lambda}^T h(p)e^{-j2\pi f p} \, dp \right] d\lambda \qquad (8.70)$$

The region of integration for the integral I is area 2 in Figure 8.16 prior to the change of variable $p = t - \lambda$, and the region specified in Figure 8.4 after this change.

The results for the Laplace transform case follow in an analogous manner, with use of the substitution $s = j2\pi f$.

APPENDIX 3: PROOF OF THEOREM 8.3

If $x, h \in L[0, \infty]$, then for all $\varepsilon > 0$ there exists numbers λ_ε and p_ε, such that

$$\left| \int_{\lambda_\varepsilon}^\infty x(\lambda)e^{-j2\pi f \lambda} \, d\lambda \right| \leqslant \int_{\lambda_\varepsilon}^\infty |x(\lambda)e^{-j2\pi f \lambda}| \, d\lambda \leqslant \int_{\lambda_\varepsilon}^\infty |x(\lambda)| d\lambda \leqslant \varepsilon \qquad (8.71)$$

$$\left| \int_{p_\varepsilon}^\infty h(p)e^{-j2\pi f p} \, dp \right| \leqslant \int_{p_\varepsilon}^\infty |h(p)e^{-j2\pi f p}| \, d\lambda \leqslant \int_{p_\varepsilon}^\infty |h(p)| dp \leqslant \varepsilon \qquad (8.72)$$

It then follows that the region over which there is a significant contribution to the integral I is as illustrated in Figure 8.17. The illustration is for the case where $\lambda_\varepsilon + p_\varepsilon > T$. Clearly, as T increases, the region over which there is a significant contribution to the integral I decreases, and for $T > \lambda_\varepsilon + p_\varepsilon$, it is expected that the value of the integral I can be made arbitrarily small, and this is shown below.

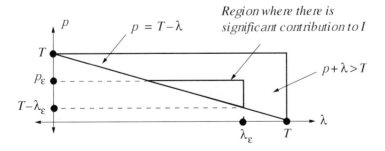

Figure 8.17 Illustration of area of integration to establish a bound on the integral I. The illustration is for the case where $\lambda_\varepsilon + p_\varepsilon > T$.

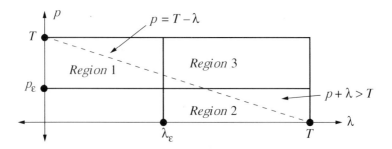

Figure 8.18 Illustration of area of integration to establish a bound on the integral I. The illustration is for the case where $\lambda_\varepsilon + p_\varepsilon < T$.

The case where $T > \lambda_\varepsilon + p_\varepsilon$ is illustrated in Figure 8.18. Using the regions 1, 2, and 3 defined in this figure, the following bound on the error integral I can be established:

$$|I(T, f)| \leq \int_0^{\lambda_\varepsilon} |x(\lambda)e^{-j2\pi f\lambda}| \, d\lambda \int_{p_\varepsilon}^T |h(p)e^{-j2\pi fp}| \, dp$$

$$+ \int_{\lambda_\varepsilon}^T |x(\lambda)e^{-j2\pi f\lambda}| \, d\lambda \int_0^{p_\varepsilon} |h(p)e^{-j2\pi fp}| \, dp \qquad (8.73)$$

$$+ \int_{\lambda_\varepsilon}^T |x(\lambda)e^{-j2\pi f\lambda}| \, d\lambda \int_{p_\varepsilon}^T |h(p)e^{-j2\pi fp}| \, dp$$

From the definitions for λ_ε and p_ε, it follows that

$$|I(T, f)| \leq \varepsilon \int_0^{\lambda_\varepsilon} |x(\lambda)| \, d\lambda + \varepsilon \int_0^{p_\varepsilon} |h(p)| \, dp + \varepsilon^2$$

$$\leq \varepsilon \left[\varepsilon + \int_0^\infty |x(\lambda)| \, d\lambda + \int_0^\infty |h(p)| \, dp \right] \qquad (8.74)$$

Thus, for any $\varepsilon > 0$, there exists a T, such that the error bound for the absolute difference between Y and $\tilde{Y}$, as given by $|I|$, is less than $k\varepsilon$ for some fixed k, which is independent of T.

When $x \in L[0, \infty]$, the results for the Laplace transform case follow in an analogous manner, with use of the substitution $s = j2\pi f$. When x is locally integrable, x does not have exponential increase, and $\text{Re}[s] > 0$, then it is the case that $x(t)e^{-\text{Re}[s]} \in L[0, \infty]$. Use of the above approach, in an analogous manner, yields the required proof.

APPENDIX 4: PROOF OF THEOREM 8.4

The graphs of $x(\lambda)$ and $h(t - \lambda)$ are shown in Figure 8.19 for the case where h and x are windowed, such that they are zero outside the interval $[0, T]$. It then follows that

$$y(t) = \int_0^t x(\lambda) h(t - \lambda) \, d\lambda = \begin{cases} \int_0^t x(\lambda) h(t - \lambda) \, d\lambda & 0 \leq t < T \\ \int_{t-T}^T x(\lambda) h(t - \lambda) \, d\lambda & T \leq t < 2T \\ 0 & \text{elsewhere} \end{cases} \quad (8.75)$$

Hence, assuming both $x, h \in L[0, T]$, it follows that

$$\tilde{Y}(2T, f) = \int_0^{2T} y(t) e^{-j2\pi f t} \, dt$$

$$= \int_0^T \left[\int_0^t x(\lambda) h(t - \lambda) \, d\lambda \right] e^{-j2\pi f t} \, dt \quad (8.76)$$

$$+ \int_T^{2T} \left[\int_{t-T}^T x(\lambda) h(t - \lambda) \, d\lambda \right] e^{-j2\pi f t} \, dt$$

The region of integration comprises areas 1 and 2 shown in Figure 8.16. Changing the order of integration yields,

$$\tilde{Y}(2T, f) = \int_0^T \left[\int_\lambda^{\lambda + T} h(t - \lambda) e^{-j2\pi f(t - \lambda)} \, dt \right] x(\lambda) e^{-j2\pi f \lambda} \, d\lambda \quad (8.77)$$

A change of variable $p = t - \lambda$ in the inner integral yields,

$$\tilde{Y}(2T, f) = \int_0^T x(\lambda) e^{-j2\pi f \lambda} \left[\int_0^T h(p) e^{-j2\pi f p} \, dp \right] d\lambda = X(T, f) H(T, f) \quad (8.78)$$

which is the required result.

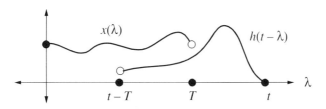

Figure 8.19 Graph of windowed input signal and shifted windowed impulse response function.

The results for the Laplace transform case follow in an analogous manner with use of the substitution $s = j2\pi f$.

APPENDIX 5: PROOF OF THEOREM 8.6

Consider the countable case, where a subscript, rather than an argument, is used, that is, $x_i(t)$ rather than $x(\gamma, t)$. First, for the case where $h \in L[0, \infty]$, the region of integration for $I_i(T, f)$ can be approximated by the region shown in Figure 8.10, that is,

$$I_i(T, f) \approx \int_0^{t_h} \int_{T-p}^{T} x_i(\lambda) h(p) e^{-j2\pi f(p+\lambda)} \, d\lambda \, dp \qquad (8.79)$$

It then follows that an upper bound on $|I_i(T, f)|$ is given by

$$|I_i(T, f)| \leq \int_0^{t_h} \int_{T-p}^{T} |x_i(\lambda)| |h(p)| \, d\lambda \, dp \leq k_I \int_0^{t_h} \int_{T-t_h}^{T} |x_i(\lambda)| |h(p)| \, d\lambda \, dp$$

$$\leq k_I \sup_X \int_0^{\infty} |h(p)| \, dp \qquad (8.80)$$

where k_I is of the order of unity and

$$\sup_X = \sup \left\{ \int_{T-t_h}^{T} |x_i(\lambda)| \, d\lambda, \, i \in \mathbf{Z}^+, \, T > t_h, \, T \in \mathbf{R}^+ \right\} \qquad (8.81)$$

As per Theorem 2.13, it follows, if the average power on $[T - t_h, T]$ does not increase with T, that $\sup_X$ is finite and $|I_i(T, f)|$ does not increase with T. It then follows that

$$\lim_{T \to \infty} \frac{|I_i(T, f)|^2}{T} = \lim_{T \to \infty} G_{I_i}(T, f) = 0 \qquad (8.82)$$

This result then implies the following for f fixed: First, if $|X_i(T, f)|$ is constant with respect to T, or is such that $\lim_{T \to \infty} |X_i(T, f)|/\sqrt{T} = 0$, then

$$\lim_{T \to \infty} |H(T, f)| |G_{X_i I_i}(T, f)| = \lim_{T \to \infty} \frac{|H(T, f)| |X_i(T, f)| |I_i^*(T, f)|}{T} = 0 \qquad (8.83)$$

For this case, both $G_{X_i}(T, f)$ and $G_{\tilde{y}}(T, f)$ converge to zero as $T \to \infty$. Second, if $|X_i(T, f)|$ is such, that $\lim_{T \to \infty} |X_i(T, f)|/\sqrt{T}$ is finite and nonzero, then both

$G_{X_i}(T, f)$ and $G_{\tilde{Y}_i}(T, f)$ are bounded as $T \to \infty$ and

$$\lim_{T \to \infty} |H(T, f)| |G_{X_i I_i}(T, f)| = \lim_{T \to \infty} \frac{|H(T, f)| |X_i(T, f)| |I_i^*(T, f)|}{T} = 0 \quad (8.84)$$

Third, if $|X_i(T, f)|$ is such, that $\lim_{T \to \infty} |X_i(T, f)|/\sqrt{T}$ is infinite, then both $G_{X_i}(T, f)$ and $G_{\tilde{Y}_i}(T, f)$ are unbounded as $T \to \infty$. It is then the case that

$$\lim_{T \to \infty} \frac{|H(T, f)| |G_{X_i I_i}(T, f)|}{G_{\tilde{Y}_i}(T, f)} = \lim_{T \to \infty} \frac{|I_i(T, f)|}{|H(T, f)| |X_i(T, f)|} = 0 \quad (8.85)$$

When these results hold for all signals in the ensemble, and this is guaranteed by the assumptions made, the following results hold:

$$\lim_{T \to \infty} G_I(T, f) = 0 \quad (8.86)$$

If, $\lim_{T \to \infty} G_{\tilde{Y}}(T, f) < \infty$ or $\lim_{T \to \infty} G_{\tilde{Y}}(T, f) = 0$, then

$$\lim_{T \to \infty} |H(T, f)| |G_{XI}(T, f)| = 0 \quad (8.87)$$

If $\lim_{T \to \infty} G_{\tilde{Y}}(T, f) = \infty$, then

$$\lim_{T \to \infty} \frac{|H(T, f)| |G_{XI}(T, f)|}{G_{\tilde{Y}}(T, f)} = 0 \quad (8.88)$$

$$\lim_{T \to \infty} G_Y(T, f) = \lim_{T \to \infty} G_{\tilde{Y}}(T, f) = \lim_{T \to \infty} |H(T, f)|^2 G_X(T, f) \quad (8.89)$$

APPENDIX 6: PROOF OF THEOREM 8.7

The proof of the first result follows directly from Theorems 4.6 and 8.6. The first form of the second result follows directly from Theorem 4.6. The second form of the second result follows from the definition of the cross power spectral density and Theorem 8.6. To show this, consider the countable case and the notation $P[x_i(\gamma_i, t), x_j(\gamma_j, t)] = p_{ij}(\gamma_i, \gamma_j)$. By definition

$$G_{Y_i Y_j}(f) = \lim_{T \to \infty} \frac{1}{T} \sum_{\gamma_i = 1}^{\infty} \sum_{\gamma_j = 1}^{\infty} p_{ij}(\gamma_i, \gamma_j) Y_i(\gamma_i, T, f) Y_j^*(\gamma_j, T, f)$$

$$= \lim_{T \to \infty} \frac{1}{T} \sum_{\gamma_i = 1}^{\infty} \sum_{\gamma_j = 1}^{\infty} p_{ij}(\gamma_i, \gamma_j) \begin{bmatrix} [X_i(\gamma_i, T, f) H_i(T, f) - I_i(\gamma_i, T, f)] \\ \times [X_j^*(\gamma_j, T, f) H_j^*(T, f) - I_j^*(\gamma_j, T, f)] \end{bmatrix}$$

$$(8.90)$$

Using definitions for the cross power spectral density, it follows that

$$G_{Y_iY_j}(f) = \lim_{T \to \infty} \Big[G_{X_iX_j}(T, f)H_i(T, f)H_j^*(T, f) - G_{X_iI_j}(T, f)H_i(T, f)$$

$$- G_{X_jI_i}^*(T, f)H_j^*(T, f) + G_{I_iI_j}(T, f) \Big] \quad (8.91)$$

It follows, from a similar argument to that used in the proof of Theorem 8.6, that the relative magnitude of the cross power spectral density terms $|G_{X_iI_j}||H_i|$, $|G_{X_jI_i}^*||H_j^*|$, and $G_{I_iI_j}$, with respect to $G_{\tilde{Y}_i\tilde{Y}_j}$ and $G_{Y_iY_j}$, become increasingly small as $T \to \infty$. Hence,

$$G_{Y_iY_j}(f) = H_i(f)H_j^*(f)G_{X_iX_j}(f) \quad (8.92)$$

which is the required result.

To show the cross power spectral density result, note that for each outcome $\zeta = (\gamma_1, \ldots, \gamma_N)$, the following individual power spectral densities and cross power spectral density can be defined,

$$G_{Z_u}(\zeta, T, f) = \frac{|Z_u(\zeta, T, f)|^2}{T} = \frac{1}{T}\sum_{i=1}^{N}\sum_{j=1}^{N} w_{ui}w_{uj}^* Y_i(\gamma_i, T, f)Y_j^*(\gamma_j, T, f) \quad (8.93)$$

$$G_{Z_v}(\zeta, T, f) = \frac{|Z_v(\zeta, T, f)|^2}{T} = \frac{1}{T}\sum_{i=1}^{N}\sum_{j=1}^{N} w_{vi}w_{vj}^* Y_i(\gamma_i, T, f)Y_j^*(\gamma_j, T, f) \quad (8.94)$$

$$G_{Z_uZ_v}(\zeta, T, f) = \frac{Z_u(\zeta, T, f)Z_v^*(\zeta, T, f)}{T}$$

$$= \frac{1}{T}\sum_{i=1}^{N}\sum_{j=1}^{N} w_{ui}w_{vj}^* Y_i(\gamma_i, T, f)Y_j^*(\gamma_j, T, f) \quad (8.95)$$

As each of these power spectral densities have the same probability and the same form, it follows, by analogy with $G_{Z_u}(T, f)$, that the cross power spectral density between Z_u and Z_v, on the infinite interval, is given by

$$G_{Z_uZ_v}(f) = \sum_{i=1}^{N} w_{ui}w_{vi}^* G_{Y_i}(f) + \sum_{\substack{i=1 \\ j \neq i}}^{N}\sum_{j=1}^{N} w_{ui}w_{vj}^* G_{Y_iY_j}(f)$$

$$= \sum_{i=1}^{N} w_{ui}w_{vi}^* |H_i(f)|^2 G_{X_i}(f) + \sum_{\substack{i=1 \\ j \neq i}}^{N}\sum_{j=1}^{N} w_{ui}w_{vj}^* H_i(f)H_j^*(f) G_{X_iX_j}(f)$$

$$(8.96)$$

APPENDIX 7: PROOF OF THEOREM 8.8

The proof of the result for $G_{Z_u}(T, f)$ follows directly from Theorems 4.6 and 8.6, and the cross power spectral density relationships given in Theorem 8.7.

To show the cross power spectral density result, note that for each outcome $\zeta = (\gamma_1, \ldots, \gamma_N)$, the following individual power spectral densities and cross power density can be defined:

$$G_{Z_u}(\zeta, T, f) = \frac{|Z_u(\zeta, T, f)|^2}{T} = \frac{1}{T} \sum_{i=1}^{N} \sum_{j=1}^{N} Y_{ui}(\gamma_i, T, f) Y_{uj}^*(\gamma_j, T, f) \quad (8.97)$$

$$G_{Z_v}(\zeta, T, f) = \frac{|Z_v(\zeta, T, f)|^2}{T} = \frac{1}{T} \sum_{i=1}^{N} \sum_{j=1}^{N} Y_{vi}(\gamma_i, T, f) Y_{vj}^*(\gamma_j, T, f) \quad (8.98)$$

$$G_{Z_u Z_v}(\zeta, T, f) = \frac{Z_u(\zeta, T, f) Z_v^*(\zeta, T, f)}{T}$$

$$= \frac{1}{T} \sum_{i=1}^{N} \sum_{j=1}^{N} Y_{ui}(\gamma_i, T, f) Y_{vj}^*(\gamma_j, T, f) \quad (8.99)$$

As each of these power spectral densities have the same probability and the same form, it follows, by analogy with $G_{Z_u}(T, f)$, that the cross power spectral density between Z_u and Z_v, on the infinite interval, is given by

$$G_{Z_u Z_v}(f) = \sum_{i=1}^{N} G_{Y_{ui} Y_{vi}}(f) + \sum_{i=1}^{N} \sum_{\substack{j=1 \\ j \neq i}}^{N} G_{Y_{ui} Y_{vj}}(f)$$

$$= \sum_{i=1}^{N} H_{ui}(f) H_{vi}^*(f) G_{X_i}(f) + \sum_{i=1}^{N} \sum_{\substack{j=1 \\ j \neq i}}^{N} H_{ui}(f) H_{vj}^*(f) G_{X_i X_j}(f)$$

$$(8.100)$$

9

Principles of Low Noise Electronic Design

9.1 INTRODUCTION

This chapter details noise models and signal theory, such that the effect of noise in linear electronic systems can be ascertained. The results are directly applicable to nonlinear systems that can be approximated around an operating point by an affine function.

An introductory section is included at the start of the chapter to provide an insight into the nature of Gaussian white noise — the most common form of noise encountered in electronics. This is followed by a description of the standard types of noise encountered in electronics and noise models for standard electronic components. The central result of the chapter is a systematic explanation of the theory underpinning the standard method of characterizing noise in electronic systems, namely, through an input equivalent noise source or sources. Further, the noise equivalent bandwidth of a system is defined. This method of characterizing a system, simplifies noise analysis — especially when a signal to noise ratio characterization is required. Finally, the input equivalent noise of a passive network is discussed which is a generalization of Nyquist's theorem. General references for noise in electronics include Ambrozy (1982), Buckingham (1983), Engberg (1995), Fish (1993), Leach (1994), Motchenbacher (1993), and van der Ziel (1986).

9.1.1 Notation and Assumptions

When dealing with noise processes in linear time invariant systems, an infinite timescale is often assumed so power spectral densities, consistent with previous notation, should be written in the form $G_\infty(f)$. However, for notational

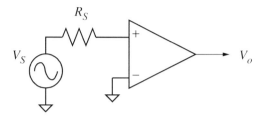

Figure 9.1 Schematic diagram of signal source and amplifier.

convenience, the subscript is removed and power spectral densities are written as $G(f)$. Further, the systems are assumed to be such that the fundamental results, as given by Theorems 8.1 and 8.6, are valid.

9.1.2 The Effect of Noise

In electronic devices, noise is a consequence of charge movement at an atomic level which is random in character. This random behaviour leads, at a macro level, to unwanted variations in signals. To illustrate this, consider a signal V_S, from a signal source, assumed to be sinusoidal and with a resistance R_S, which is amplified by a low noise amplifier as illustrated in Figure 9.1. The equivalent noise signal at the amplifier input for the case of a 1 kΩ source resistance, and where the noise from this resistance dominates other sources of noise, is shown in Figure 9.2. A sample rate of 2.048 kSamples/sec has been used, and 200 samples are displayed. The specific details of the amplifier are described in Howard (1999b). In particular, the amplifier bandwidth is 30 kHz.

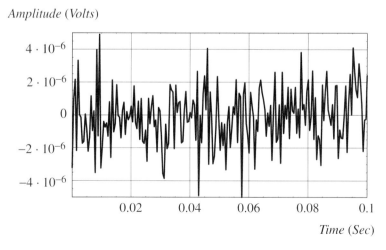

Figure 9.2 Time record of equivalent noise at amplifier input.

258 PRINCIPLES OF LOW NOISE ELECTRONIC DESIGN

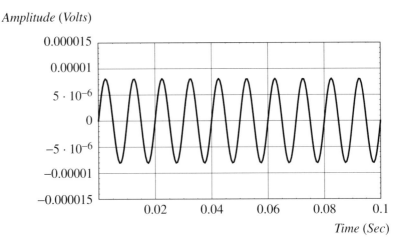

Figure 9.3 Sinusoid of 100 Hz whose amplitude is consistent with a signal-to-noise ratio of 10.

In Figure 9.3 a 100 Hz sine wave is displayed, whose amplitude is consistent with a signal-to-noise ratio of 10 assuming the noise waveform of Figure 9.2. The addition of this 100 Hz sinusoid, and the noise signal of Figure 9.2, is shown in Figure 9.4 to illustrate the effect of noise corrupting the integrity of a signal.

For completeness, in Figure 9.5, the power spectral density of the noise referenced to the amplifier input is shown. In this figure, the power spectral

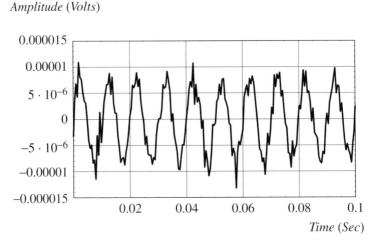

Figure 9.4 100 Hz sinusoidal signal plus noise due to the source resistance and amplifier. The signal-to-noise ratio is 10.

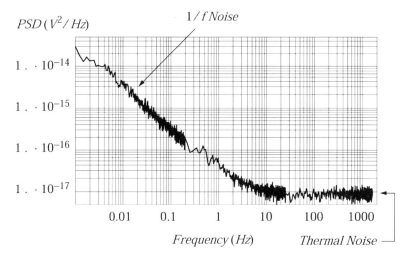

Figure 9.5 Power spectral density of amplifier noise referenced to the amplifier input.

density has a $1/f$ form at low frequencies, and at higher frequencies is constant. For frequencies greater than 10 Hz, the thermal noise from the resistor dominates the overall noise.

9.2 GAUSSIAN WHITE NOISE

Gaussian white noise, by which is meant noise whose amplitude distribution at a set time has a Gaussian density function and whose power spectral density is flat, that is, white, is the most common type of noise encountered in electronics. The following section gives a description of a model which gives rise to such noise. Since the model is consistent with many physical noise processes it provides insight into why Gaussian white noise is ubiquitous.

9.2.1 A Model for Gaussian White Noise

In many instances, a measured noise waveform is a consequence of the weighted sum of waveforms from a large number of independent random processes. For example, the observed randomly varying voltage across a resistor is due to the independent random thermal motion of many electrons. In such cases, the observed waveform z, can be modelled according to

$$z(t) = \sum_{i=1}^{M} w_i z_i(t) \qquad z_i \in E_i \tag{9.1}$$

where w_i is the weighting factor for the ith waveform z_i, which is from the ith

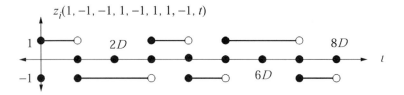

Figure 9.6 One waveform from a binary digital random process on the interval [0, 8D].

ensemble E_i defining the ith random process Z_i. Here, z is one waveform from a random process Z which is defined as the weighted summation of the random processes $Z_1, \ldots, Z_M$. Consider the case, where all the random processes $Z_1, \ldots, Z_M$ are identical, but independent, signalling random processes and are defined, on the interval $[0, ND]$, by the ensemble

$$E_i = \left\{ z_i(\gamma_1, \ldots, \gamma_N, t) = \sum_{k=1}^{N} \gamma_k \phi(t - (k-1)D) \quad \begin{array}{c} \gamma_k \in \{-1, 1\} \\ P[\gamma_k = \pm 1] = 0.5 \end{array} \right\} \quad (9.2)$$

where the pulse function ϕ is defined according to

$$\phi(t) = \begin{cases} 1 & 0 \leqslant t < D \\ 0 & \text{elsewhere} \end{cases} \qquad \Phi(f) = D \operatorname{sinc}(fD) e^{-j\pi fD} \quad (9.3)$$

All waveforms in the ensemble have equal probability, and are binary digital information signals. One waveform from the ensemble is illustrated in Figure 9.6.

One outcome of the random process Z, as defined by Eq. (9.1), has the form illustrated in Figure 9.7 for the case of equal weightings, $w_i = 1$, $D = 1$, and $M = 500$. The following subsections show, as the number of waveforms M, increases, that the amplitude density function approaches that of a Gaussian function, and that over a restricted frequency range the power spectral density is flat or "white".

9.2.2 Gaussian Amplitude Distribution

The following, details the reasons why, as the number of waveforms, M, comprising the random process increases, the amplitude density function approaches that of a Gaussian function.

The waveform defined by the sum of M equally weighted independent binary digital waveforms, as per Eq. (9.1), has the following properties: (1) the amplitudes of the waveform during the intervals $[iD, (i+1)D)$, and $[jD, (j+1)D)$, are independent for $i \neq j$; (2) the amplitude A, in any interval $[iD, (i+1)D]$ is, for the case where M is even, from the set

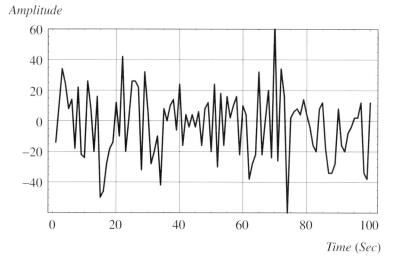

Figure 9.7 Sum of 500 equally weighted, independent, binary digital waveforms where $D = 1$. Linear interpolation has been used between the values of the function at integer values of time.

$S_A = \{-M, -M+2, \ldots, 0, \ldots, M-2, M\}$, and M is assumed to be even in subsequent analysis; (3) at a specific time, the amplitude A, is a consequence of k ones, and m negative ones where $k + m = M$. Thus, $A \in S_A$ is such that $A = k - m$. Given A and M, it then follows that

$$k = (M + A)/2 \qquad m = (M - A)/2 \tag{9.4}$$

Hence, $P[A]$ equals the probability of $k = 0.5(A + M)$ successes in M outcomes of a Bernoulli trial. For the case where the probability of success is p, and the probability of failure is q, it follows that (Papoulis 2002 p. 53)

$$P[A] = \frac{M!(p^k)q^{M-k}}{k!(M-k)!} = \frac{M! p^{0.5(A+M)} q^{0.5(M-A)}}{[0.5(M+A)]![0.5(M-A)]!} \tag{9.5}$$

To show that $P[A]$ can be approximated by a Gaussian function, consider the DeMoivre–Laplace theorem (Papoulis 2002 p. 105, Feller 1957 p. 168f):

Consider M trials of a Bernoulli random process, where the probability of success is p, and the probability of failure is q. With the definitions $\sigma = \sqrt{Mpq}$ and $\mu = Mp$, and the assumption $\sigma^2 \gg 1$, the probability of k successes in M trials can be approximated according to:

$$P[k \text{ out of } M \text{ trials}] = \frac{M!}{k!(M-k)!} p^k q^{M-k} \approx \frac{e^{-(k-\mu)^2/2\sigma^2}}{\sqrt{2\pi}\sigma} \tag{9.6}$$

where a bound on the relative error in this approximation is:

$$\left|\frac{(k-\mu)^3}{6\sigma^2} - \frac{(k-\mu)}{2\sigma}\right| \quad k \neq \mu \tag{9.7}$$

For the case being considered, where $k = 0.5(A + M)$, and $p = q = 0.5$, it follows that $\sigma = 0.5\sqrt{M}$, $\mu = 0.5M$, and $k - \mu = 0.5A$. Thus, for $0.25M \gg 1$, the amplitude distribution in any interval $[jD, (j+1)D]$, can be approximated by the Gaussian form:

$$P(A) = P\left[\frac{A+M}{2} \text{ out of } M \text{ trials}\right] \approx \frac{2e^{-A^2/2M}}{\sqrt{2\pi M}} \tag{9.8}$$

where a bound on the relative error is

$$\left|\frac{A^3}{12M} - \frac{A}{2\sqrt{M}}\right| \tag{9.9}$$

Note, with the assumptions made, the mean of A is zero, and the rms value of A is $\sqrt{M}$. The factor of 2 in Eq. (9.8) arises from the fact that A only takes on even values. Consistent with this result, many noise sources have a Gaussian amplitude distribution, and the term Gaussian noise is widely used.

Confirmation, and illustration of this result is shown in Figure 9.8, where the probability of an amplitude obtained from 1000 repetitions of 100 trials of a Bernoulli process (possible outcomes are from the set $\{-100, -98, \ldots, 0, \ldots, 100\}$) is shown. The smooth curve is the Gaussian probability density function as per Eq. (9.8) with $M = 100$.

9.2.3 White Power Spectral Density

The power spectral density of the individual random processes comprising Z are zero mean signaling random processes, as defined by the ensemble of Eq. (9.2). It then follows, from Theorem 5.1, that the power spectral density of each of these random processes, on the interval $[0, ND]$, is

$$G_i(ND, f) = r|\Phi(f)|^2 = \frac{1}{r}\text{sinc}^2\left(\frac{f}{r}\right) \tag{9.10}$$

where, $r = 1/D$, and Φ is the Fourier transform of the pulse function ϕ. As Z is the sum of independent random processes with zero means, it follows, from Theorem 4.6, that the power spectral density of Z is the sum of the weighted

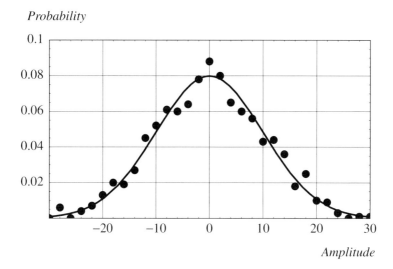

Figure 9.8 Probability of an amplitude from the set $\{-100, -98, \ldots, 98, 100\}$ arising from 1000 repetitions of 100 trials of a Bernoulli process. The probabilities agree with the Gaussian form, as defined in the text.

individual power spectral densities, that is,

$$G_Z(ND, f) = \sum_{i=1}^{M} |w_i|^2 G_i(ND, f) = r|\Phi(f)|^2 \sum_{i=1}^{M} |w_i|^2$$

$$= \frac{1}{r} \operatorname{sinc}^2(f/r) \sum_{i=1}^{M} |w_i|^2 \qquad (9.11)$$

This power spectral density is shown in Figure 9.9 for the normalized case of $M = r = 1$, and $w_1 = 1$. For frequencies lower than $r/4$, the power spectral density is approximately constant at a level of M/r, and it is this constant level that is typically observed from noise sources arising from electron movement. This is the case because, first, the dominant source of electron movement is, typically, thermal energy, and electron thermal movement is correlated over an extremely short time interval. Second, a consequence of this very short correlation time, is that the rate r, used for modelling purposes, is much higher than the bandwidth of practical electronic devices. Thus, the common case is where the noise power spectral density, appears flat for all measurable frequencies, and the phrase "white Gaussian noise" is appropriate, and is commonly used.

Note, for processes whose correlation time is very short compared with the response time of the measurement system (for example, rise time), the power spectral density will be constant within the bandwidth of the measurement

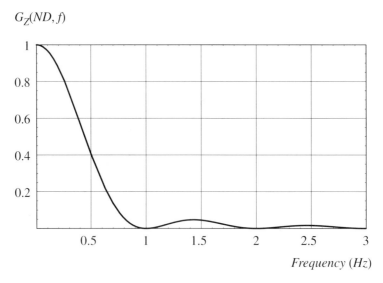

Figure 9.9 Normalized power spectral density as defined by the case where $r = M = w_1 = 1$.

system and, consistent with Eq. 9.11, this constancy is independent of the pulse shape.

9.3 STANDARD NOISE SOURCES

The noise sources commonly encountered in electronics are thermal noise, shot noise, and $1/f$ noise. These are discussed briefly below.

9.3.1 Thermal Noise

Thermal noise is associated with the random movement of electrons, due to the electrons thermal energy. As a consequence of such electron movement, there is a net movement of charge, during any interval of time, through an elemental section of a resistive material as illustrated in Figure 9.10. Such a net movement of charge, is consistent with a current flow, and as the elemental section has a defined resistance, the current flow generates an elemental voltage dV. The sum of the elemental voltages, each of which has a random magnitude, is a random time varying voltage.

Consistent with such a description, equivalent noise models for a resistor are shown in Figure 9.11. In this figure, v and i, respectively, are randomly varying voltage and current sources. These sources are related via Thevenin's and Norton's equivalence statements, namely $v(t) = Ri(t)$, and $i(t) = v(t)/R$.

Statistical arguments (for example, Reif, 1965 pp. 589–594, Bell, 1960 ch. 3) can be used to show that the power spectral density of the random processes,

STANDARD NOISE SOURCES 265

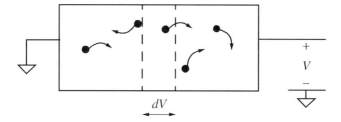

Figure 9.10 Illustration of electron movement in a resistive material.

which give rise to v and i, respectively, are:

$$G_V(f) = \frac{2h|f|R}{e^{h|f|/kT} - 1} \qquad V^2/\text{Hz} \qquad (9.12)$$

$$G_I(f) = \frac{2h|f|}{R(e^{h|f|/kT} - 1)} \qquad A^2/\text{Hz} \qquad (9.13)$$

where T is the absolute temperature, k is Boltzmann's constant (1.38×10^{-23} J/K), h is Planck's constant (6.62×10^{-34} J.sec) and R is the resistance of the material. For frequencies, such that $|f| < 0.1kT/h \approx 10^{12}$ Hz (assuming $T = 300K$) a Taylor series expansion for the exponential term in these equations, namely,

$$e^{h|f|/kT} \approx 1 + h|f|/kT \qquad (9.14)$$

is valid, and the following approximations hold:

$$G_V(f) \approx 2kTR \quad V^2/\text{Hz} \qquad G_I(f) \approx \frac{2kT}{R} \quad A^2/\text{Hz} \qquad (9.15)$$

These equations were derived using the equipartition theorem, and statistical arguments, by Nyquist in 1928 (Nyquist 1928; Kittel 1958 p. 141; Reif 1965 p. 589; Freeman 1958 p. 117) and are denoted as Nyquist's theorem. A

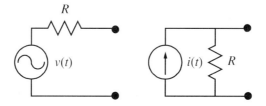

Figure 9.11 Equivalent noise models for a resistor.

derivation of these results, based on electron movement, is given in Buckingham 1983 pp. 39–41. Further, these equations are the ones that are nearly always used in analysis. Note that the power spectral density is "white", that is, it has a constant level independent of the frequency.

One point to note: In analysis, the Norton, rather than the Thevenin equivalent noise model for a resistor best facilitates analysis.

9.3.2 Shot Noise

As shown in Section 5.5, shot noise is associated with charge carriers crossing a barrier, such as that inherent in a PN junction, at random times, but with a constant average rate. As detailed in Section 5.5.1 the power spectral density, for all but high frequencies, is given by

$$G(f) \approx q\bar{I} + \bar{I}^2 \delta(f) \qquad A^2/\text{Hz} \tag{9.16}$$

where q is the electronic charge (1.6×10^{-19} C), and $\bar{I}$ is the mean current. Note that, apart from the impulse at DC, the power spectral density is "white". In electronic circuits the mean current is associated with circuit bias. As variations away from the bias state are of interest in analogue electronics, it is usual to approximate the power spectral density in such circuits, according to

$$G(f) \approx q\bar{I} \qquad A^2/\text{Hz} \tag{9.17}$$

9.3.3 1/f Noise

As discussed in Section 6.5, the power spectral density of a $1/f$ random process has a power spectral density given by

$$G(f) = \frac{k}{f^\alpha} \tag{9.18}$$

where k is a constant, and α determines the slope. Typically, α is close to unity. At low frequencies, $1/f$ noise often dominates other noise sources, and this is well illustrated in Figure 9.5.

9.4 NOISE MODELS FOR STANDARD ELECTRONIC DEVICES

9.4.1 Passive Components

In an ideal capacitor with an ideal dielectric, all charge is bound, such that interatomic movement of charge is not possible. Accordingly, an ideal capacitor is noiseless. An ideal inductor is made from material with zero resistance, and in such a material the voltage created by the thermal motion of electrons is zero. Hence, ideal inductors are noiseless. As discussed above, resistors exhibit thermal noise, and have either of the noise models shown in Figure 9.11. Fish (1993 ch. 6) gives a more detailed analysis of noise in passive components.

9.4.2 Active Components

The small signal equivalent noise model for a diode, is shown in Figure 9.12 (Fish, 1993 pp. 126–127). In this figure I_D is the mean diode current, and the power spectral density of the small signal equivalent noise source i, is given by

$$G_D(f) = q|I_D| \quad A^2/\text{Hz} \qquad (9.19)$$

Note, the model of Figure 9.12(c) is also applicable to standard nonavalanche photodetectors, when they are operated with reverse bias.

The small signal noise equivalent model for a PNP or NPN BJT transistor, operating in the forward active region, is shown in Figure 9.13 (Fish, 1993 p. 128). The sources i_{BB}, i_B, and i_C in this figure, respectively model the thermal noise in the base due to the base spreading resistance r_b, which is typically in the range of 10–500 Ohms (Gray, 2001 p. 32; Fish, 1993 pp. 128–139), the shot noise of the base current and the collector current shot noise (see Edwards, 2000). The respective power spectral densities of these noise sources are

$$G_{BB}(f) = 2kT/r_b \quad A^2/\text{Hz} \qquad (9.20)$$

$$G_B(f) = qI_B \quad A^2/\text{Hz} \qquad G_C(f) = qI_C \quad A^2/\text{Hz} \qquad (9.21)$$

In analysis, it is usual to neglect r_o as, typically, it is in parallel with a much lower value load resistance.

The small signal noise equivalent model for a NMOS or PMOS MOSFET, with the source connected to the substrate, and a N or P channel JFET, when they are operating in the saturation region, is shown in Figure 9.14 (for

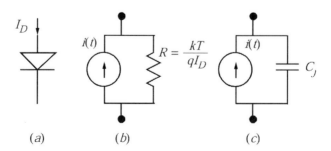

Figure 9.12 (a) Diode symbol. (b) Noise equivalent model for a diode under forward bias. (c) Noise equivalent model for a diode under reverse bias. I_D is the DC current flow and C_j is the junction capacitance.

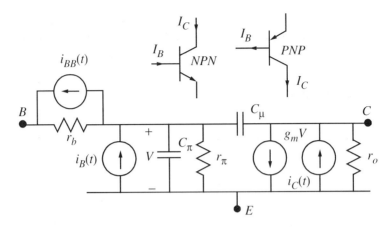

Figure 9.13 Small signal equivalent noise model for a NPN or PNP BJT operating in the forward active region.

example, Fish, 1993 p. 140; Levinzon, 2000; Howard, 1987). In this figure, the noise sources i_G and i_D, respectively, account for the noise at the gate, which is due to the gate leakage current and the induced noise in the gate due to thermal noise in the channel, and the thermal noise in the channel. The respective power spectral densities of these sources are

$$G_G(f) = q|I_G| + 2kT\theta(2\pi f C_{gs})^2/g_m \qquad A^2/\text{Hz} \qquad (9.22)$$

$$G_D(f) = 2kTPg_m \qquad A^2/\text{Hz} \qquad (9.23)$$

In these equations, θ is a constant with a value of around 0.25 for JFETs, and 0.1 for MOSFETS (Fish, 1993 p. 141). P is a constant with a theoretical value of 0.7, but practical values can be higher (Howard, 1987; Muoi, 1984; Ogawa, 1981). I_G is the gate leakage current which, typically, is in the pA range. As with a BJT, it is usual to neglect r_o in analysis.

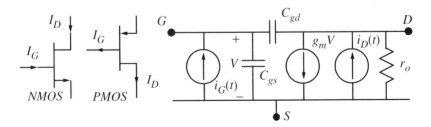

Figure 9.14 Small signal equivalent noise model for a PMOS or NMOS MOSFET, or a N or P channel JFET, operating in the saturation region.

9.5 NOISE ANALYSIS FOR LINEAR TIME INVARIANT SYSTEMS

The following discussion relates to analysis of noise in linear time invariant systems — linear electronic systems are an important subset of such systems.

9.5.1 Background and Assumptions

A schematic diagram of a linear system is shown in Figure 9.15. With the assumption that the results of Theorems 8.1 and 8.6 are valid, the relationship between the input and output power spectral densities, on the infinite interval $[0, \infty]$, or a sufficiently long interval relative to the impulse response time of the system, is given by

$$G_Y(f) = |H(f)|^2 G_X(f) \qquad (9.24)$$

In this diagram, the input random process X is defined by the ensemble E_X, and the output random process Y is defined by the ensemble E_Y.

9.5.1.1 Transfer Functions and Notation
Analysis of electronic circuits is usually performed through use of Laplace transforms (for example, Chua, 1987 ch. 10). Such analysis yields a relationship, assuming appropriate excitation, between the Laplace transform of the ith and jth node voltage or current, of the form $V_j(s)/V_i(s) = L_{ij}(s)$. If the time domain input at the ith node, $v_i(t)$, is an "impulse," then $V_i(s) = 1$ and, hence, the output signal $v_j(t)$ is the impulse response, whose Laplace transform is given by $L_{ij}(s)$. In the subsequent text, the following notation will be used: L_{ij} is denoted the Laplace transfer function, while H_{ij}, which is the Fourier transform of the impulse response, is simply denoted the transfer function. From the definitions for the Laplace and Fourier transform, it follows that the relationship between these transfer functions is

$$H_{ij}(f) = L_{ij}(j2\pi f) \qquad (9.25)$$

The Fourier transform H_{ij}, is guaranteed to exist if the impulse response h_{ij}, is such that $h_{ij} \in L[0, \infty]$. Similarly, the Laplace transform L_{ij}, will exist, with a region of convergence including the imaginary axis, when $h_{ij} \in L[0, \infty]$.

Finally, in circuit analysis, it is usual to omit the argument s from Laplace transformed functions. To distinguish between a time function, and its associated Laplace transform, capital letters are used for the latter, while lowercase letters are used for the former.

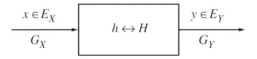

Figure 9.15 Schematic diagram of a linear system.

9.5.2 Input Equivalent Noise — Individual Case

The definition of the input equivalent noise of a linear system, is fundamental to low noise amplifier design. The following is a brief summary: When all components in a linear circuit have been replaced by their equivalent circuit models, including appropriate models for noise sources, the circuit, as illustrated in Figure 9.16, results.

In this figure w_0 and w_M respectively, are the input and output signals of the circuit, and $w_1, \ldots, w_N$ are signals from the ensembles defining the N noise sources in the circuit. The Laplace transform of these signals are, respectively, denoted by $W_0, W_1, \ldots, W_N, W_M$. The transfer function between the source and the output, denoted H_{0M}, is defined according to

$$H_{0M}(f) = L_{0M}(j2\pi f) = \left. \frac{W_M(j2\pi f)}{W_0(j2\pi f)} \right|_{w_0 = \delta,\ w_1 = \cdots = w_N = 0} \qquad (9.26)$$

where, δ denotes the Dirac delta function, and it is assumed that $w_M \in L[0, \infty]$, when $w_0 = \delta$, such that, the results of Theorem 8.3 are valid. Similarly, the transfer functions $H_{1M}, \ldots, H_{NM}$ are defined as the transfer functions that relate the noise sources $w_1, \ldots, w_N$ to the amplifier output, and are defined as

$$H_{iM}(f) = L_{iM}(j2\pi f) = \left. \frac{W_M(j2\pi f)}{W_i(j2\pi f)} \right|_{w_i = \delta,\ w_0 = w_1 = \cdots = w_{i-1} = w_{i+1} = w_N = 0} \qquad (9.27)$$

It is usual, when quantifying the noise performance of an amplifier, to refer the noise to the amplifier input in order that it is independent of the amplifier gain. To achieve this, it is necessary to define an input equivalent noise source for each of the noise sources in the amplifier. By definition, the input equivalent noise source, denoted w_{ei}, for the ith noise source w_i, is the equivalent noise source at the amplifier input that produces the same level of output noise as w_i. That is, by definition, w_{ei} guarantees the equivalence of the circuits shown in Figures 9.17 and 9.18, as far as the output noise is concerned.

Assume, for the circuit shown in Figure 9.17, that either, or both the source w_0, and the ith noise source w_i, have zero mean, and the source is independent of the ith noise source. It then follows, from Theorem 8.7, that the output

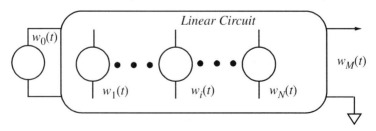

Figure 9.16 Schematic diagram of a linear system with N noise sources.

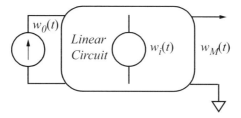

Figure 9.17 Noise model for ith noise source.

power spectral density, G_M, due to w_0 and w_i is

$$G_M(f) = |H_{0M}(f)|^2 G_0(f) + |H_{iM}(f)|^2 G_i(f) \qquad (9.28)$$

where G_0 and G_i, respectively, are the power spectral densities of w_0 and w_i. For the circuit shown in Figure 9.18, the output power spectral density due to the noise sources w_0 and w_{ei}, is

$$G_M(f) = |H_{0M}(f)|^2 G_0(f) + |H_{0M}(f)|^2 G_{ei}(f) \qquad (9.29)$$

where G_{ei} is the power spectral density of the input equivalent source w_{ei}. A comparison of Eqs. (9.28) and (9.29) shows that these two circuits are equivalent, in terms of the output power spectral density, when

$$|H_{0M}(f)|^2 G_{ei}(f) = |H_{iM}(f)|^2 G_i(f) \qquad (9.30)$$

Thus, the power spectral density of the input equivalent noise source associated with the ith noise source is

$$G_{ei}(f) = \frac{|H_{iM}(f)|^2}{|H_{0M}(f)|^2} G_i(f) = |H_{iM}^{eq}(f)|^2 G_i(f) \qquad (9.31)$$

where $H_{iM}^{eq}(f) = H_{iM}(f)/H_{0M}(f)$ is the transfer function between the ith noise source, w_i, and the associated input equivalent noise source w_{ei}.

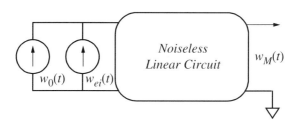

Figure 9.18 Equivalent noise model, as far as the output node is concerned, for the ith noise source.

9.5.3 Input Equivalent Noise—General Case

For the general case of determining the input equivalent noise of all the N noise sources, the approach is to, first, establish the input equivalent signal and, then, evaluate its power spectral density. The details are as follows: the N noise signals generate an output signal according to

$$w_M(t) = \int_0^t w_1(\lambda) h_{1M}(t - \lambda) d\lambda + \cdots + \int_0^t w_N(\lambda) h_{NM}(t - \lambda) d\lambda \quad (9.32)$$

where, $h_{1M}, \ldots, h_{NM}$ are the impulse responses of the systems between w_i and w_m for $i \in \{1, \ldots, N\}$. From Theorem 8.3 it follows that

$$W_M(s) = W_1(s) L_{1M}(s) + \cdots + W_N(s) L_{NM}(s) \quad \text{Re}[s] > 0 \quad (9.33)$$

where, L_{iM} is the Laplace transform of h_{iM}, and the noise signals are assumed not to have exponential increase, which is the usual case. An equivalent input signal, w_{eq}, whose Laplace transform is W_{eq}, will result in an output signal with the same Laplace transform when

$$W_{eq}(s) L_{0M}(s) = W_1(s) L_{1M}(s) + \cdots + W_N(s) L_{NM}(s) \quad \text{Re}[s] > 0 \quad (9.34)$$

Thus, provided $L_{0M}(s) \neq 0$, it is the case that

$$W_{eq}(s) = \sum_{i=1}^N W_{ei}(s) = \sum_{i=1}^N \frac{W_i(s) L_{iM}(s)}{L_{0M}(s)} \quad \text{Re}[s] > 0 \quad (9.35)$$

where W_{ei} is the Laplace transform of the ith input equivalent signal associated with w_i. Consistent with this result, an equivalent model for the input equivalent noise is as shown in Figure 9.19. The ith transfer function in this figure, from Eq. (9.35), is given by

$$H_{iM}^{eq}(f) = \frac{L_{iM}(j2\pi f)}{L_{0M}(j2\pi f)} = \frac{H_{iM}(f)}{H_{0M}(f)} \quad H_{0M}(f) \neq 0 \quad (9.36)$$

where $L_{iM}(s)$ and $L_{0M}(s)$ are validly defined when $\text{Re}[s] = 0$, as assumed in Eqs. (9.26) and (9.27). The following theorem states the power spectral density of the input equivalent noise random process.

THEOREM 9.1 POWER SPECTRAL DENSITY OF INPUT EQUIVALENT NOISE *For independent noise sources with zero means, the amplifier input equivalent power spectral density, denoted $G_{eq}(f)$, is the sum of the individual input equivalent power spectral densities, that is,*

$$G_{eq}(f) = \sum_{i=1}^N G_{ei}(f) = \sum_{i=1}^N |H_{iM}^{eq}(f)|^2 G_i(f) = \sum_{i=1}^N \left| \frac{H_{iM}(f)}{H_{0M}(f)} \right|^2 G_i(f) \quad (9.37)$$

where G_i and G_{ei}, respectively, are the power spectral density, and the input equivalent power spectral density, of the ith noise source. For the general case,

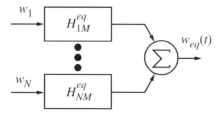

Figure 9.19 Equivalent model for input equivalent noise source.

the input equivalent power spectral density is given by

$$G_{eq}(f) = \sum_{i=1}^{N} \left| \frac{H_{iM}(f)}{H_{0M}(f)} \right|^2 G_i(f) + \sum_{i=1}^{N} \sum_{\substack{j=1 \\ j \neq i}}^{N} \frac{H_{iM}(f) H_{jM}^*(f)}{|H_{0M}(f)|^2} G_{ij}(f) \quad (9.38)$$

where G_{ij} is the cross power spectral density between the ith and jth noise sources.

Proof. These results follow directly from the model shown in Figure 9.19 and Theorem 8.7.

9.5.4 Notation

When analysing a linear circuit arising from several transistor stages, it is convenient to label the sources according to which node they are between, as is indicated in Figure 9.20. In this figure I_{13} is a noise current source between nodes 1 and 3, I_1 is a current source between node 1 and ground and so on. For the circuit arising from a single transistor stage amplifier, it is more convenient to label the noise sources according to their origin, as is illustrated in the following example.

9.5.5 Example: Input Equivalent Noise of a Common Emitter Amplifier

To illustrate the theory related to input equivalent noise characterization of a circuit, consider the Common Emitter (CE) amplifier shown in Figure 9.21. The small signal equivalent noise model for such a structure, is shown in Figure 9.22. The noise current sources i_S, i_{BB}, i_B, i_C defined in this figure are independent and have zero means. Their respective power spectral densities are:

$$G_S(f) = \frac{2kT}{R_S} \qquad G_{BB}(f) = \frac{2kT}{r_b} \qquad A^2/\text{Hz} \qquad (9.39)$$

$$G_B(f) = qI_B \qquad G_C(f) = qI_C + \frac{2kT}{R_C} \qquad A^2/\text{Hz} \qquad (9.40)$$

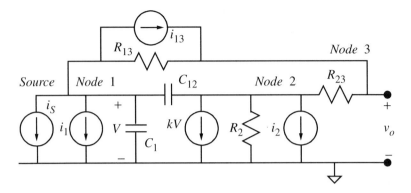

Figure 9.20 Notation for labelling noise sources in a circuit.

The amplifier voltage transfer function, L_o, is

$$L_o(s) = \left.\frac{V_o(s)}{V_S(s)}\right|_{v_S=\delta,\, i_S=i_{BB}=i_B=i_C=0}$$

$$= \frac{-g_m R_C}{1 + \dfrac{R_{Sb}}{r_\pi}} \cdot \frac{1 - sC_\mu/g_m}{1 + s(r_\pi//R_{Sb})\left[C_\pi + C_\mu\left(1 + g_m R_C + \dfrac{R_C}{r_\pi//R_{Sb}}\right)\right] + s^2 D_2} \quad (9.41)$$

where $D_2 = (r_\pi//R_{Sb})R_C C_\pi C_\mu$, and $R_{Sb} = R_S + r_b$. Using the parameter values tabulated in Table 9.1, the normalized magnitude, $|H_o(f)| = |L_o(j2\pi f)|$, of this transfer function, is plotted in Figure 9.23. The low frequency gain is 37.5, and the 3 dB bandwidth is 58 MHz.

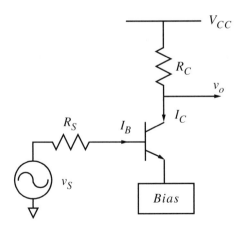

Figure 9.21 Schematic diagram of a common emitter amplifier.

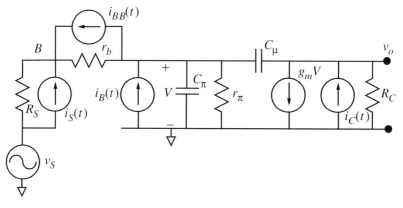

Figure 9.22 Small signal equivalent noise model for a common emitter amplifier.

The small signal input equivalent noise model for the common emitter amplifier, is shown in Figure 9.24, where the power spectral density of the input equivalent noise source v_{eq}, from Theorem 9.1, is given by

$$G_{eq}(f) = G_S(f)\frac{|H_S(f)|^2}{|H_o(f)|^2} + G_{BB}(f)\frac{|H_{BB}(f)|^2}{|H_o(f)|^2} + G_B(f)\frac{|H_B(f)|^2}{|H_o(f)|^2} + G_C(f)\frac{|H_C(f)|^2}{|H_o(f)|^2}$$

(9.42)

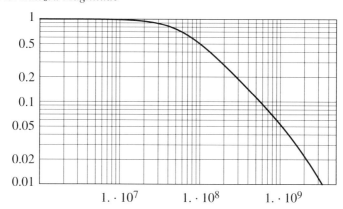

Figure 9.23 Normalized magnitude of transfer function of common emitter amplifier for parameter values listed in Table 9.1. The low frequency gain is 37.5 and the 3 dB bandwidth is 58 MHz.

TABLE 9.1 Parameters for BJT common emitter amplifier

Parameter	Value (Rounded)
$V_T = kT/q$	0.0259
I_C	1 mA
$\beta = I_C/I_B$	100
$g_m = I_C/V_T$	0.039
$r_\pi = \beta/g_m$	2600 Ohms
r_b	50 Ohms
R_S	50 Ohms
R_C	1000 Ohms
f_T	1.5 GHz
C_μ	0.5 pF
$C_\pi = g_m/2\pi f_T - C_\mu$	4 pF

Here, $H_S(f) = L_S(j2\pi f)$, $L_S(s) = V_o(s)/I_S(s)|_{i_S = \delta, v_S = i_{BB} = i_B = i_C = 0}$, and similarly for the other transfer functions. Standard circuit analysis yields the following results:

$$\frac{|H_S(f)|^2}{|H_o(f)|^2} = R_S^2 \qquad \frac{|H_{BB}(f)|^2}{|H_o(f)|^2} = r_b^2 \qquad \frac{|H_B(f)|^2}{|H_o(f)|^2} = (R_S + r_b)^2 \qquad (9.43)$$

$$\frac{|H_C(f)|^2}{|H_o(f)|^2} = \frac{\left(1 + \frac{R_{Sb}}{r_\pi}\right)^2}{g_m^2} \frac{1 + 4\pi^2 f^2 (r_\pi \| R_{Sb})^2 (C_\pi + C_\mu)^2}{1 + 4\pi^2 f^2 C_\mu^2/g_m^2} \qquad (9.44)$$

where $R_{Sb} = R_S + r_b$. It then follows that the power spectral density G_{eq} of the input equivalent voltage noise source v_{eq}, is

$$G_{eq}(f) = 2kTR_S + 2kTr_b$$

$$+ qI_B R_{Sb}^2 \left[1 + \frac{r_\pi \left(1 + \frac{R_{Sb}}{r_\pi}\right)^2}{g_m R_{Sb}^2} \frac{1 + 4\pi^2 f^2 (r_\pi \| R_{Sb})^2 (C_\pi + C_\mu)^2}{1 + 4\pi^2 f^2 C_\mu^2/g_m^2} \right] \qquad (9.45)$$

$$+ \frac{2kT}{R_C} \frac{\left(1 + \frac{R_{Sb}}{r_\pi}\right)^2}{g_m^2} \frac{1 + 4\pi^2 f^2 (r_\pi \| R_{Sb})^2 (C_\pi + C_\mu)^2}{1 + 4\pi^2 f^2 C_\mu^2/g_m^2} \qquad V^2/Hz$$

where the fact that $I_C = \beta I_B = g_m r_\pi I_B$ has been used to combine the power spectral density of the base and collector shot noise.

Clearly, such an analytical expression facilitates low noise design. For example, low noise performance is consistent with a low source resistance, R_S; low base spreading resistance, r_b; and low base current, I_B. However, with

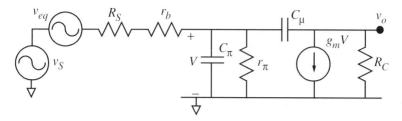

Figure 9.24 Equivalent noise model for CE amplifier as far as the output node is concerned.

respect to the base current note, as $g_m = I_C/V_T$, with $V_T = kT/q$ being the thermal voltage, that $r_\pi/g_m = V_T^2/\beta I_B^2$ and hence, an optimum base current exists to minimize the third term in the expression for G_{eq} (Hullett, 1977).

Using the parameter values given in Table 9.1, the power spectral density, as defined by Eq. (9.45), is shown in Figure 9.25. Also shown is the output power spectral density given by

$$G_o(f) = |H_o(f)|^2 G_{eq}(f)$$

$$= [2kTR_{Sb} + qI_B R_{Sb}^2]\left[\frac{g_m^2 R_C^2}{(1 + R_{Sb}/r_\pi)^2}\frac{1 + 4\pi^2 f^2 C_\mu^2/g_m}{D(f)}\right] \quad (9.46)$$

$$+ \left[qI_C + \frac{2kT}{R_C}\right]\frac{R_C^2[1 + 4\pi^2 f^2 (R_{Sb}\|r_\pi)^2(C_\pi + C_\mu)^2]}{D(f)} \quad V^2/\text{Hz}$$

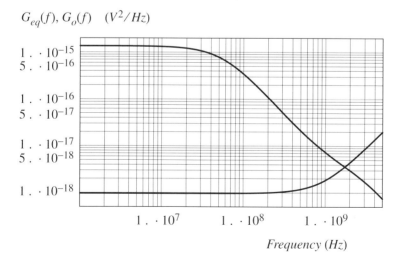

Figure 9.25 Input equivalent (lower trace) and output (upper trace) power spectral density of common emitter amplifier for parameter values listed in Table 9.1.

where

$$D(f) = [1 - 4\pi^2 f^2 R_C (R_{Sb} \| r_\pi) C_\pi C_\mu]^2$$
$$+ 4\pi^2 f^2 (R_{Sb} \| r_\pi)^2 \left[C_\pi + C_\mu \left(1 + g_m R_C + \frac{R_C}{R_{Sb} \| r_\pi} \right) \right]^2 \quad (9.47)$$

One interesting example of the usefulness of input equivalent noise characterization can be found in Howard (1999a) which details a novel structure for an optoelectronic receiver, that potentially, has half the input equivalent noise level of a standard optoelectronic receiver.

9.6 INPUT EQUIVALENT CURRENT AND VOLTAGE SOURCES

In many instances it is convenient to be able to characterize the noise of a structure by an equivalent noise source at its input, which is insensitive to the way the structure is driven, that is, insensitive to the nature of the source impedance characteristics. Such a characterization is possible through use of an input equivalent current source I_{eq}, and an input equivalent voltage source V_{eq} (Haus, 1960; Lam, 1992; Netzer, 1981; Gray, 2001 p. 768), as per the model shown in Figure 9.26.

In this model Z_{11} is the input impedance of the linear circuit and $Z_S = 1/Y_S$ is the source impedance. With the Norton equivalent model for the source, I_S is the source current and I_{SS} is a current source to account for the noise associated with the source impedance. With the Thevenin equivalent model, V_S is the source voltage ($V_S = I_S Z_S$) and V_{SS} is a voltage source to account for the noise associated with the source impedance ($V_{SS} = I_{SS} Z_S$).

An alternative model often used with differential input circuits such as operational amplifier circuits [see, for example, Trofimenkoff (1989)], is shown in Figure 9.27. It is important to note that the two current sources I_{eq}, are 100% correlated.

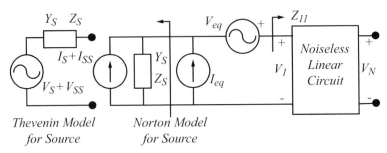

Figure 9.26 Equivalent input noise model for a linear time invariant circuit where an input equivalent voltage source and an input equivalent current source have been used to characterize the circuit noise.

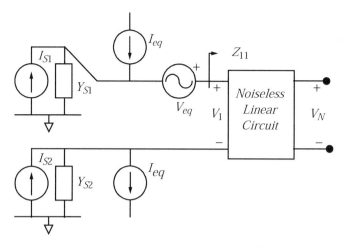

Figure 9.27 An alternative input equivalent model for differential input circuits including operational amplifier circuits. Norton models have been used for the two sources.

The following theorem states the input equivalent voltage and current defined in this model.

THEOREM 9.2. INPUT EQUIVALENT SOURCE CURRENT AND SOURCE VOLTAGE
The model shown in Figure 9.26 is valid, and the input source voltage or input source current that replace the noise sources in the circuit as far as the output node is concerned, are such that

$$\begin{aligned} V_S &= V_{SS} + V_{eq} + Z_S I_{eq} & \text{Thevenin} \\ I_S &= I_{SS} + I_{eq} + Y_S V_{eq} & \text{Norton} \end{aligned} \qquad (9.48)$$

where the input equivalent current I_{eq} and input equivalent voltage V_{eq} are defined according to

$$V_{eq}(s) = \frac{V_N^{sc}(s)}{L_{1Nsc}(s)} \qquad I_{eq}(s) = \frac{V_N^{oc}(s)}{L_{1Noc}^I(s)} \qquad (9.49)$$

Here V_N^{sc} and V_N^{oc}, respectively, are the output noise voltage when the input is short circuited and open circuited, that is,

$$V_N^{sc}(s) = V_N(s)|_{Z_S = V_S = V_{SS} = 0} \qquad V_N^{oc}(s) = V_N(s)|_{Y_S = I_S = I_{SS} = 0} \qquad (9.50)$$

and the transfer functions L_{1Nsc} and L_{1Noc}^I, are defined according to

$$L_{1Nsc}(s) = \frac{V_N(s)}{V_S(s)}\bigg|_{V_S = 1, V_{SS} = Z_S = I_i = 0} \qquad L_{1Noc}^I(s) = \frac{V_N(s)}{I_S(s)}\bigg|_{I_S = 1, I_{SS} = Y_S = I_i = 0} \qquad (9.51)$$

In these definitions, I_i is the Laplace transform of the current sources connected to the ith node in the linear circuit, to account for the various noise sources connected to that node.

Proof. The proof of this theorem is given in Appendix 1.

9.6.0.1 Notes Consistent with Eqs. (9.49)–(9.51), V_{eq} is the input equivalent voltage noise when the input is shoft circuited, that is, when $V_S = V_{SS} = Z_S = 0$, assuming a Thevenin equivalent model for the source. Similarly, I_{eq} is the input equivalent current noise when the input is open circuited, that is, when $I_S = I_{SS} = Y_S = 0$, assuming a Norton equivalent model for the source.

9.6.1 Input Equivalent Noise Power Spectral Density

THEOREM 9.3. INPUT EQUIVALENT NOISE POWER SPECTRAL DENSITY *The power spectral density of the voltage V_S, replacing the noise sources V_{SS}, I_{eq}, and V_{eq}, assuming V_{SS} is independent of both I_{eq} and V_{eq}, and all noise sources have zero means, is given by*

$$G_{V_S}(f) = G_{V_{SS}}(f) + G_{V_{eq}}(f) + |Z_S|^2 G_{I_{eq}}(f) + 2Re[Z_S G_{I_{eq}V_{eq}}(f)] \qquad V^2/Hz$$
(9.52)

where $G_{V_{SS}}$, $G_{V_{eq}}$, and $G_{I_{eq}}$ are, respectively, the power spectral density of V_{SS}, V_{eq}, and I_{eq}, and where $G_{I_{eq}V_{eq}}$ is the cross power spectral density between I_{eq} and V_{eq}. According to Eq. (9.25), all the impedances are evaluated with arguments of $j2\pi f$.

With the assumptions noted above, the power spectral density of the current I_S replacing I_{SS}, I_{eq}, and V_{eq}, is given by

$$G_{I_S}(f) = G_{I_{SS}}(f) + G_{I_{eq}}(f) + |Y_S|^2 G_{V_{eq}}(f) + 2Re[Y_S^* G_{I_{eq}V_{eq}}(f)] \qquad A^2/Hz$$
(9.53)

where $G_{I_{SS}}$ is the power spectral density of I_{SS}.

Proof. The proof of these results for G_{V_S} and G_{I_S} follows from Eq. (9.48) and Theorem 8.8.

The results specified in this theorem can be simplified, under certain conditions, as outlined in the following two subsections.

9.6.1.1 Case 1: Input Equivalent Voltage Dominates Noise If $G_{V_{eq}}(f) \gg |Z_S|^2 G_{I_{eq}}(f)$ which is consistent with $|Re[Z_S G_{I_{eq}V_{eq}}(f)]| \ll G_{V_{eq}}(f)$, it follows that

$$G_{V_S}(f) \approx G_{V_{SS}} + G_{V_{eq}}(f) \qquad (9.54)$$

INPUT EQUIVALENT CURRENT AND VOLTAGE SOURCES

that is, the input equivalent noise is determined by the source noise and the input equivalent voltage noise of the circuit.

9.6.1.2 Case 2: Input Equivalent Current Dominates Noise
If $G_{Ieq}(f) \gg |Y_S|^2 G_{Veq}(f)$ which is consistent with $|\text{Re}[Y_S^* G_{IeqVeq}(f)]| \ll G_{Ieq}(f)$, it follows that

$$G_{Is}(f) = G_{Iss}(f) + G_{Ieq}(f) \tag{9.55}$$

that is, the input equivalent noise is determined by the source noise and the input equivalent current noise of the circuit.

9.6.2 Example: Input Equivalent Results for Common Emitter Amplifier

Consider the common emitter amplifier shown in Figure 9.21, and its small signal equivalent model shown in Figure 9.22. Analysis for the input equivalent current and voltage yields,

$$\begin{aligned} V_{eq}(s) = &- I_{BB}(s) r_b + I_B(s) r_b \\ &- I_C(s) \frac{1 + r_b/r_\pi}{g_m} \frac{1 + s(r_b \| r_\pi)(C_\pi + C_\mu)}{1 - sC_\mu/g_m} \end{aligned} \tag{9.56}$$

$$I_{eq}(s) = I_B(s) - I_C(s) \frac{1}{g_m r_\pi} \frac{1 + sr_\pi(C_\pi + C_\mu)}{1 - sC_\mu/g_m} \tag{9.57}$$

Hence, with independent and zero mean noise sources the input equivalent power spectral densities are

$$\begin{aligned} G_{Veq}(f) = &\, G_{BB}(f) r_b^2 + G_B(f) r_b^2 \\ &+ G_C(f) \left[\frac{1 + r_b/r_\pi}{g_m} \right]^2 \frac{1 + 4\pi^2 f^2 (r_b \| r_\pi)^2 (C_\pi + C_\mu)^2}{1 + 4\pi^2 f^2 C_\mu^2 / g_m^2} \end{aligned} \tag{9.58}$$

$$G_{Ieq}(f) = G_B(f) + G_C(f) \left[\frac{1}{g_m r_\pi} \right]^2 \frac{1 + 4\pi^2 f^2 r_\pi^2 (C_\pi + C_\mu)^2}{1 + 4\pi^2 f^2 C_\mu^2 / g_m^2} \tag{9.59}$$

$$\begin{aligned} G_{IeqVeq}(f) = &\, G_B(f) r_b + G_C(f) \left[\frac{1 + r_b/r_\pi}{g_m r_\pi} \right] \\ &\times \frac{(1 + j2\pi f r_\pi (C_\pi + C_\mu))(1 - j2\pi f (r_b \| r_\pi)(C_\pi + C_\mu))}{1 + 4\pi^2 f^2 C_\mu^2 / g_m^2} \end{aligned} \tag{9.60}$$

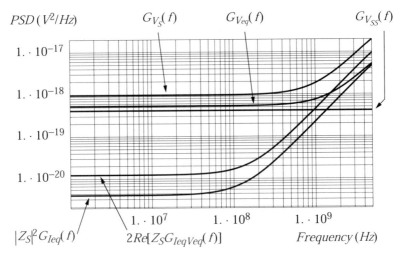

Figure 9.28 Power spectral density of input equivalent voltage source for a common emitter amplifier as well as the constituent components of this power spectral density.

These results follow from Theorem 8.8. Substitution of Eqs. (9.58)–(9.60) into Eq. (9.52) yields,

$$G_{V_S}(f) = 2kTR_S + 2kTr_b + qI_B R_{Sb}^2$$
$$+ \left[\frac{2kT}{R_C} + qI_C\right] \frac{\left(1 + \frac{R_{Sb}}{r_\pi}\right)^2}{g_m^2} \frac{1 + 4\pi^2 f^2 (r_\pi \| R_{Sb})^2 (C_\pi + C_\mu)^2}{1 + 4\pi^2 f^2 C_\mu^2/g_m^2} \quad V^2/\text{Hz}$$

(9.61)

which can easily be shown to be equivalent to the form given in Eq. (9.45), for the case where the input equivalent noise was directly evaluated. In Figure 9.28, $G_{V_S}(f)$ is plotted along with its constituent components as given by Eq. (9.52). Note, for frequencies within the amplifier bandwidth (58 MHz), the input equivalent power spectral density is dominated by the power spectral density of the input equivalent voltage source and the power spectral density of the source resistance. For frequencies significantly higher than the amplifier bandwidth the noise due to the cross power spectral density dominates.

9.7 TRANSFERRING NOISE SOURCES

Consider a cascade of N stages, as shown in Figure 9.29, where it is required to replace the kth noise source by an equivalent one at the $(k+1)$th stage, such that the noise power spectral density at the output node N, is unchanged. The following theorem states the required result.

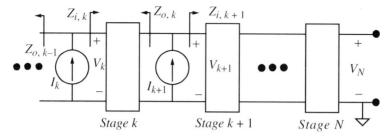

Figure 9.29 Cascade of N stages.

THEOREM 9.4. TRANSFERRING NOISE SOURCES VIA SHORT CIRCUIT CURRENT GAIN *The two circuits shown in Figure 9.30 are equivalent, as far as the output node voltage is concerned, provided*

$$I_{k+1}(s) = L_{sc}(s)I_k(s) \qquad L_{sc}(s) = \left. \frac{I_{sc}(s)}{I_k(s)} \right|_{I_k=1, Z_{i,k+1}=0} \qquad (9.62)$$

where L_{sc} is the short circuit current transfer function (commonly called the short circuit current gain) of the kth stage.

Proof. The proof of this theorem is given in Appendix 2.

The theory pertaining to, and application of, this result has been detailed in Moustakas (1981).

9.7.1 Example—Output Noise of Common Emitter Amplifier

Consider the small signal equivalent noise model for the common emitter amplifier shown in Figure 9.22, and the requirement to replace the input noise sources i_S, i_B, and i_{BB} with an equivalent noise source at the output and in parallel with the noise source i_C. To achieve this, the small signal equivalent

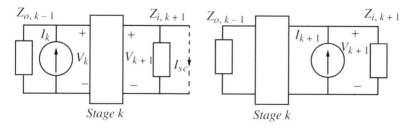

Figure 9.30 Two equivalent circuits as far as the (k + 1)th node voltage is concerned. Equivalence is when $I_{k+1} = L_{sc} I_k$ where $L_{sc} = I_{sc}/I_k$.

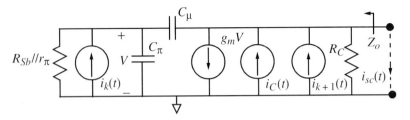

Figure 9.31 Small signal equivalent model for common emitter amplifier where i_{k+1} replaces i_k. In this model $R_{Sb} = R_S + r_b$.

model shown in Figure 9.22, can be redrawn as shown in Figure 9.31 where the noise current i_k is such that

$$i_k(t) = \frac{R_S i_S(s)}{R_S + r_b} + \frac{r_b i_{BB}(t)}{R_S + r_b} + i_B(t) \qquad G_k(f) = \frac{2kT}{R_S + r_b} + qI_B \qquad (9.63)$$

Analysis of this model yields the short circuit current transfer function,

$$L_{sc}(s) = \left.\frac{I_{sc}(s)}{I_k(s)}\right|_{I_k = 1} = \frac{-g_m(R_{Sb} \| r_\pi)(1 - sC_\mu/g_m)}{1 + s(R_{Sb} \| r_\pi)(C_\pi + C_\mu)} \qquad (9.64)$$

and hence,

$$I_{k+1}(s) = \frac{-g_m(R_{Sb} \| r_\pi)(1 - sC_\mu/g_m)I_k(s)}{1 + s(R_{Sb} \| r_\pi)(C_\pi + C_\mu)} \qquad (9.65)$$

Thus, the noise power spectral density associated with the current sources i_C and i_{k+1} at the output node are

$$G_C(f) + G_{k+1}(f) = qI_C + \frac{2kT}{R_C} + \left[\frac{2kT}{R_S + r_b} + qI_B\right] \\ \times \left[\frac{g_m^2(R_{Sb}\|r_\pi)^2(1 + 4\pi^2 f^2 C_\mu^2/g_m^2)}{1 + 4\pi^2 f^2 (R_{Sb}\|r_\pi)^2(C_\pi + C_\mu)^2}\right] \quad A^2/Hz \qquad (9.66)$$

This result can be used to, first, infer the input equivalent noise power spectral density via the transfer function of Eq. (9.44). The result stated in Eq. (9.45) readily follows. Second, as the output impedance, Z_o of the common emitter stage is

$$Z_o(s) = \frac{R_C[1+s(R_{Sb}\|r_\pi)(C_\pi+C_\mu)]}{1+s(R_{Sb}\|r_\pi)\left[C_\pi+C_\mu\left(1+g_mR_C+\dfrac{R_C}{R_{Sb}\|r_\pi}\right)\right]+s^2R_C(R_{Sb}\|r_\pi)C_\pi C_\mu}$$

(9.67)

it then follows that the output power spectral density is

$$\begin{aligned}G_o(f) &= |Z_o(j2\pi f)|^2[G_C(f) + G_{k+1}(f)] \\ &= \left[\frac{2kT}{R_{Sb}} + qI_B\right]\frac{g_m^2 R_C^2(R_{Sb}\|r_\pi)^2(1+4\pi^2 f^2 C_\mu^2/g_m^2)}{D(f)} \\ &+ \left[qI_C + \frac{2kT}{R_C}\right]\frac{R_C^2[1+4\pi^2 f^2(R_{Sb}\|r_\pi)^2(C_\pi+C_\mu)^2]}{D(f)}\end{aligned}$$

(9.68)

where $D(f)$ is defined in Eq. (9.47). This equation agrees with the result previously derived and stated in Eq. (9.46). The output power spectral density G_o is plotted in Figure 9.25.

9.8 RESULTS FOR LOW NOISE DESIGN

The principles outlined in preceding sections can be used to show, for example, the following results: (1) at low frequencies and with a low source impedance a cascade of common emitter/source stages will yield a lower level of noise than a common gate/base or common collector/drain cascade (Moustakis, 1981); (2) at low frequencies and for low source impedances a BJT will yield, in general, lower noise than a JFET, but the reverse is true for high source impedances (Leach, 1994); (3) at low frequencies and for low source impedances, paralleling transistors will reduce the input equivalent noise (Hallgren, 1988); and (4) transformer coupling is effective in reducing the input equivalent noise when the source impedance is low (Lepaisant, 1992).

9.9 NOISE EQUIVALENT BANDWIDTH

If a system has been characterized by a noise equivalent bandwidth, the calculation of the noise power at the output of the system, as required, for example, when the signal-to-noise ratio of a system is being evaluated, is greatly simplified.

DEFINITION: NOISE EQUIVALENT BANDWIDTH The noise equivalent bandwidth B_N, of a real system with transfer function $H(f)$ and with a gain given by

$|H(f_o)|$, is the bandwidth of an ideal filter with a gain of $|H(f_o)|$, which yields the same level of output power as the system being considered. By definition, and for a low pass transfer function, B_N is defined such that,

$$\int_0^\infty G_{IN}(f)|H(f)|^2\,df = |H(f_o)|^2 \int_0^{B_N} G_{IN}(f)\,df \tag{9.69}$$

where G_{IN} is the power spectral density of the input noise. The definition is readily generalized for the bandpass case.

This definition arises from the evenness of $|H(f)|$ and the power spectral density of real signals, as well as the definition of the output power, namely,

$$\bar{P} = \int_{-\infty}^\infty G_{IN}(f)|H(f)|^2\,df = |H(f_o)|^2 \int_{-B_N}^{B_N} G_{IN}(f)\,df \tag{9.70}$$

9.9.1 Examples

For the common case of white noise, where $G_{IN}(f) = G_{IN}(0)$, an explicit expression for B_N is readily obtained,

$$B_N = \frac{1}{|H(f_o)|^2} \int_0^\infty |H(f)|^2\,df \tag{9.71}$$

and the output noise power is

$$\bar{P} = 2|H(f_o)|^2 G_{IN}(0) B_N \tag{9.72}$$

If, for example, H has a single pole form $H(f) = H_o/(1 + jf/f_{3dB})$, then it follows that $B_N = \pi f_{3dB}/2 = 1.57 f_{3dB}$.

For the case where $G_{IN}(f) = kf^2$, which occurs, for example, in high speed optoelectronic receiver amplifiers [see, for example, Jain (1985)], it follows that

$$B_N = \left[\frac{3}{|H(f_o)|^2} \int_0^\infty f^2 |H(f)|^2\,df\right]^{1/3} \tag{9.73}$$

and the output noise power is

$$\bar{P} = \frac{2k|H(f_o)|^2 B_N^3}{3} \tag{9.74}$$

If, for example, H has a Gaussian form with a 3-dB bandwidth of f_{3dB} Hz, that is, $H(f) = H_o e^{-\ln(2)f^2/2f_{3dB}^2}$, then the noise equivalent bandwidth is

$$B_N = \frac{[3\sqrt{\pi}/\sqrt{2}]^{1/3} f_{3dB}}{\sqrt{\ln(2)}} = 1.87 f_{3dB} \qquad (9.75)$$

9.9.2 Signal-to-Noise Ratio of Common Emitter Amplifier

Consider the Common Emitter amplifier whose power spectral density is shown in Figure 9.25 and whose transfer function is shown in Figure 9.23. This transfer function can be approximated by a single pole form $H(f) = H_o/(1 + jf/f_{3dB})$, where $H_o = 37.5$ and $f_{3dB} = 5.8 \times 10^7$. As is evident in Figure 9.25, the input power spectral density is approximately flat up until well beyond the amplifier bandwidth, and consistent with Eq. (9.71), the noise equivalent bandwidth is approximately 91 MHz. With an input equivalent power spectral density level close to $10^{-18} V^2/Hz$, it follows from Eq. (9.72), that the output rms noise level is 0.506 mV consistent with an equivalent rms input noise level of 13.5 μV. With a 1 mV rms input signal, the signal-to-noise ratio is 5490 or 37.4 dB.

9.10 POWER SPECTRAL DENSITY OF A PASSIVE NETWORK

The well-known result, which is a generalization of Nyquist's theorem, is that the power spectral density of noise measured across the terminals of a passive network, is given by

$$G_V(f) = 2kT \, \text{Re}[Z_{in}^F(f)] = 2kT R_{in}^F(f) \qquad V^2/Hz \qquad (9.76)$$

where the input impedance at the same two terminals is Z_{in}, $Z_{in}^F(f) = Z_{in}(j2\pi f) = R_{in}(j2\pi f) + X_{in}(j2\pi f)$, $R_{in}(j2\pi f)$ is real while $X_{in}(j2\pi f)$ is imaginary, and $R_{in}^F(f) = R_{in}(j2\pi f)$. This result is consistent with the models for a passive network shown in Figure 9.32, where G_V is the power spectral density of the voltage source v and the power spectral densities of the sources i and i_R

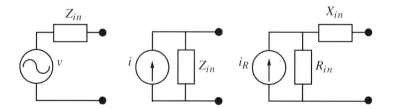

Figure 9.32 Equivalent models for a passive network.

are given by

$$G_I(f) = \frac{2kTR_{in}^F(f)}{|Z_{in}^F(f)|^2} \qquad G_{I_R}(f) = \frac{2kT}{R_{in}^F(f)} \qquad (9.77)$$

A proof of Eq. (9.76) usually uses an argument based on conservation of energy as opposed to direct circuit analysis (Williams, 1937; Helmstrom, 1991 pp. 427–429; Papoulis, 2002 p. 452). A partial proof of Eq. (9.76), based on direct circuit analysis, is given in the following subsection.

Suitable references for the extension of the generalized Nyquist result, as per Eq. 9.76, to nonlinear circuits are Coram (2000) and Weiss (1998, 2000).

9.10.1 Implications of Nyquist's Theorem

Given resistive elements in a passive circuit generate zero mean independent noise waveforms, it follows from Theorem 8.7, that Eq. (9.76) is consistent with the following result:

$$G_V(f) = \sum_{i=1}^{N} \frac{2kT}{R_i} |H_i(f)|^2 = \sum_{i=1}^{N} 2kT \operatorname{Re}[Y_i^F(f)] |H_i(f)|^2 \qquad (9.78)$$

where $H_i(f) = L_i(j2\pi f)$, with L_i being the Laplace transfer function V/I_i relating the output voltage V to the ith current source I_i, associated with the noise generated by the ith resistance R_i. Further, $Y_i^F(f) = Y_i(j2\pi f)$ is the admittance of R_i and associated lossless circuitry between the two nodes that R_i is between. Equating Eqs. (9.76) and (9.78) yields,

$$\operatorname{Re}[Z_{in}^F(f)] = \sum_{i=1}^{N} \operatorname{Re}[Y_i^F(f)] |H_i(f)|^2 \qquad (9.79)$$

A conjecture associated with this result is the generalization,

$$Z_{in}(s) = \sum_{i=1}^{N} Y_i(s) L_i(s) L_i(-s) \qquad (9.80)$$

In fact the correct generalization is

$$Z_{in}(s) = \sum_{i=1}^{N} Y_i(-s) L_i(s) L_i(-s) \qquad (9.81)$$

A partial proof of this conjecture is given in Appendix 3. It is based on proving the result for the general ladder structure shown in Figure 9.33, where the current sources account for the noise of the resistive elements in the adjacent admittance.

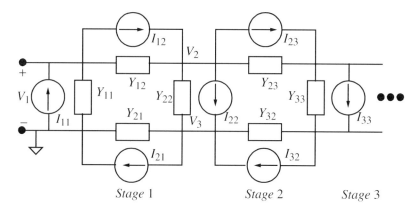

Figure 9.33 Structure of a passive ladder network.

For a N stage passive ladder network defined by Figure 9.33, the input impedance can be written as

$$Z_{in}(s) = Y_{11}(-s)L_{11}(s)L_{11}(-s) + \sum_{i=1}^{N} Y_{i,i+1}(-s)L_{i,i+1}(s)L_{i,i+1}(-s)$$
$$+ Y_{i+1,i}(-s)L_{i+1,i}(s)L_{i+1,i}(-s) \qquad (9.82)$$
$$+ Y_{i+1,i+1}(-s)L_{i+1,i+1}(s)L_{i+1,i+1}(-s)$$

where

$$L_{ij}(s) = \left.\frac{V_1(s)}{I_{ij}(s)}\right|_{I_{ij}=1, I_{pq}=0, p \neq i, q \neq j} \qquad (9.83)$$

9.10.2 Example

Consider the determination of the noise at the input node of a three stage, doubly terminated lossless ladder (van Valkenburg, 1982 pp. 399f) shown in Figure 9.34. Analysis yields the input impedance,

$$Z_{in}(s) = \left.\frac{V_1(s)}{I_1(s)}\right|_{I_1=1, I_{R1}=I_{R4}=0} = \frac{N(s)}{D(s)} \qquad (9.84)$$

where

$$N(s) = R^2 + 3RLs + 6R^2CLs^2 + 4RCL^2s^3 + 5R^2C^2L^2s^4 + RC^2L^3s^5 + R^2C^3L^3s^6 \qquad (9.85)$$

$$D(s) = 2R + 3(L + R^2C)s + 9RCLs^2 + 4CL(L + R^2C)s^3 + 6RC^2L^2s^4 + C^2L^2(L + R^2C)s^5 + RC^3L^3s^6 \qquad (9.86)$$

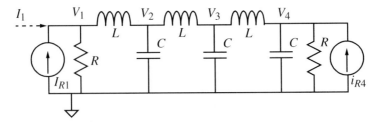

Figure 9.34 Three stage doubly terminated lossless ladder.

Hence, the input equivalent noise at node 1 has the power spectral density,

$$G_1(f) = 2kT\,\text{Re}[Z_{IN}(j2\pi f)] = 2kT\,\text{Re}\left[\frac{N(j2\pi f)D(-j2\pi f)}{|D(j2\pi f)|^2}\right] \quad V^2/\text{Hz}$$

(9.87)

As a check, the transfer functions for the noise current sources I_{R1} and I_{R4} are,

$$\left.\frac{V_1(s)}{I_{R1}(s)}\right|_{I_{R1}=1, I_1 = I_{R4}=0} = \frac{N(s)}{D(s)} \qquad \left.\frac{V_1(s)}{I_{R4}(s)}\right|_{I_{R4}=1, I_1 = I_{R1}=0} = \frac{R^2}{D(s)} \quad (9.88)$$

and direct analysis yields,

$$G_1(f) = \frac{2kT}{R}\left[\left|\frac{V_1(j2\pi f)}{I_{R1}(j2\pi f)}\right|^2 + \left|\frac{V_1(j2\pi f)}{I_{R4}(j2\pi f)}\right|^2\right]$$

$$= 2kT\,\frac{1}{|D(j2\pi f)|^2}\left[\frac{N(j2\pi f)N(-j2\pi f) + R^4}{R}\right] \quad V^2/\text{Hz}$$

(9.89)

Comparing Eq. (9.87) with Eq. (9.89) yields the requirement

$$\frac{N(j2\pi f)D(-j2\pi f) + N(-j2\pi f)D(j2\pi f)}{2} = \frac{N(j2\pi f)N(-j2\pi f) + R^4}{R} \quad (9.90)$$

which can be readily verified. G_1 is plotted in Figure 9.35 for the case where $R = 50$, $C = 10^{-9}$, and $L = 10^{-7}$. At low frequencies, the power spectral density is that of $R\|R$; at high frequencies, when the inductor impedance becomes high, the power spectral density is that of the resistor R.

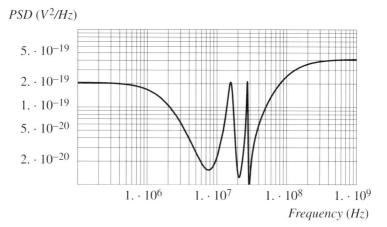

Figure 9.35 Power spectral density at input node of passive network of Figure 9.34 when $R = 50$, $C = 10^{-9}$ and $L = 10^{-7}$.

Finally, for the noiseless case where $R = \infty$, it should be the case that $\text{Re}[Z_{IN}(j2\pi f)] = 0$. To check this, note that when $R = \infty$,

$$Z_{in}(s) = \left.\frac{V_1(s)}{I_1(s)}\right|_{I_1 = 1, R = \infty} = \frac{1 + 6CLs^2 + 5C^2L^2s^4 + C^3L^3s^6}{sC(3 + 4CLs^2 + C^2L^2s^4)} \quad (9.91)$$

Clearly, this transfer function is purely imaginary, when $s = j2\pi f$, as required.

APPENDIX 1: PROOF OF THEOREM 9.2

The linear circuit, as per Figure 9.26, is assumed to have N nodes, where node 1 is the input node and node N is the output node. When the input node is open circuited, the circuit is assumed to be characterized by the N node equations according to

$$\begin{bmatrix} Y_{11} & \cdots & Y_{1N} \\ \vdots & & \vdots \\ Y_{N1} & \cdots & Y_{NN} \end{bmatrix} \begin{bmatrix} V_1 \\ \vdots \\ V_N \end{bmatrix} = \begin{bmatrix} I_1 \\ \vdots \\ I_N \end{bmatrix} \quad \text{or} \quad YV = I \quad (9.92)$$

where V_i is the Laplace transform of the ith node voltage, Y_{ij} is the admittance parameter for the $i - j$th node, and I_i is the Laplace transform of the current sources connected to the ith node to account for the various noise sources connected to that node. The following inverse matrix is assumed to exist:

$$\begin{bmatrix} Z_{11} & \cdots & Z_{1N} \\ \vdots & & \vdots \\ Z_{N1} & \cdots & Z_{NN} \end{bmatrix} = \begin{bmatrix} Y_{11} & \cdots & Y_{1N} \\ \vdots & & \vdots \\ Y_{N1} & \cdots & Y_{NN} \end{bmatrix}^{-1} \quad (9.93)$$

The elements of the inverse matrix are defined by $Z_{ij} = \Delta_{ji}/\Delta$, where Δ is the determinant of Y and Δ_{ji} is the cofactor of Y_{ji} (Anton, 1991 pp. 86–94). Further, Z_{11} is the input impedance at node 1.

For the case where the input node is driven by a Norton source, as per Figure 9.26, the node equations for the overall circuit can be written as

$$\begin{bmatrix} Y_{11} + Y_S & Y_{12} & \cdots & Y_{1N} \\ Y_{21} & Y_{22} & \cdots & Y_{2N} \\ \vdots & \vdots & & \vdots \\ Y_{N1} & Y_{N2} & \cdots & Y_{NN} \end{bmatrix} \begin{bmatrix} V_1 \\ V_2 \\ \vdots \\ V_N \end{bmatrix}$$

$$= \left(\begin{bmatrix} Y_S & 0 & \cdots & 0 \\ 0 & 0 & \cdots & 0 \\ \vdots & \vdots & & \vdots \\ 0 & \cdots & & 0 \end{bmatrix} + \begin{bmatrix} Y_{11} & Y_{12} & \cdots & Y_{1N} \\ Y_{21} & Y_{22} & \cdots & Y_{2N} \\ \vdots & \vdots & & \vdots \\ Y_{N1} & Y_{N2} & \cdots & Y_{NN} \end{bmatrix} \right) \begin{bmatrix} V_1 \\ V_2 \\ \vdots \\ V_N \end{bmatrix} = \begin{bmatrix} I_S + I_{SS} + I_1 \\ I_2 \\ \vdots \\ I_N \end{bmatrix}$$

$$(9.94)$$

and hence, after appropriate manipulation,

$$\begin{bmatrix} V_1 \\ V_2 \\ \vdots \\ V_N \end{bmatrix} = \begin{bmatrix} \dfrac{Z_{11}}{1 + Y_S Z_{11}} & \cdots & \dfrac{Z_{1N}}{1 + Y_S Z_{11}} \\ Z_{21} - \dfrac{Y_S Z_{21} Z_{11}}{1 + Y_S Z_{11}} & \cdots & Z_{2N} - \dfrac{Y_S Z_{21} Z_{1N}}{1 + Y_S Z_{11}} \\ \vdots & & \vdots \\ Z_{N1} - \dfrac{Y_S Z_{N1} Z_{11}}{1 + Y_S Z_{11}} & \cdots & Z_{NN} - \dfrac{Y_S Z_{N1} Z_{1N}}{1 + Y_S Z_{11}} \end{bmatrix} \begin{bmatrix} I_S + I_{SS} + I_1 \\ I_2 \\ \vdots \\ I_N \end{bmatrix} \quad (9.95)$$

It then follows that the transfer function between the ith noise source I_i and the output signal V_N is

$$L_i(s) = \dfrac{V_N(s)}{I_i(s)} \bigg|_{I_i=1, I_S=I_{SS}=I_j=0, j \neq i} = Z_{Ni} - \dfrac{Y_S Z_{N1} Z_{1i}}{1 + Y_S Z_{11}} \quad (9.96)$$

APPENDIX 1: PROOF OF THEOREM 9.2

which can be written in the following form:

$$L_i(s) = \frac{Z_{Ni}}{1 + Y_S Z_{11}} + \frac{Y_S(Z_{11}Z_{Ni} - Z_{1i}Z_{N1})}{1 + Y_S Z_{11}} \qquad (9.97)$$

Thus, the output voltage is

$$V_N(s) = \frac{Z_{N1}(I_S + I_{SS})}{1 + Y_S Z_{11}} + \sum_{i=1}^{N}\left[\frac{Z_{Ni}}{1 + Y_S Z_{11}} + \frac{Y_S(Z_{11}Z_{Ni} - Z_{1i}Z_{N1})}{1 + Y_S Z_{11}}\right]I_i \qquad (9.98)$$

To refer this output noise to the input, and hence, establish the validity of input equivalent current and voltage noise sources, the transfer function between the source current I_S and the output node voltage V_N is required. From this equation it readily follows that

$$L^I_{1N}(s) = \frac{V_N(s)}{I_S(s)}\bigg|_{I_S=1, I_{SS}=I_i=0} = \frac{Z_{N1}}{1 + Y_S Z_{11}} \qquad (9.99)$$

Further, as $I_S = Y_S V_S$ the transfer function between the source voltage V_S and the output voltage V_N can be defined according to

$$L_{1N}(s) = \frac{V_N(s)}{V_S(s)}\bigg|_{V_S=1, V_{SS}=I_i=0} = \frac{Z_{N1} Y_S}{1 + Y_S Z_{11}} \qquad (9.100)$$

Hence, the input equivalent source current I_S that generates the same output noise as the noise sources I_{SS} and $I_1, \ldots, I_N$ is,

$$I_S(s) = I_{SS} + \sum_{i=1}^{N}\left[\frac{Z_{Ni}}{Z_{N1}} + \frac{Y_S(Z_{11}Z_{Ni} - Z_{1i}Z_{N1})}{Z_{N1}}\right]I_i \qquad (9.101)$$

which can be written as

$$I_S = I_{SS} + I_{eq} + Y_S V_{eq} \qquad (9.102)$$

where

$$I_{eq} = \sum_{i=1}^{N}\left[\frac{Z_{Ni}}{Z_{N1}}\right]I_i \qquad V_{eq} = \sum_{i=1}^{N}\left[\frac{Z_{11}Z_{Ni} - Z_{1i}Z_{N1}}{Z_{N1}}\right]I_i \qquad (9.103)$$

As I_{eq} and V_{eq} are independent of the source impedance, the model is justified.

The results $V_S = Z_S I_S$ and $V_{SS} = Z_S I_{SS}$ yields the input source voltage that generates the same output voltage as the noise sources I_{SS} and $I_1, \ldots, I_N$,

$$V_S = V_{SS} + V_{eq} + Z_S I_{eq} \qquad (9.104)$$

Finally, from Eqs. (9.98) and (9.103), I_{eq} and V_{eq} can be written as

$$V_{eq}(s) = \frac{Z_{11}V_N^{sc}(s)}{Z_{N1}} \qquad I_{eq}(s) = \frac{V_N^{oc}(s)}{Z_{N1}} \qquad (9.105)$$

where V_N^{sc} and V_N^{oc}, respectively, are the output voltage when the input is short circuited and open circuited, that is,

$$V_N^{sc}(s) = V_N(s)|_{I_S = I_{SS} = 0, Y_S = \infty} = \sum_{i=1}^{N} \left[\frac{Z_{11}Z_{Ni} - Z_{1i}Z_{N1}}{Z_{11}}\right] I_i$$

$$V_N^{oc}(s) = V_N(s)|_{Y_S = I_S = I_{SS} = 0} = \sum_{i=1}^{N} Z_{Ni} I_i \qquad (9.106)$$

From Eqs. (9.99) and (9.100) it follows that the transfer functions L_{1Nsc} and L_{1Noc}^I can be defined according to

$$L_{1Nsc}(s) = \frac{V_N(s)}{V_S(s)}\bigg|_{V_S = 1, V_{SS} = I_i = 0, Y_S = \infty} = \frac{Z_{N1}}{Z_{11}} \qquad (9.107)$$

$$L_{1Noc}^I(s) = \frac{V_N(s)}{I_S(s)}\bigg|_{I_S = 1, I_{SS} = Y_S = I_i = 0} = Z_{N1} \qquad (9.108)$$

Thus, I_{eq} and V_{eq} can be written as

$$V_{eq}(s) = \frac{V_N^{sc}(s)}{L_{1Nsc}(s)} \qquad I_{eq}(s) = \frac{V_N^{oc}(s)}{L_{1Noc}^I(s)} \qquad (9.109)$$

which is the last required result.

APPENDIX 2: PROOF OF THEOREM 9.4

First, consider the Thevenin equivalent model shown in Figure 9.36, which arises from looking in at the output of the kth stage of the circuit shown in Figure 9.29. In this figure V_{koc} is the open circuit voltage at the output of the kth node when this node is open circuited, that is, when $Z_{i,k+1} = \infty$. Clearly, from Figure 9.36 it follows that

$$V_{k+1} = \frac{Z_{i,k+1} V_{koc}}{Z_{o,k} + Z_{i,k+1}} \qquad (9.110)$$

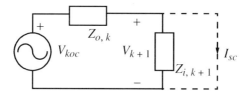

Figure 9.36 Thevenin equivalent circuit at output of kth stage.

Second, consider the case where the open circuit voltage, V_{koc}, is a consequence solely of the current source I_k, shown in Figure 9.29. The transfer function between V_{koc} and I_k is denoted L_k, that is,

$$L_k = \left. \frac{V_{koc}}{I_k} \right|_{I_k=1, I_{k-1}=I_{k-2}=\cdots=0} \tag{9.111}$$

It then follows that the voltage V_{k+1}, due to the current source I_k, is

$$V_{k+1} = \frac{Z_{i,k+1} L_k I_k}{Z_{o,k} + Z_{i,k+1}} \tag{9.112}$$

Third, a current I_{k+1}, acting alone, generates a voltage V_{k+1}, where

$$V_{k+1} = [Z_{o,k} \| Z_{i,k+1}] I_{k+1} = \frac{Z_{o,k} Z_{i,k+1} I_{k+1}}{Z_{o,k} + Z_{i,k+1}} \tag{9.113}$$

Thus, the current I_{k+1} generates the same voltage V_{k+1} as the current I_k, when

$$\frac{Z_{o,k} Z_{i,k+1} I_{k+1}}{Z_{o,k} + Z_{i,k+1}} = \frac{Z_{i,k+1} L_k I_k}{Z_{o,k} + Z_{i,k+1}} \Rightarrow I_{k+1} = \frac{L_k I_k}{Z_{o,k}} \tag{9.114}$$

As $V_{koc} = L_k I_k$, it follows that

$$I_{k+1} = \frac{V_{koc}}{Z_{o,k}} = I_{sc} = L_{sc} I_k \tag{9.115}$$

where I_{sc} is the short circuit current that flows in the circuit, as illustrated in Figure 9.36, and

$$L_{sc} = \left. \frac{I_{sc}}{I_k} \right|_{I_k=1, Z_{i,k+1}=0} \tag{9.116}$$

APPENDIX 3: PROOF OF CONJECTURE FOR LADDER STRUCTURE

Kirchoff's current law applied to the nodes of the first stage in Figure 9.33, yields the following matrix of equations:

$$\begin{bmatrix} Y_{11}+Y_{12} & -Y_{12} & 0 \\ -Y_{12} & Y_{12}+Y_{2T} & -Y_{2T} \\ 0 & -Y_{2T} & Y_{21}+Y_{2T} \end{bmatrix} \begin{bmatrix} V_1 \\ V_2 \\ V_3 \end{bmatrix} = \begin{bmatrix} I_{11}-I_{12} \\ I_{12}-I_{22}-I_{23} \\ I_{22}-I_{21}+I_{32} \end{bmatrix} \quad (9.117)$$

where Y_{2T} is equal to the admittance Y_{22} plus the admittance of the network to the right of Y_{22}. Solving this equation yields the following transfer functions:

$$Z_{in} = L_{11} = \left.\frac{V_1}{I_{11}}\right|_{I_{12}=I_{21}=I_{22}=I_{23}=I_{32}=0} = \frac{Y_{12}Y_{21}+Y_{12}Y_{2T}+Y_{21}Y_{2T}}{\Delta} \quad (9.118)$$

$$L_{12} = \left.\frac{V_1}{I_{12}}\right|_{I_{11}=I_{21}=I_{22}=I_{23}=I_{32}=0} = \frac{-Y_{21}Y_{2T}}{\Delta} \quad (9.119)$$

$$L_{21} = \left.\frac{V_1}{I_{21}}\right|_{I_{11}=I_{12}=I_{22}=I_{23}=I_{32}=0} = \frac{-Y_{12}Y_{2T}}{\Delta} \quad (9.120)$$

$$L_{22} = \left.\frac{V_1}{I_{22}}\right|_{I_{11}=I_{12}=I_{21}=I_{23}=I_{32}=0} = \frac{-Y_{12}Y_{21}}{\Delta} \quad (9.121)$$

where Δ is the determinant of the matrix and is given by

$$\Delta = Y_{11}(Y_{12}Y_{21}+Y_{12}Y_{2T}+Y_{21}Y_{2T}) + Y_{12}Y_{21}Y_{2T} \quad (9.122)$$

It then follows that the input impedance Z_{in}, defined in Eq. (9.118), can be written according to

$$Z_{in} = \frac{Y_{11}^*|Y_{12}Y_{21}+Y_{12}Y_{2T}+Y_{21}Y_{2T}|^2}{|\Delta|^2} + \frac{Y_{12}^*Y_{21}^*Y_{2T}^*(Y_{12}Y_{21}+Y_{12}Y_{2T}+Y_{21}Y_{2T})}{|\Delta|^2}$$

$$(9.123)$$

where, for convenience, the notation $|W(s)|^2 = W(s)W(-s)$ and $W^*(s) = W(-s)$ has been used, and is used in subsequent analysis. Using Eqs. (9.118)–(9.121) in this equation, the input impedance can be written in the form

$$Z_{in} = Y_{11}^*|L_{11}|^2 + Y_{12}^2|L_{12}|^2 + Y_{21}^*|L_{21}|^2 + Y_{2T}^*|L_{22}|^2 \quad (9.124)$$

This is the required result for a circuit consisting of the four stage 1 admittances Y_{11}, Y_{12}, Y_{21}, and Y_{2T}. This result needs to be extended to account

APPENDIX 3: PROOF OF CONJECTURE FOR LADDER STRUCTURE 297

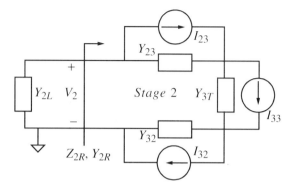

Figure 9.37 Definition of Z_{2R} and Y_{2L} for second stage of ladder network.

for the admittances in stage 2. To do this, first, note that $Y_{2T} = Y_{22} + Y_{2R}$, where Y_{2R} is the admittance to the right of Y_{22}, as per Figure 9.37. Hence, the input impedance can be written as

$$Z_{in} = Y_{11}^*|L_{11}|^2 + Y_{12}^*|L_{12}|^2 + Y_{21}^*|L_{21}|^2 + Y_{22}^*|L_{22}|^2 + Y_{2R}^*|L_{22}|^2 \quad (9.125)$$

Second, the last term in this equation can be written as

$$Y_{2R}^*|L_{22}|^2 = |L_{22}|^2|Y_{2R}|^2 Z_{2R} \quad (9.126)$$

where $Z_{2R} = 1/Y_{2R}$. Using the result of Eq. (9.124), it follows that Z_{2R} can immediately be written as

$$Z_{2R} = Y_{23}^*|F_{23}|^2 + Y_{32}^*|F_{32}|^2 + Y_{3T}^*|F_{33}|^2 \quad (9.127)$$

where the transfer functions F_{23}, F_{32}, and F_{33} are defined according to the following equations where the voltage and current definitions are as per Figure 9.37:

$$F_{23} = \left.\frac{V_{2oc}}{I_{23}}\right|_{Y_{2L}=I_{32}=I_{33}=0} \qquad L_{23} = \left.\frac{V_1}{I_{23}}\right|_{I_{32}=I_{33}=0} \quad (9.128)$$

$$F_{32} = \left.\frac{V_{2oc}}{I_{32}}\right|_{Y_{2L}=I_{23}=I_{33}=0} \qquad L_{32} = \left.\frac{V_1}{I_{32}}\right|_{I_{23}=I_{33}=0} \quad (9.129)$$

$$F_{33} = \left.\frac{V_{2oc}}{I_{33}}\right|_{Y_{2L}=I_{23}=I_{32}=0} \qquad L_{33} = \left.\frac{V_1}{I_{33}}\right|_{I_{23}=I_{32}=0} \quad (9.130)$$

In these equations, V_{2oc} is the open circuit voltage defined by $V_{2oc} = V_2|_{Y_{2L}=0}$.

Thus, substitution of Eqs. (9.127) and (9.126) into Eq. (9.125) yields,

$$Z_{in} = Y_{11}^*|L_{11}|^2 + Y_{12}^*|L_{12}|^2 + Y_{21}^*|L_{21}|^2 + Y_{22}^*|L_{22}|^2 \\ + |L_{22}|^2|Y_{2R}|^2[Y_{23}^*|F_{23}|^2 + Y_{32}^*|F_{32}|^2 + Y_{3T}^*|F_{33}|^2] \quad (9.131)$$

The goal is to rewrite the last three terms in this expression, in terms of the transfer functions from the current sources to the node 1 voltage V_1. To do this, consider the first of these three terms $|L_{22}|^2|Y_{2R}|^2 Y_{23}^*|F_{23}|^2$. First, the transfer function F_{23} is the relationship between the open circuit voltage V_{2oc} and the current source I_{23}, that is, $V_{2oc} = F_{23}I_{23}$. By considering the Thevenin equivalent circuit shown in Figure 9.38, it follows that

$$V_2 = V_{2oc}\left[\frac{Y_{2R}}{Y_{2R} + Y_{2L}}\right] \quad (9.132)$$

and hence, the current I_{23} will generate a voltage V_2 given by

$$V_2 = F_{23}\left[\frac{Y_{2R}}{Y_{2R} + Y_{2L}}\right]I_{23} \quad (9.133)$$

Now, the voltage V_2 could also have been generated by a current I_{22}, according to the relationship $V_2 = I_{22}/(Y_{2R} + Y_{2L})$. Hence, the relationship between the equivalent current I_{22} and I_{23}, is

$$I_{22} = Y_{2R}F_{23}I_{23} \quad (9.134)$$

From the definitions used for stage 1 of the network, a current I_{22} will generate a voltage V_1 at the left end of the network, according to $V_1 = L_{22}I_{22}$.

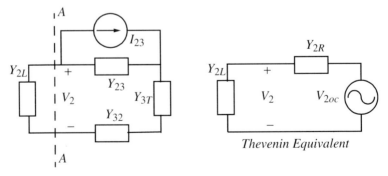

Figure 9.38 Thevenin equivalent model for the circuit to the right of the line A − A.

APPENDIX 3: PROOF OF CONJECTURE FOR LADDER STRUCTURE

Hence, the output voltage generated by I_{23} is given by

$$V_1 = L_{22}Y_{2R}F_{23}I_{23} = L_{23}I_{23} \qquad (9.135)$$

where by definition, $L_{23} = L_{22}Y_{2R}F_{23}$ is the transfer function relating I_{23} to V_1. Similar definitions can be made for L_{32} and L_{33}, and then, using Eq. (9.135), Eq. (9.131) can be written in the required form:

$$\begin{aligned}Z_{in} = & Y_{11}^*|L_{11}|^2 + Y_{12}^*|L_{12}|^2 + Y_{21}^*|L_{21}|^2 + Y_{22}^*|L_{22}|^2 \\ & + Y_{23}^*|L_{23}|^2 + Y_{32}^*|L_{32}|^2 + Y_{3T}^*|L_{33}|^2 \end{aligned} \qquad (9.136)$$

In a similar manner, the admittance Y_{3T} can be expanded, and the above argument repeated to obtain a further expansion for the impedance Z_{in}. The general result for a N stage ladder network is,

$$Z_{in} = Y_{11}^*|L_{11}|^2 + \sum_{i=1}^{N} Y_{i,i+1}^*|L_{i,i+1}|^2 + Y_{i+1,i}^*|L_{i+1,i}|^2 + Y_{i+1,i+1}^*|L_{i+1,i+1}|^2$$

$$(9.137)$$

Notation

COMMON MATHEMATICAL NOTATION

$\lfloor \; \rfloor$	Greatest integer function
$*$	Conjugation operator
$\forall$	For all
$\exists$	There exists
δ	Dirac Delta function
$\emptyset$	The empty set
χ	Characteristic function
Inf	Infimum
f_X	Probability density function for a random variable X
j	Imaginary unit number: $j^2 = -1$
sinc(x)	Sinc function: $\sin(\pi x)/\pi x$
sup	Supremum
$x \in A$	x is an element of A
$x, X, X(T, f)$	The signal x, its Fourier transform and the Fourier transform evaluated on the interval $[0, T]$
u	Unit step function
$A \subseteq B$	A is a subset of B
$A \times B$	Cartesian product of the sets A and B
$\mathbf{C}$	The set of complex numbers
E_X	Ensemble associated with the random process X
$G(T, f), G_\infty(f)$	Power spectral density evaluated on $[0, T]$ and $[0, \infty]$
I	The interval I
Im	Imaginary part of
L	Set of Lebesgue integrable functions on $(-\infty, \infty)$
$L[\alpha, \beta]$	Set of Lebesgue integrable functions on $[\alpha, \beta]$
M	Measure operator
$\mathbf{N}$	The set of positive integers $\{1, 2, \ldots\}$

NOTATION

P	Probability operator
$\bar{P}(T)$, $\bar{P}_\infty$	Average power evaluated on $[0, T]$ and $[0, \infty]$
P_X	Probability space associated with the random process X
$\mathbf{Q}$	The set of rational numbers
$\mathbf{R}$, $\mathbf{R}^+$	The set of real numbers, the set of real numbers greater than zero
$\bar{R}(T, \tau)$, $\bar{R}_\infty(\tau)$	Time averaged autocorrelation function evaluated on $[0, T]$ and on $[0, \infty]$
$R(T, t, \tau)$	Autocorrelation function evaluated on $[0, T]$
Re	Real part of
S^c	Complement of the set S
S_X	Index set identifying outcome of a random process X
$\mathbf{Z}$, $\mathbf{Z}^+$	The set of integers, the set of positive integers $\{1, 2, \ldots\}$

DEFINITION OF ACRONYMS

a.e.	almost everywhere
c.c.	countable case
iff	if and only if
rms	root mean square
s.t.	such that
u.c.	uncountable case
CE	common emitter
DAC	digital to analogue converter
FM	frequency modulation
FSK	frequency shift keyed
MIMO	multiple input–multiple output
NBHD	neighbourhood
PSD	power spectral density
QAM	quadrature amplitude modulation
RP	random process
SNR	signal to noise ratio

PARAMETER VALUES

q-Electronic Charge	1.6×10^{-19} C
T-temperature	300 degree Kelvin (room temp.)
k-Boltzmann's constant	1.38×10^{-23} J/degree Kelvin
$V_T = kT/q$-Thermal Voltage	0.0257 J/C (room temp.)

References

Ambrozy, A. 1982, 'Electronic Noise', McGraw Hill.

Anton, H. and Rorres, C. 1991, 'Elementary Linear Algebra', Wiley.

Apostol, T. M. 1974, 'Mathematical Analysis', Addison Wesley.

Ball, D. G. 1973, 'An Introduction to Real Analysis', Pergamon Press.

Bell, D. A. 1960, 'Electrical Noise: Fundamentals and Physical Mechanism', D. Van Nostrand Co.

Bell, D. A. 1980, 'A survey of 1/f noise in electrical conductors', Journal of Physics C: Solid State Physics, Vol. 13, pp. 4425–4437.

Brown, J. W. and Churchill, R. V. 1995, 'Complex Variables and Applications', McGraw Hill.

Buckingham, M. J. 1983, 'Noise in Electronic Devices and Systems', Ellis Horwood.

Burk, F. 1998, 'Lebesgue Measure and Integration: An Introduction', Wiley.

Carlson, A. B. 1986, 'Communication Systems', McGraw Hill.

Champeney, D. C. 1987, 'A Handbook of Fourier Theorems', Cambridge University Press.

Chua, L. O., Desoer, C. A. and Kuh, E. S. 1987, 'Linear and Nonlinear Circuits', McGraw Hill.

Coram, G. J. Anderson, B. D. O. and Wyatt, J. L. 2000, 'Limits to the fluctuation-dissipation theorem for nonlinear circuits', IEEE Transactions on Circuits and Systems—I: Fundamental Theory and Applications, Vol. 47, pp. 1323–1329.

Davenport, W. B. and Root, W. L. 1958, 'An Introduction to the Theory of Random Signals and Noise', McGraw Hill.

Debnath, L. and Mikusinski, P. 1999, 'Introduction to Hilbert Spaces with Applications', Academic Press.

Demir, A. and Sangiovanni-Vincentelli, A. 1998, 'Analysis and Simulation of Noise in Nonlinear Electronic Circuits and Systems', Kluwer Academic Publishers.

Edwards, P. J., McDonald, R. and Cheng, W. N. 2000, 'Semiconductor junction noise revisited', Unsolved Problems of Noise and Fluctuations, Adelaide, Australia, 12–15

July 1999, American Institute of Physics, AIP Conference Proceedings, Vol. 511, pp. 415–421.

Engberg, J. and Larsen, T. 1995, 'Noise Theory of Linear and Nonlinear Circuits', Wiley.

Epp, S. S. 1995, 'Discrete Mathematics with Applications', Brooks/Cole Publishing Co.

Feller, W. 1957, 'An Introduction to Probability Theory and its Applications', Wiley.

Fish, P. J. 1993, 'Electronic Noise and Low Noise Design', Macmillan.

Franco, S. 2002, 'Design with Operational Amplifiers and Analog Integrated Circuits', McGraw Hill.

Franks, L. E. 1969, 'Signal Theory', Prentice Hall.

Freeman, J. J. 1958, 'Principles of Noise', Wiley.

Gabel, R. A. and Roberts, R.A. 1987, 'Signals and Linear Systems', Wiley.

Gardner, C. W. 1990, 'Handbook of Stochastic Methods for Physics, Chemistry and the Natural Sciences', Springer-Verlag.

Gardner, W. A., 1988, 'Statistical Spectral Analysis: A Non-probabilistic Theory', Prentice Hall.

Gillespie, D. T. 1996, 'The mathematics of Brownian motion and Johnson noise', American Journal of Physics, Vol. 64, pp. 225–240.

Gradshteyn, I. S. and Ryzhik, I. M. 1980, 'Tables of Integrals, Series, and Products', Academic Press.

Gray, M. R., and Stockham, T. G. 1993, 'Dithered quantizers', IEEE Transactions on Information Theory, Vol. 39, pp. 805–812.

Gray, P. R., Hurst, P. J., Lewis, S. H. and Meyer, R. G. 2001, 'Analysis and Design of Analog Integrated Circuits', Wiley.

Grimmett, G. R. and Stirzaker, D. R. 1992, 'Probability and Random Processes', Oxford University Press.

Hallgren, R. B. 1988, 'Paralleled transconductance ultralow-noise preamplifier', Review of Scientific Instruments, Vol. 59, pp. 2070–2074.

Hanggi, P., Inchiosa, M. E., Fogliatti, D. and Bulsara, A. R. 2000, 'Nonlinear stochastic resonance: The saga of anomalous output-input gain', Physical Review E, Vol. 62, pp. 6155–6163.

Haus, H. A. et al. 1960, 'Representation of noise in linear twoports', Proceedings of the IRE, Vol. 48, pp. 69–74.

Helstrom, C. W. 1991, 'Probability and Stochastic Processes for Engineers', Macmillan.

Higgins, J. R. 1996, 'Sampling Theory in Fourier and Signal Analysis', Oxford University Press.

Hirschman, I. I. 1962, 'Infinite Series', Holt, Rinehart and Winston.

Hooge, F. N., Kleinpenning, T. G. M., and Vandamme, L. K. J. 1981, 'Experimental studies on 1/f noise', Reports on Progress in Physics, Vol. 44, pp. 479–532.

Hooge, F. N, 1997, '40 years of 1/f noise modelling', Proceedings of the 14th International Conference on Noise in Physical Systems and 1/f Fluctuations', World Scientific, pp. 3–10.

Howard, R. M., Jeffery, R. D. and Hullett, J. L. 1987, 'On the noise of high-transimpedance amplifiers for long-wavelength pulse OTDRs', Optical and Quantum Electronics, Vol. 19, pp. 123–129.

Howard, R. M. 1999a, 'Dual sensing receiver structure for halving the amplifier contribution to the input equivalent noise of an opto-electronic receiver', IEE Proceedings Optoelectronics, Vol. 146, pp. 189–200.

Howard, R. M. 1999b, 'Low noise amplifier design and low noise amplifiers for characterizing the low frequency noise of IR detectors', Proceedings of the Conference on Optoelectronic and Microelectronic Materials and Devices, Perth, Australia, 14–16 December 1998, IEEE, pp. 179–182.

Howard, R. M. 2000, 'The representative ensemble and its application to 1/f type random processes', Unsolved Problems of Noise and Fluctuations, Adelaide, Australia, 12–15 July 1999, American Institute of Physics, AIP Conference Proceedings Vol. 511, pp. 124–129.

Hullett, J. L., Muoi, T. V. and Moustakis, S. 1977, 'High-speed optical preamplifiers', Electronics Letters, Vol. 13, pp. 688–690.

Jain, P. K. and Gupta, V. P. 1986, 'Lebesgue Measure and Integration', Wiley Eastern Limited.

Jain, V. K., Kumar, P. and Gupta, S. N. 1985, 'Design of an optimum optical receiver', Journal of Optical Communications, Vol. 6, pp. 106–112.

Jantsch, O. 1987, 'Flicker (1/f) noise generated by a random walk of electrons in interfaces', IEEE Transactions on Electron Devices', Vol. 34, pp. 1100–1115.

Jenkins, G. M. and Watts, D. G. 1968, 'Spectral Analysis and its Applications', Holden-Day.

Kaulakys, B. and Meskauskas, T. 1998, 'Modelling 1/f noise', Physical Review E, Vol. 58, pp. 7013–7019.

Keshner, M. S. 1982, '1/f noise', Proceedings of the IEEE, Vol. 70, pp. 212–218.

Kittel, C. 1958, 'Elementary Statistical Physics', Wiley.

Knopp, K. 1956, 'Infinite Sequences and Series', Dover.

Kreyszig, E. 1978, 'Introductory Functional Analysis with Applications', Wiley.

Lam, V. M. T., Poole, C, R. and Yip, P. C. L. 1992, 'Exact noise figure of a noisy twoport with feedback', IEE Proceedings-G, Vol. 139, pp. 473–476.

Larson, R. J. and Marx, M. L. 1986, 'An Introduction to Mathematical Statistics and its Applications', Prentice Hall.

Leach, W. M. 1994, 'Fundamentals of Low-Noise Analog Circuit Design', Proceedings of the IEEE, Vol. 82, pp. 1515–1538.

Lepaisant, J., Lam Chok Sing, M. and Bloyet, D. 1992, 'Low-noise preamplifier with input and feedback transformers for low source resistance sensors', Review of Scientific Instruments', Vol. 63, pp. 2089–2094.

Levinzon, F. A., 2000, 'Noise of the JFET amplifier', IEEE Transactions on Circuits and Systems I: Fundamental Theory and Applications', Vol. 47, pp. 981–985.

Lipschutz, S. 1998, 'Theory and Problems of Set Theory and Related Topics', McGraw Hill.

Luchinsky, D. G., Mannella, R., McClintock, P. V. E. and Stocks, N. G. 1999, 'Stochastic resonance in electrical circuits: I. Conventional stochastic resonance', IEEE Transactions on Circuits and Systems II: Analog and Digital Signal Processing, Vol. 46, pp. 1205–1214.

Marsden, J. E. and Hoffman, M. J. 1987, Basic Complex Analysis', W. H. Freeman and Co.

Marsden, J. E. and Hoffman, M. J. 1993, 'Elementary Classical Analysis', W. H. Freeman and Co.

McGillem, C. D. and Cooper, G. R. 1991, 'Continuous and Discrete Signal and System Analysis', Saunders.

Motchenbacher, C. D. and Connelly, J. A. 1993, 'Low-Noise Electronic System Design', Wiley.

Moustakas, S. and Hullett, J. L. 1981, 'Noise modelling for broadband amplifier design', IEE Proceedings Part G, Vol. 128, pp. 67–76.

Muoi, T. V. 1984, 'Receiver design for high speed optical fibre systems', Journal of Lightwave Technology, Vol. 2, pp. 243–267.

Netzer, Y. 1981, 'The design of low-noise amplifiers', Proceedings of the IEEE, Vol. 69, pp. 728–741.

Nyquist, H. 1928, 'Thermal agitation of electric charge in conductors', Physical Review, Vol. 32, pp. 110–113.

Ogawa, K. 1981, 'Noise caused by GaAs MESFETs in optical receivers', The Bell System Technical Journal, Vol. 60, pp. 923–928.

Papoulis, A. 1977, 'Signal Analysis', McGraw Hill.

Papoulis, A. and Pillai, S. U. 2002, 'Probability, Random Variables, and Stochastic Processes', McGraw Hill.

Parzen, E. 1960, 'Modern Probability Theory and Its Applications', Wiley.

Parzen, E. 1962, 'Stochastic Processes', Holden-Day Inc.

Peebles, P. Z. 1993, 'Probability, Random Variables, and Random Signal Principles', McGraw-Hill.

Polimeni, A. D. and Straight, H. J. 1990, 'Foundations of Discrete Mathematics', Brooks/Cole Publishing Co.

Potzick, J. 1999, 'Noise averaging and measurement resolution (or 'A little noise is a good thing')', Review of Scientific Instruments, Vol. 70, pp. 2038–2040.

Priestley, M. B. 1981, 'Spectral Analysis and Time Series, Vol. 1: Univariate Series', Academic Press.

Proakis, J. G. 1995, 'Digital Communications', McGraw Hill.

Reif, F. 1965, 'Fundamentals of Statistical and Thermal Physics', McGraw Hill.

Rensen, N. and Howard, R. M. 1999, 'A systematic approach for determining the spectral imperfections introduced by a DSP-DAC, Proceedings of the 5th International Symposium on DSP for Communication Systems, Perth, Australia, 1–4 February 1999, pp. 122–127.

Rice, S. O. 1944, 'Mathematical analysis of random noise', Bell System Technical Journal, Vol. 23, pp. 282–332.

Smith, D., Eggen, M. and St. Andre, R. 1990, 'A Transition to Advanced Mathematics', Brooks/Cole Publishing Co.

Spiegel, M. R. 1968, 'Mathematical Handbook of Formulas and Tables', McGraw Hill.

Spivak, M. 1994, 'Calculus', Publish or Perish Inc.

Sprecher, D. A. 1970, 'Elements of Real Analysis', Dover.

Stephany, J. F. 1998, 'A theory of 1/f noise', Journal of Applied Physics, Vol. 83, pp. 3139–3143.

Taylor, H. M. and Karlin, S. 1998, 'An Introduction to Stochastic Modelling', Academic Press.

Thomas, J. B. 1969, 'An Introduction to Statistical Communication Theory', Wiley.

Titchmarsh, E. C. 1939, 'The Theory of Functions', Oxford University Press.

Titchmarsh, E. C. 1948, 'Introduction to the Theory of Fourier Integrals', Oxford University Press.

Tolstov, G. P. 1962, 'Fourier Series', Dover.

Trofimenkoff, F. N. 1989, 'Noise performance of operational amplifier circuits', IEEE Transactions on Education, Vol. 32, pp. 12–16.

Tunaley, J. K. E. 1976, 'A Theory of 1/f current noise based on a random walk model', Journal of Statistical Physics, Vol. 15, pp. 149–156.

van den Elzen, H. C. 1970, 'Calculating power spectral densities for data signals', Proceedings of the IEEE, Vol. 58. pp. 942–943.

van der Ziel, A. 1986, 'Noise in Solid State Devices and Circuits', Wiley.

van Valkenburg, M. E. 1982, 'Analog Filter Design', Holt Saunders.

Weiss, L. and Mathis, W. 1998, 'A thermodynamical approach to noise in non-linear networks', International Journal of Circuit Theory and Applications, Vol. 26, pp. 147–165.

Weiss, L. and Mathis, W. 2000, 'A unified description of thermal noise and shot noise in nonlinear resistors', Unsolved Problems of Noise and Fluctuations, Adelaide, Australia, 12–15 July 1999, American Institute of Physics, AIP Conference Proceedings Vol. 511, pp. 89–100.

Williams, F. C. 1937, 'Thermal fluctuations in complex networks', Journal of Institute of Electrical Engineers, Vol. 81, pp. 751–760.

Index

1/f noise, 198, 266
1/f noise and bounded random walks, 198

Absolute continuity, 18, 22, 23
Absolute continuity and differentiability, 23
Absolute convergence, 34
Almost everywhere, 24
Approximation
   continuous, 30
   step, 32
Autocorrelation
   definition, 79
   existence conditions, 81
   infinite interval case, 82
   notation, 80
   power spectral density relationship, 81
   random process case, 80
   single waveform case, 79
   time averaged, 79

Bounded power spectral density, 75, 92
Bounded random walk, 196
Bounded variation, 18
Boundedness, 17
Boundedness and integrability, 29, 35
Boundedness and local integrability, 35
Brownian motion, 192

Cartesian product, 4
CE amplifier
   input noise, 273, 281
   noise model, 273
   output noise, 277, 283
   signal to noise ratio, 287
Characteristic function, 4

Communication signals, 140
Complex numbers, 6
Conjugation operator, 8
Continuity
   absolute, 18
   left, 13
   piecewise, 13
   point, 14
   pointwise, 14
   right, 13
   uniform, 14
Continuity and absolute continuity, 23
Continuity and boundedness, 19
Continuity and Dirichlet point, 42
Continuity and discontinuities, 20
Continuity and maxima-minima, 20
Continuity and path length, 20
Continuous approximation, 30
Convergence, 36
   dominated, 36
   Fourier series, 39
   in mean, 36, 37
   monotone, 37
   pointwise, 36
   uniform, 36
Correlogram, 80
Countable set, 6
Cross power spectral density, 102
   bound, 106
   definition, 104

DAC, 152
Differentiability, 15
   piecewise, 16
Differentiability and Dirichlet point, 41

Dirac Delta function, 43
Dirichlet point, 40
Dirichlet value, 41
Disjoint signaling random process, 207
Disjoint signals, 8
Dominated convergence theorem, 37

Energy, 13
Ensemble, 44
Equivalent disjoint random process, 207
Equivalent low pass signal, 187

Fourier series, 38
Fourier theory, 38
Fourier transform, 39
Frequency modulation
   binary FSK, 215
   raised cosine pulse shaping, 218
Fubini-Tonelli theorem, 34
Function definition, 7

Gaussian white noise, 259
Generalized signaling process, 166

Impulse response, 230
Impulsive power spectral density, 76
   graphing, 122
Infimum, 4
Information signal, 138, 141
Input equivalent noise, 270, 278
   CE amplifier, 273, 281
   input equivalent current, 278
   input equivalent voltage, 278
   power spectral density, 280
Integers, 5
Integrability, 35
Integrated spectrum, 77
Integration
   Lebesgue, 25
   local, 27, 35
   Riemann, 26
Interchanging integration order, 34
Interchanging summation order, 34
Interval, 6
Inverse Fourier transform, 42

Jitter, 155

Lebesgue integration, 25
Linear system
   impulse response, 230
   input-output relationship, 232
   power signal case, 234
   power spectral density of output, 238
   transforms of output signal, 232
   widowed case, 234
Linear system theory, 229
Locally integrable, 27
Low noise design, 285

Measurable functions, 25
Measurable set, 24
Measure, 23
Memoryless system, 10, 11
Memoryless transformation of a random
   process, 206
Monotone convergence theorem, 37
Multiple input-multiple output system, 243

Natural numbers, 5
Neighbourhood, 6
Noise
   1/f, 198, 266
   CE amplifier, 273, 281, 283
   doubly terminated lossless ladder, 289
   effect of, 257
   oscillator, 241
   passive network, 287
   shot, 160, 266
   thermal, 264
Noise analysis
   input equivalent noise, 270
   linear time invariant system, 269
Noise equivalent bandwidth, 285
Noise model
   BJT, 267
   diode, 267
   JFET and MOSFET, 267
   resistor, 264
Non-stationary random processes, 71
Nyquist's theorem, 265
   generalized, 287

Operator definition, 7
Ordered pair, 4
Orthogonal set, 9
Orthogonality, 8
Oscillator noise, 241

Parseval's theorem, 44
Partition, 4, 6
Passive network, 287
Piecewise continuity, 13
Piecewise smoothness, 16
Piecewise smoothness and absolute continuity,
   23
   bounded variation, 22
   Dirichlet point, 42

discontinuities, 21
piecewise continuity, 22
Pointwise continuity, 14
Pointwise convergence, 36
Power
  relative, 59
  signal, 13
Power and power spectral density, 65
Power spectral density, 60
  approximate, 99
  autocorrelation relationship, 78, 81
  bounded, 75, 92
  continuity, 63
  continuous form, 62
  cross, 102
  definition, 63, 67, 69
  discrete approximation, 72
  existence criteria, 73
  graphing, 122
  impulsive, 74
  infinite interval, 67, 69
  input equivalent current, 280
  input equivalent noise, 272
  input equivalent voltage, 280
  integrability, 66
  interpretation, 64
  multiple input–multiple output system, 244
  nonlinear transformation, 209
  nonzero mean case, 121
  output of linear system, 238
  periodic component case, 119
  properties, 65
  random process, 67
  required characteristics, 60
  resolution, 66
  simplification, 98
  single sided, 72
  single waveform, 60
  symmetry, 66
  via signal decomposition, 95
Power spectral density example, 96, 98, 102, 122
  amplitude signaling through memoryless nonlinearity, 211
  binary FSK, 215
  bipolar signaling, 148
  CE amplifier input noise, 273, 281
  CE amplifier output noise, 277, 283
  digital random process, 70, 122
  doubly terminated lossless ladder, 289
  FM with pulse shaping, 218
  jittered pulse train, 159
  oscillator noise, 241
  periodic pulse train, 115
  quadrature amplitude modulatin, 191

RZ signaling, 146
shot noise with dead time, 165
sinusoid, 65
spectral narrowing, 213
Power spectral density of
  bounded random walk, 196
  DAC quantization error, 152
  electrons crossing a barrier, 161
  equivalent low pass signal, 188
  generalized signaling process, 167
  infinite sum of periodic signals, 118
  infinite sum of random processes, 108
  input equivalent noise sources, 280
  jittered signal, 158
  linear system output, 238
  multiple input–multiple output system, 244
  passive network, 287
  periodic random process, 117
  periodic signal, 112
  quadrature amplitude modulated signal, 186
  random walk, 193
  sampled random process, 184
  sampled signal, 181
  shot noise, 161
  shot noise with dead time, 164
  signaling random processes, 141
  sum of N random processes, 112
  sum of two random processes, 107
Probability space, 44

Quadrature amplitude modulation, 185
Quantization, 152

Raised cosine pulse, 218
Raised cosine spectrum, 148
Random processes, 44
  signaling, 138
Random walk, 192
  bounded, 196
Rational numbers, 5
Real numbers, 5
Relative power measures, 59
Riemann integration, 26
Riemann sum, 179
Riemann-Lebesgue theorem, 40

Sample autocorrelation function, 80
Sampling, 179
Schottky's formula, 163
Schwarz inequality, 29
Set theory, 3
Shot noise, 160, 266
Shot noise with dead time, 163

Signal
    classifiction, 35
    decomposition, 9
    definition, 7
    disjointness, 8
    energy, 13
    orethogonality, 8
    power, 13
Signal path length, 17
Signal to noise ratio
    CE amplifier, 287
    DAC, 155
Signaling invariant system, 210
Signaling random process, 138
    definition, 139
    disjoint, 207
    generalized, 166
Simple function, 30
Single-sided power spectral density, 72
Spectral distribution function, 77
Spectral issues, 146
Spectral spread, 213

Square law device, 212
Step approximation, 32
Supremum, 4
System
    definition, 7
    linear, 9
    memoryless, 10, 11

Thermal noise, 264
Transferring noise sources, 282

Uncountable set, 6
Uniform and pointwise continuity, 14
Uniform continuity, 14
Uniform convergence, 36

White noise, 259
Wiener Khintchine relations, 83
Wiener process, 192

Zero measure, 24, 27